BIOACTIVE COMPOUNDS AND NUTRACEUTICALS FROM PLANT SOURCES

Extraction Technology, Analytical Techniques, and Potential Health Prospects

BIOACTIVE COMPOUNDS AND NUTRACEUTICALS FROM PLANT SOURCES

Extraction Technology, Analytical Techniques, and Potential Health Prospects

Edited by

Charanjit S. Riar, PhD

Parmjit S. Panesar, PhD

AAP APPLE ACADEMIC PRESS

First edition published 2024

Apple Academic Press Inc.
1265 Goldenrod Circle, NE,
Palm Bay, FL 32905 USA

760 Laurentian Drive, Unit 19,
Burlington, ON L7N 0A4, CANADA

CRC Press
2385 NW Executive Center Drive,
Suite 320, Boca Raton FL 33431

4 Park Square, Milton Park,
Abingdon, Oxon, OX14 4RN UK

Library and Archives Canada Cataloguing in Publication

Title: Bioactive compounds and nutraceuticals from plant sources : extraction technology, analytical techniques, and potential health prospects / edited by Charanjit Singh Riar, PhD, Parmjit S. Panesar, PhD.
Names: Riar, Charanjit Singh, editor. | Panesar, P. S. (Parmjit Singh), editor.
Description: First edition. | Includes bibliographical references and index.
Identifiers: Canadiana (print) 20240289188 | Canadiana (ebook) 20240289234 | ISBN 9781774915004 (hardcover) | ISBN 9781774915011 (softcover) | ISBN 9781003455172 (ebook)
Subjects: LCSH: Plant bioactive compounds. | LCSH: Bioactive compounds—Analysis. | LCSH: Functional foods—Analysis.
Classification: LCC QK898.B54 B56 2024 | DDC 580—dc23

Library of Congress Cataloging-in-Publication Data

CIP data on file with US Library of Congress

ISBN: 978-1-77491-500-4 (hbk)
ISBN: 978-1-77491-501-1 (pbk)
ISBN: 978-1-00345-517-2 (ebk)

About the Editors

Charanjit S. Riar, PhD
Sant Longowal Institute of Engineering and Technology, Sangrur, Punjab, India

Charanjit S. Riar, PhD, has completed his graduation and post-graduation in Food Technology from Guru Nanak Dev University, Amritsar (India), and PhD in Food Technology from Punjab Technical University, Jalandhar, India. Dr. Riar started his career as a lecturer in the Department of Food Technology, Sant Longowal Institute of Engineering and Technology, Longowal, Sangrur (Punjab), and is presently working as a Professor in the same Department. He has 25 years of teaching and research experience. Dr. Riar has published 108 research papers in reputed national and international journals, having 2,642 citations and 'h' index of 31. He has authored/edited 4 books, presented more than 65 papers in conferences/seminars, and written 20 book chapters. Dr. Riar is a reviewer of national/international publishing houses such as Elsevier, Springer, SAGE, Taylor & Francis, etc. He is on the panel of experts of a number of reputed national universities/institutions in India. Dr. Riar has guided 45 BTech, 20 MTech and guided/under guidance, nine PhD scholars. His areas of research are starch modification, biodegradable film, bioactive compounds (nutraceuticals) extraction and analysis, seed germination, and characteristics like pasting, rheological, thermal, morphological, molecular, and functional characteristics of food and its products.

Parmjit S. Panesar, PhD
Food Biotechnology Research Laboratory, Department of Food Engineering and Technology, Sant Longowal Institute of Engineering and Technology, Longowal, Sangrur, Punjab, India

Parmjit S. Panesar, PhD, is currently working as a Professor and Dean (Planning and Development) at Sant Longowal Institute of Engineering and

Technology (Deemed to be University: Established by Government of India), Longowal, Punjab. Prof. Panesar has more than 24 years of teaching and research experience and has also served in administrative positions like Dean (Research and Consultancy) and Head of the Department of Food Engineering and Technology, SLIET Longowal. In 2005, he was awarded a BOYSCAST (Better Opportunities for Young Scientists in Chosen Areas of Science and Technology) fellowship by the Department of Science and Technology (DST), Government of India, to carry out advanced research at Chembiotech labs, University of Birmingham Research Park, UK. He was also involved in USA and Australian-sponsored projects during his stay in the UK. In 1999, Prof. Panesar was awarded a Young Scientist Fellowship by the Punjab State Council for Science and Technology, Punjab.

Prof. Panesar has successfully completed seven research projects funded by DBT, CSIR, MHRD, AICTE, New Delhi, and three projects are in progress. He has published more than 160 international/national scientific papers, 50 book reviews in peer-reviewed journals, 40 chapters, and has authored/edited 19 books. He has supervised 14 PhD students and more than 30 MTech students. He is a member of the editorial advisory boards of various national/international journals, including the *International Journal of Biological Macromolecules, Journal of Food Science and Technology*, and *Carbohydrate Polymer Technologies and Applications*. He is now serving as a member of the National Level Scientific Panel on "Alcoholic Beverages and Water and Non-alcoholic Beverages" constituted by FSSAI and has acted as a member of the international collective of experts of the Foundation for Science and Technology (FCT), Portugal. He has also received advanced training from different institutes/labs, including the International Center for Genetic Engineering and Biotechnology, Italy, and Chembiotech Labs, University of Birmingham Research Park, Birmingham, United Kingdom.

In recognition of his work, Prof. Panesar was elected as 'Fellow 2018' by The Biotech Research Society of India (BRSI) and 'Fellow 2021' by the 'International Society for Energy, Environment, and Sustainability (ISEES).' He has also been selected for the award of the prestigious INSA Teachers Award (2020) by Indian National Science Academy (INSA). Prof. P. S. Panesar was also listed in the most coveted list (2020, 2021) of World Ranking of Top 2% Scientists published by Stanford University, USA. He has served as the National Vice President of the AFSTI during 2019. Prof. Panesar has also served as a member of the Board of Studies (BOS) of various universities, such as the Central University of Punjab, Punjabi

University, Patiala, Punjab Technical University, Jalandhar, HP University, Shimla, Chandigarh University, etc.

He is a life member of the International Forum on Industrial Bioprocesses (IFIBiop), Biotechnology Research Society of India, Association of Microbiologists of India, Association of Food Scientists, Indian Science Congress Association, and Technologists, Indian Society for Technical Education, and Punjab Academy of Sciences. His research is focused on the area of value addition of food industry byproducts, prebiotics, carbohydrate polymers, circular economy, green extraction of bioactives, and nanoencapsulation. He has visited several countries like the UK, USA, Canada, Switzerland, New Zealand, Australia, Germany, France, Singapore, Malaysia, China, Iran, Thailand, etc.

Contents

Contributors

Farhana Mehraj Allai
Department of Food Technology, Islamic University of Science and Technology, Awantipora, India

Farhan Mohiuddin Bhat
CSK Himachal Pradesh Agricultural University, Palampur, Himachal Pradesh, India

Hanuman Bobade
Department of Food Science and Technology, Punjab Agricultural University, Ludhiana, Punjab, India

Lucía Ortega Cabello
Biological Systems Department, Metropolitan Autonomous University, Xochimilco Campus, Mexico City, Mexico

Amit Baran Das
Department of Food Engineering and Technology, Tezpur University, Tezpur, Assam, India

Dipak Das
Department of Food Engineering and Technology, Sant Longowal Institute of Engineering and Technology, Longowal, Punjab, India

Ren-You Gan
Key Laboratory of Coarse Cereal Processing (Ministry of Agriculture and Rural Affairs), Sichuan Engineering and Technology Research Center of Coarse Cereal Industrialization, School of Food and Biological Engineering, Chengdu University, Chengdu, China; Research Center for Plants and Human Health, Institute of Urban Agriculture, Chinese Academy of Agricultural Sciences, Chengdu, China

Khalid Gul
Department of Food Process Engineering, National Institute of Technology, Rourkela, Odisha, India

Antima Gupta
Department of Food Science and Technology, Punjab Agricultural University, Ludhiana, Punjab, India

Vivek Gupta
Department of Food Engineering and Technology, Tezpur University, Tezpur, Assam, India

Romee Jan
University of Kashmir, Hazratbal, Srinagar, Kashmir, India

María Eugenia Mendoza Jasso
Biological Systems Department, Metropolitan Autonomous University, Xochimilco Campus, Mexico City, Mexico; Pharmacy Department, Chemistry Faculty, National Autonomous University of Mexico, Mexico City, Mexico

Brahmeet Kaur
Department of Food Engineering and Technology, Sant Longowal Institute of Engineering and Technology, Longowal, Punjab, India

Ravneet Kaur
Department of Food Engineering and Technology, SLIET, Longowal, Punjab, India

José de Jesús Lira-Ricárdez
Biological Systems Department, Metropolitan Autonomous University, Xochimilco Campus, Mexico City, Mexico

Nisar Ahmad Mir
Department of Food Technology, School of Engineering and Technology, Islamic University of Science and Technology, Awantipora, Jammu & Kashmir, India

Gaurav Panesar
Department of Food Engineering and Technology, Sant Longowal Institute of Engineering and Technology, Longowal, Punjab, India

Parmjit S. Panesar
Food Biotechnology Research Laboratory, Department of Food Engineering and Technology, Sant Longowal Institute of Engineering and Technology, Longowal, Punjab, India

Prem Prakash
Department of Food Engineering and Technology, SLIET, Longowal, Punjab, India

Kamlesh Prasad
Department of Food Engineering and Technology, SLIET, Longowal, Punjab, India

Lokeswari Ramireddy
Department of Food Process Engineering, National Institute of Technology, Rourkela, Odisha, India

Charanjit S. Riar
Sant Longowal Institute of Engineering and Technology, Sangrur, Punjab, India

Rajan Sharma
Department of Food Science and Technology, Punjab Agricultural University, Ludhiana, Punjab, India

Savita Sharma
Department of Food Science and Technology, Punjab Agricultural University, Ludhiana, Punjab, India

Seema Sharma
Jaipur National University, Jaipur, Rajasthan, India

Mohd. Aaqib Sheikh
Department of Food Engineering and Technology, Sant Longowal Institute of Engineering and Technology, Longowal, Punjab, India

Shubhra Shekhar
Department of Food Process Engineering, National Institute of Technology, Rourkela, Odisha, India

Sukhcharn Singh
Department of Food Process Engineering, National Institute of Technology, Rourkela, Odisha, India

Pooja Singha
Department of Food Engineering and Technology, Tezpur University, Tezpur, Assam, India

Sakshi Sukhija
Amity University, Noida, Uttar Pradesh, India

Rahul Thakur
Department of Food Engineering and Technology, Tezpur University, Tezpur, Assam, India

Ranjana Verma
CSK Himachal Pradesh Agricultural University, Palampur, Himachal Pradesh, India

Basharat Yousuf
Department of Food Science and Technology, University of Kashmir, Hazratbal, Srinagar, Jammu and Kashmir, India

Zainul A. Zakaria
School of Chemical and Energy Engineering, Faculty of Engineering, Universiti Teknologi Malaysia, UTM Johor Bahru, Johor, Malaysia

Abbreviations

ACSO	s-allyl-l-cysteine sulfoxide
AFE	accelerated fluid extraction
AM	amylose
AP	amylopectin
APCI	atmosphere pressure chemical ionization
ASE	accelerated solvent extraction
AX	arabinoxylan
BOD	biochemical oxygen demand
CE/CEP	capillary electrophoresis
CO	coconut oil
COD	chemical oxygen demand
DAD	diode array detector
DADS	diallyl disulfide
DATS	diallyl trisulfide
DFs	dietary fibers
DP	degree of polymerization
DPPH	2,2-diphenyl-1-picrylhydrazyl
DSC	differential scanning calorimetry
EAAE	enzyme-assisted aqueous extraction
EACP	enzyme-assisted cold pressing
EAE	enzyme-assisted extraction
EI-MS	electron impact-mass spectrometer
ELISA	enzyme-linked immunosorbent assays
ELSD/ESD	evaporative light scattering detector
EPA	Environmental Protection Agency
ESE	enhanced solvent extraction
FD	fluorescence detector
FTIR	Fourier transform infrared
FTMS	Fourier transform mass spectrometry
GABA	gamma-aminobutyric acid
GAE	gallic acid equivalent

GC	gas chromatography
GC-MS	gas chromatography-mass spectroscopy
GHz	gigahertz
GLC	gas-liquid chromatography
HAT	hypoxanthine aminopterin thymidine
HDL	high-density lipoprotein
HIV	human immunodeficiency virus
HLWE	hot liquid water extraction
HPLC	high-performance liquid chromatography
HPLC-ESI-MS	high-performance liquid chromatography-mass spectrometry
HPP	high-pressure processing
HPPLC	high-pressure performance liquid chromatography
HPTLC	high-performance thin liquid chromatography
HR/MAS	high-resolution magic angle spinning
HSL	hormone-sensitive lipase
HSPE	high-pressure solvent extraction
HTLC	high-temperature liquid chromatography
HTWE	high-temperature water extraction
HWE	hot water extraction
IMS	ion mobility spectrometry
KBr	potassium bromide
kHz	kilohertz
LC	liquid chromatography
MAE	microwave-assisted extraction
MCFAs	medium-chain fatty acids
MCSO	methyl-L-cysteine sulfoxide
MCTs	medium-chain triglycerides
MMTC	methanethiosulfonic acid S-methyl ester
MS	mass spectrometry
NIR	near-infrared spectroscopy
NMR	nuclear magnetic resonance
ORAC	oxygen radical absorbance capacity
OT	oolong tea
OTPP	OT-polymerized polyphenols
PDA	photodiode array detector
PDAC	pancreatic ductal adenocarcinoma prevention
PEF	pulse electric field

PEFAE	pulsed electric field assisted extraction
PFE	pressurized fluid extraction
PHWE	pressurized hot water extraction
PLE	pressurized liquid extraction
PO	palm oil
PSE	pressurized solvent extraction
RT	retention time
RW®	refractance window
SCFE/SFE	supercritical fluid extraction
SCWE/SWE	subcritical water extraction
SDS	sodium dodecyl sulfate
SF	supercritical fluids
SFC	supercritical fluid chromatography
SHWE	superheated water extraction
SVOCs	semi-volatile organic compounds
TA	thermal analysis
TBARS	thiobarbituric acid-reactive species
TFC	total flavonoid content
TGA	thermogravimetric analysis
TLC	thin-layer chromatography
TPC	total phenolic content
TPC	total polyphenols content
UA	ursolic acid
UAE	ultrasonic-assisted extraction
UHPLC	ultrahigh-performance liquid chromatography
UPLC	ultra-pressure liquid chromatography
USA	United States of America
UV	ultraviolet
UVD	UV detectors
VOCs	volatile organic compounds
WHO	World Health Organization

Preface

Bioactive compounds are extracted from natural sources, and they have beneficial effects on human health. Natural products from medicinal plants, either as pure compounds or as standardized extracts, provide unlimited opportunities for new product development and drug development due to the availability of natural biochemical compounds of unmatched diversity. Due to an increasing demand for functional biochemical diversity in screening programs and seeking therapeutic drugs from natural products, interest, particularly in edible plants, has grown throughout the world. Botanicals, herbal preparations, and health food development for medicinal usage contain various types of bioactive compounds. There is an enhanced concern for the quality and safety of food products and increased preference for natural products. Since extraction is the crucial first step in the analysis of bioactive compounds, there arises the necessity to extract the desired chemical components from the naturally available materials for further separation and characterization. The focus of this book is to elaborate in detail the insight of the extraction technology and analytical techniques of bioactive compounds from various plant food sources.

The book, which contains nine chapters, is divided into two sections. The first section comprises the technological and analytical aspects of bioactives from fruits and vegetable sources, and the second section is focused on bioactives from food grain sources. The inclusion of techniques and key adoptions in the extraction and characterization of active ingredients from food preparations in each specific chapter enhances the reader's knowledge and helps in understanding the perilous aspects of extraction, which are considered as the desirable precondition in the analysis of medicinal plants. Technological aspects of the extraction of bioactive compounds are considered a promising aspect both for researchers and industry, giving first-hand information about the dispensation and quality control attributes required during the handling of numerous sensitive, expensive, and medicinal compounds available in low quantity in natural resources. On the other hand, analytical techniques are the sole physicochemical aspects of profiling different functional ingredients of foods and nutraceutical products and identifying the most popular and specific ones. Various new extraction and analytical techniques have been developed, but no sole method is regarded as a standard for extracting bioactive compounds from plant sources. From established processes to emerging techniques, many methods of extraction are now available and used commercially and have the advantage of being effective in target extraction. Various analytical techniques used for qualitative and quantity estimation in the various natural food sources, such as GCMS, NMR, LCMS, chromatographic analysis, spectroscopy, along with other available analytical and techniques, are discussed. The researchers can get full-scale information about the extraction and analysis of bioactive compounds from natural sources. The most important aspect of bioactive compounds is their extraction and analysis, which has been covered and handled in depth with recent trends in extraction and analytical techniques.

—*Editors*

PART I

Technology of Bioactive Compounds and Nutraceuticals from Fruits and Vegetables Sources

CHAPTER 1

Technological and Analytical Aspects of Bioactive Compounds and Nutraceuticals from Plant (Vegetable) Sources

PREM PRAKASH,[1] RAVNEET KAUR,[1] SHUBHRA SHEKHAR,[2] and KAMLESH PRASAD[1]

[1]*Department of Food Engineering and Technology, SLIET, Longowal, Punjab, India*

[2]*Department of Food Process Engineering, National Institute of Technology, Rourkela, Odisha, India*

ABSTRACT

Recently, there has been an increase in demand for a healthy and balanced diet. With the increase in health problems due to a sedentary lifestyle, people are focused more on foods that have therapeutic benefits. Vegetables are naturally rich source of various bioactive compounds and nutraceuticals. Different groups of phytochemicals are present in different vegetables, having high antioxidant properties and various health benefits. Therefore, this is an area of focus for researchers and industrialists. Efforts are being made for the extraction of bioactive compounds from vegetables and their by-products obtained during processing and their utilization in the development of functional foods. Various conventional and novel techniques are used for the extraction of bioactive compounds. These include solvent extraction, hydrodistillation, ultrasound-assisted extraction, microwave-assisted extraction (MAE), enzyme-assisted extraction, supercritical fluid extraction (SCFE), etc. Further, separation, and quantification of specific compounds are carried out using several analytical techniques like LC-MS,

Bioactive Compounds and Nutraceuticals from Plant Sources: Extraction Technology, Analytical Techniques, and Potential Health Prospects. Charanjit Singh Riar, Parmjit S. Panesar (Eds.)

HPLC, GC-MS, and UHPLC. This chapter focuses mainly on the bioactive compounds present in different vegetables, their extraction techniques, and analytical methods for the quantification of bioactive compounds.

1.1 INTRODUCTION

Vegetables may be defined as a plant or portion of the plant that is consumed as raw or processed either as a salad or in a staple food. From the production point of view, vegetables may be defined as a cash crop that needs intensive special care in the preharvest and postharvest periods [45]. Some vegetables are harvested in an immature stage, so special care must be taken to maintain their quality attributes from field to consumer point. Depending on the climatic conditions and geographical location, vegetables may be classified as annual, biennial, and perennial [46]. Vegetables are a rich source of micronutrients, nutraceuticals, and bioactive compounds. These compounds generally appear under nutrient (dietary fiber, vitamins, and minerals) and non-nutrient (phenolic compounds, flavonoids, bioactive peptides, etc.) categories (Table 1.1). Vegetable nutrients, which are present at the micro-level, helps in various health cure potassium helps to maintain blood pressure, folic acid decrease the risk of birth defects; vitamin A helps with eyes and skin problem, and vitamin C helps in teeth and gums problem as well as helps in iron absorption [47]. Apart from these, it also helps in cardiovascular diseases, obesity, diabetes, and cancer by preventing free-radical damage [48, 49]. Fiber present in the vegetables provides roughages for digestion, and it decreases the blood cholesterol level and reduces the risk of heart problems. Vegetables are rich in phytochemicals, and the quantity and composition of these phytochemicals vary in different groups of vegetables. For example, carrots are rich in carotenoids and contain a high amount of three flavonoids namely, quercetin, kaempferol, and luteolin [50]. Vegetables of the Cucurbitaceae family are rich sources of tocopherols, vitamin C and carotenoids [51]. Lycopene, phytoene, neurosporene, and carotene are the major phytochemicals present in tomatoes [52].

1.1.1 CLASSIFICATION OF VEGETABLES

The vegetables are harvested at the horticultural maturity stage (green pea, cabbage, spinach, tomato, okra, etc.) and physiological maturity stage (potato, onion, garlic, fruit yam, etc.). These factors are important for consumer and postharvest handling points. Some vegetables need to be pre-cooled and then stored in refrigerated condition for longer storage, e.g., tomato, green pea, carrot, radish, bean, etc. Whereas leafy vegetables can't be stored for a longer time due to being highly perishable. On the other hand, some vegetables can be stored for a longer period in normal conditions. Tubers and bulbs are an example of this group. Vegetables can be classified into the following different groups depending on the part of the plant from which they are derived:

- **Leaves:** Spinach, cabbage, lettuce, mustard leaf, fenugreek.
- **Stem:** Asparagus, celery, Knol-Khol.
- **Fruits:** Bitter guard, pointed guard, tomato, cucumber.
- **Pods:** Beans, okra, snap pea, snow pea.
- **Flowers:** Cauliflower, broccoli, squash blossom.
- **Bulbs:** Onion, garlic, leek.
- **Roots and Tubers:** Potato, sweet potato, cassava, taro radish, turnip, carrot, beetroot, yam.
- **Seeds:** Bean, pea.
- **Fungi:** Mushroom.

1.1.2 NUTRACEUTICALS IMPORTANCE OF VEGETABLES

The bioactive compounds are present in different kinds of plants in small quantities, which are secondary metabolites [33]. Generally, bioactive compounds appear under the phytochemical groups that are capable to modulate the metabolic processes and have health benefits [34]. Bioactive compounds are important for human health due to their numerous biological benefits, such as reduction in cardiovascular risk, antioxidant properties, antimutagenic properties, antiallergenic properties, anticarcinogenic properties, anti-inflammatory properties, and antimicrobial activities [35–38]. These compounds are extracted by various processes, but the selection of appropriate methods is based on their application and cost. Solid-liquid extraction (using organic solvents), supercritical fluid extraction (SCFE), high-pressure processing (HPP), microwave-assisted extraction (MAE), subcritical water extraction (SCWE), and ultrasound-assisted extraction are some methods used to extract the bioactive compounds from natural sources [39–41]. Nutraceutical foods is the term applied to describe any food materials which are prepared by incorporating bioactive components from any other food matrix with supplementary health benefits that adds to the basic nutritional composition of food [22]. It is derived from the word "nutrition" and "pharmaceutical," which was invented in the year 1989 by Stephen De Felice. He defined the term nutraceuticals as, "a food (or part of a food) that provides medicinal or health benefits, including the prevention and/or treatment of a disease" [23]. The main idea behind the nutraceuticals is to focus on prevention rather than illness according to the statement mentioned by Hippocrates, as "let food be your medicine." In the science forum, nutraceutical can be defined as, "a diet supplement that delivers a concentrated form of a presumed bioactive agent from a food, presented in a non-food matrix, and used to enhance health" [24]. Green tea, glucosamine, ginseng, echinacea, omega-3, folic acid, allicin, and spirulina are some of the important natural nutraceuticals which provide multiple therapeutic properties [25, 26]. Sometimes, nutraceuticals are also called functional foods, nutritional supplements, dietary supplements, and pharmaceutical or medicinal foods. Nowadays, humans take an interest in natural nutraceuticals due to their nutritional properties, therapeutic effects, and reduced risk from various diseases. It helps in lifestyle disorders such as obesity, mental stress, heart problem, and diabetes without any side effects. Nutraceuticals are also useful in clinical disorders like immune deficiency, indigestion, depression, inflammation, allergy, arthritis, hypertension, blood cholesterol control, and malignancies. In India, the word nutraceuticals are generally applied in foods made from herbal or botanical raw materials as well as products fortified with bioactive components extracted from fruit and vegetables which may prevent acute and chronic problem.

In 2019, the global market of nutraceuticals products was approximately 400 billion USD, and in the future, its demand is expected to cross 700 billion USD by 2027. The main reason for the rapid demand for nutraceutical products is due to innovative research and healthy products booming in the market [27]. The other factors associated with nutraceuticals are safety and efficacy, whereas pharmaceutical drugs are correlated with adverse effects [28].

1.1.3 BIOACTIVE COMPOUNDS IN VEGETABLES

1.1.3.1 LEAFY VEGETABLES

Leafy vegetables are highly perishable and include spinach, cabbage, mustard leaf, and fenugreek, etc. Leafy vegetables are mainly recognized by their nutrients and phytochemical compounds, which contribute to health-promoting activities. Spinach (*Spinacia oleracea* L.) is considered as functional food since it contains a diversified group of nutritional compounds including vitamins, minerals, and phytochemicals that support health beyond basic nutrition. It contains little amounts of carbohydrate (3.6%), protein (2.9%), and a negligible amount of fat (0.9%) [53]. The major lipids present in the spinach are oleic acid, α-linolenic acid, linoleic acid, capric acid, myristic acid, and stearic acid. It contains significant amounts of minerals (magnesium, potassium, and iron) and vitamins (vitamin A, folate, vitamin C, and vitamin K) (USDA, database). The phytochemical-derived compounds in the spinach are capable of (i) preventing oxidative damage of macromolecule; (ii) involved in metabolism, antioxidant defense, and inflammation; and (iii) restraining food ingestion by inducing the secretion of satiety hormones. The major phytochemicals present in the spinach are phenolic compounds, lutein, and carotenoids. These bioactive compounds are involved in biological activities, namely anti-obesity, anticancer, hypolipidemic, and hypoglycemic properties [54].

Purslane (*Portulaca oleracea*) is a green leafy vegetable crop. It is also famous for the names pigweed, little hogweed, and fat weed. It is generally eaten raw or in cooked form, having a slightly sour or salty taste like spinach. Purslane contains a small amount of carbohydrate (2.32%), protein (1.81%), fat (0.32%), and fiber (0.98%) [55]. Apart from basic nutrition, it is a rich source of bioactive compounds and fatty acids, namely omega-3 fatty acids, carotenes, tocopherols, quercetin, and phenolic alkaloids, which are mainly present in stems and leaves [56]. These bioactive compounds act as an antioxidant by reducing free oxygen radicals and give neuroprotective effects against neurotoxic agents [57]. It also contains minerals like K (6,920 mg/100 g), Ca (900 mg/100 g), P (360 mg/100 g), Mg (300 mg/g), Fe (14 mg/100 g), Mn (5.15 mg/100 gm), and Zn (5.79 mg/100 g) [55]. These minerals have diversified functionalities in body metabolism.

1.1.3.2 STEM VEGETABLES

Asparagus (*Asparagus officinalis* L.) is one of the most important vegetable crops because of its low-calorie value. Nowadays, it is popular all over the world, as it is a good source of bioactive phytochemicals, such as ascorbic acid, rutin, quercetin, chlorophylls, carotenoids, and other phenolic compounds. Among all the commercially consumed vegetables, asparagus is rich in antioxidant activity [58, 59]. The major group of phenolic compounds present in asparagus is flavonoids [60, 61]. Asparagus contains polyphenolic compounds like quercetin glucosyl rutinoside, quercetin-3-O-rutinoside (rutin), kaempferol-3-O-rutinoside (nicotiflorin), isorhamnetin-O-rutinoside (narcissin), dicoumaroyl glycerol, coumaroylferuloyl glycerol, diferuloyl glycerol, which was identified by the UHPLC-PDA-MS analytical instrument [62]. Asparagus is a good source of essential nutrients like protein (4.34%), fat (2.55%), and carbohydrate (5.28%).

Celery (*Apium graveolens* L) is one of the annual or perennial crops that are cultivated in the major area of Europe, Africa, and Asia. The importance of celery is increasing day by day because it provides powerful phytochemical compounds such as caffeic acid, *p*-coumaric acid, ferulic acid, apigenin, luteolin, apigenin, limonene, selinene, furocoumarin glycosides, thiamine, ascorbic acid tannin, saponin, kaempferol, phenolics, flavonoids (apigenin), alkaloids, and steroids [63]. It can prevent cardiovascular diseases and also has an anticoagulation activity of blood plasma [64]. Celery increases fertility as well as improves the hormonal imbalance, oligospermia, and immotile sperm [63]. It has antimicrobial, antioxidant, antifungal, anti-inflammatory, and antibacterial properties [65–67]. Its seeds are useful as a flavoring agent, either in the form of whole seed, or powder due to the presence of flavor substances such as β-pinene, limonene, β-selinene, and β-caryophyllene. Oil extracted from celery seeds is highly useful in the pharmaceutical industries [64]. Seeds are also helpful in bronchitis, asthenopia, asthma, chronic skin disorders, vomiting, fever, and tumors [63]. Phytochemical compounds, particularly polyphenols such as flavonoids, phenylpropanoids, phenolic acids, and tannins, are responsible for free radicals scavenging and antioxidant activities of the plant [68]. Polyphenols have several biological activities. Generally, polyphenols exhibit comparable chemical properties, which means one or more phenolic groups react with hydrogen ions and neutralize free radicals. Its root and leaves have the property to eliminate the hydroxyl ion and 2,2-diphenyl-1-picrylhydrazyl (DPPH) radicals and thus reduces the liposomal peroxidation and hence protects the plants [69].

Knol-Khol (*Brassica caulorapa*) is also known as ganth ghobi, nookal, or kholrabi. It is a rich source of antioxidants like ascorbic acid, tocopherol, and carotene and a good source of dietary fiber. Knol-Khol contains phytochemicals such as alkaloids, glycosides, steroids, flavonoids, saponin, tannin, terpenoids, phytosterols, isothiocyanates, sulforaphane, and indole-3-carbinol [70]. Antioxidants present in Knol-Khol help to eliminate the harmful free radicals from the body and thus facilitate to fight against the cancer cell, especially in curing colon cancers. Bioactive compounds like sulforaphane and isothiocyanates help in the building of protective enzymes in the body, which make the body's immune system stronger [71]. Along with other phytochemicals, sulforaphane helps to protect some specific enzymes, which helps to destroy cancer-causing agents in the body and also helps to eliminate the carcinogens out from the body before they can injure the cells. Knol-Khol also has antidiabetic activity and preventive action against hyperglycemia and oxidative stress [72].

1.1.3.3 GOURD

Bitter gourd (*Momordica charantia*) is the fruit part of the plant that contains bioactive compounds and is cultivated in tropical and subtropical regions in Asia and various parts of the world. The main bioactive compounds present are alkaloids, steroidal saponins, polypeptides, tannins, terpenoids, and aromatic volatile compounds. Among the bioactive compounds, p-insulin is parallel to insulin, and steroidal saponins (charantin) and analogous to peptides and certain alkaloids, which are beneficial in lowering down the blood glucose levels in diabetic patients. In bitter gourd, two types of saponins, namely oleanane and cucurbitane-type triterpenoids, are present [73]. But, in the case of the whole plant of bitter gourd other types of cucurbitane-type triterpenoid glycoside are also recognized, such compounds are momordicoside U, kuguaglycoside G, 3-hydroxycucurbita-5,24-dien-19-al-7,23-di-O-β-glucopyranoside, momordicine I, 3β,7 β,25-trihydroxycucurbita-5,23 (E)-dien-19-al and momordicine II [74–77]. Bitter gourd contains momordicine and charantin, which have various health-promoting effects and provide bitterness to the fruit [78, 79]. It is a good source of various types of phenolic compounds such as epicatechin, chlorogenic acid, catechin, and gentisic acid, and gallic acid [80–82]. The bitter gourd oil contains a large amount of polyunsaturated fatty acids (59.96%), which are conjugated with linolenic acid [83, 84]. α-oleostearic acid present in the oil, is useful for inhibiting the tumor cell, inhibiting the CVD, and also has anti-inflammatory action [84]. Bitter guard is useful in regulating the blood cholesterol level, thus protecting the human body from cardiovascular disorders (atherosclerosis). It also can protect from plenty of human disorders related to change in lifestyle [85, 86]. Apart from that, it contains minor amounts of vitamins, minerals, and fiber. Overall, it is effective against numerous disorders namely ulcer, diabetes, insurgence cancer, abdominal pain, kidney (stone), fever, scabies, and inflammation [77, 87–90].

The quality of proteins present in the bitter guard seeds meets the amino acids requirements laid down by the regulatory agency (FAO/WHO/UNU) for preschool children. Albumin (49.3%), globulin (29.3%), and glutelin (3.1%) are the main protein fraction (molecular weight: 45 to 55 kDa) present in the bitter guard seeds [80, 91]. Bitter guard seeds contain approximately 35 to 40% of oil and their fatty acid profile show, it contains monounsaturated fatty acid (3.33%) and saturated fatty acids (36.71%). Through GC-MS, approximately 25 different types of compounds are identified in the bitter gourd seed, some major compounds are germacrene, cis-dihydrocarveol, and trans-nerolidol apiole [27, 92]. The bitter melon seed contains mineral compounds like potassium, magnesium, calcium, sodium, and phosphorous in the highest amounts. It is the best natural source of chromium (5.65 mg/100 g) and zinc (45.45 mg/100 g) [84].

Pointed guard (*Trichosanthes dioica)* is a well-known vegetable all over the world. In India, it is known by different names such as Parval (Hindi), Potol (Bengali), and Putulika (Sanskrit). Based on the size, shape, color, and striation patterns, this fruit is categorized into four groups: (i) fruits that are 10–13 cm long having dark green color with white stripes; (ii) fruits that are 10–16 cm long having relatively thick with dark green color and very pale-green stripes; (iii) fruits that are 5–8 cm long and round having dark green color with white stripes; and (iv) fruits that are 5–8 cm long having green color, tapered, and striped at the ends [53, 79]. It is highly nutritious and contains different types of bioactive compounds namely peptides (tricho-santhin and lectin); triterpenes (cucurbitacin B, euphol, α-amyrin, β-amyrin, lupeol, taraxerol, betulin, and karounidiol); sterols, steroidal saponin, tannin, and flavonoids. Not only fruits but their different parts like seeds, leaves, and roots have the same importance since they also provide bioactive compounds (Figure 1.1). Due to the presence of bioactive compounds, these parts have promising anti-hyperglycemic, anti-hyperlipidemic, anti-HIV, cytotoxic, arsenic poisoning ameliorative, anti-inflammatory, anti-diarrheal, and varieties of pharmacological properties [93].

1.1.3.4 PODS

Okra is one of the most important vegetable crops, which contains multi-nutrients that help in human health. These nutrients are dietary fiber, vitamins, minerals (P, Ca, and Fe), protein, carbohydrates, unsaturated fatty acid (linolenic and oleic acid), and bioactive compounds [94]. The okra is 10–30 cm in length and 1–4 cm in diameter, having approximately five cavities containing ovules. The whole okra plant is known to be a multipurpose crop due to its versatile application. Its various parts like leaves, stems, buds, flowers, pods, and seeds provide different types of bioactive compounds (Figure 1.2) [95].

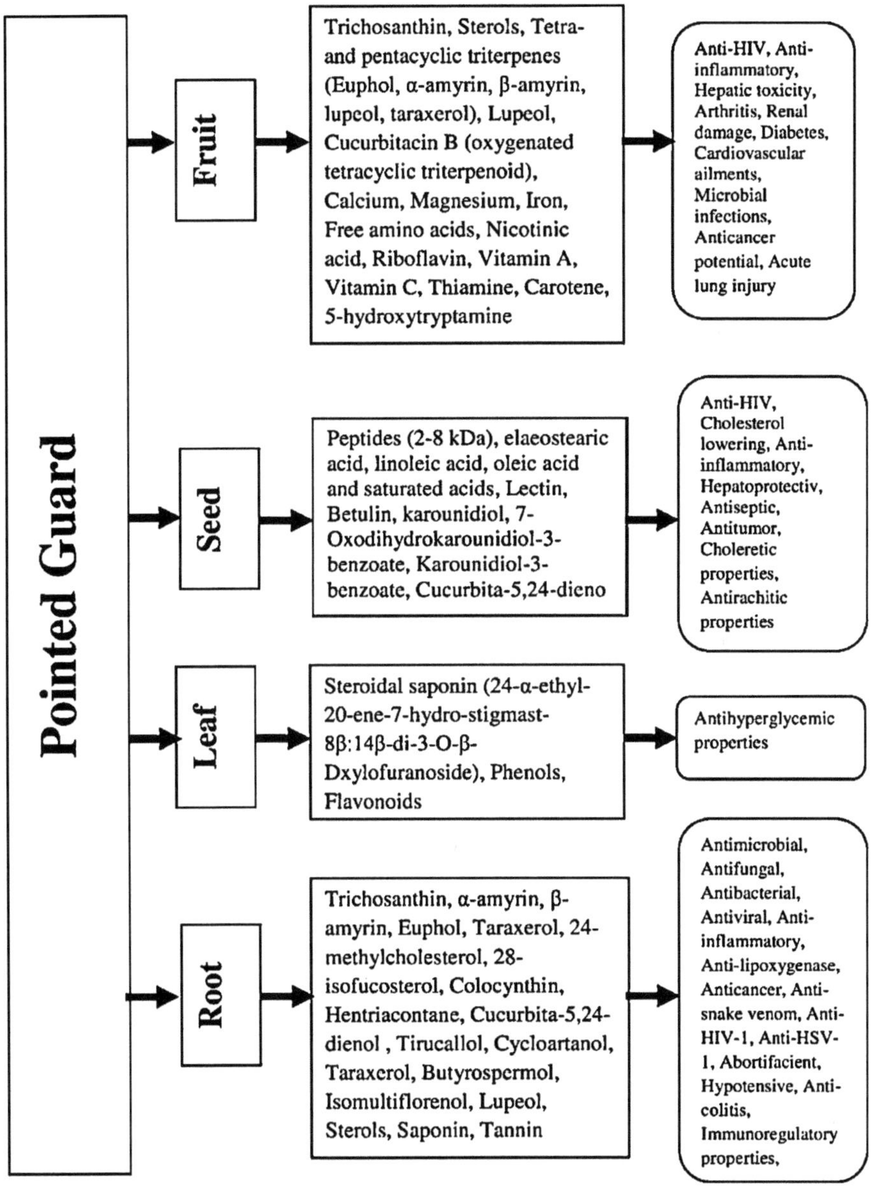

FIGURE 1.1 Bioactive compounds in different sections of the pointed guard.

Source: Adapted from Ref. [93]

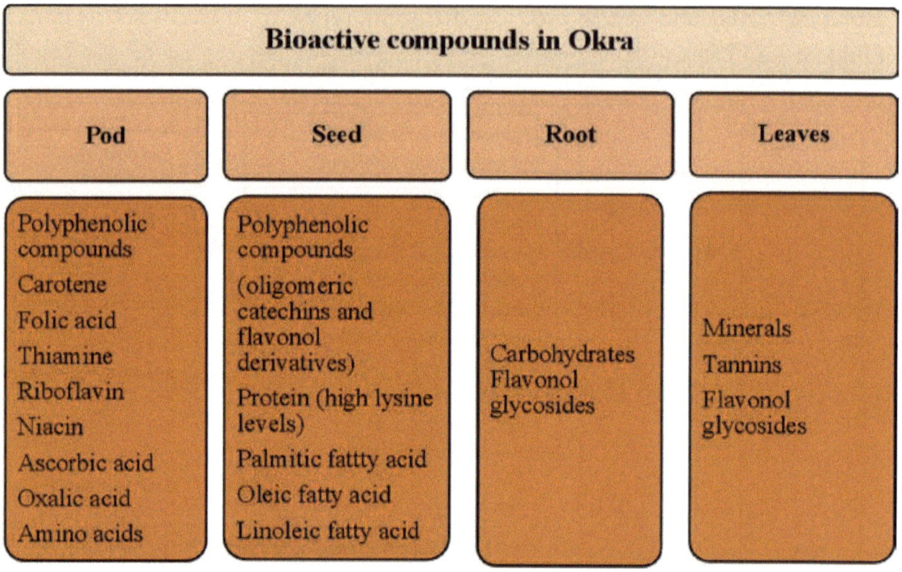

FIGURE 1.2 Bioactive compounds present in major parts of the okra plant.

Okra mucilage is mainly composed of carbohydrates, but in the case of young pods, it is constituted with polysaccharides having a molecular weight of approximately 1,70,000 along with amino acids. Furthermore, okra pods are principally constituted with 25% galactose, 27% galacturonic acid, and 22% rhamnose [96]. Dried okra pods are deficient in one of the major nutrients, namely carotene or retinol [97].

1.1.3.5 FLOWER VEGETABLES

Cauliflower (*Brassica oleracea* L. var. botrytis) is one of the popular vegetable crops and consumed all over the world. It is the richest source of phytochemical compounds, namely glucosinolates, vitamin C, phenolic compounds, and little amount of carotenoids [98–100]. Indole glucosinolates and their products formed after breakdown have anticarcinogenic properties [101]. The phytochemical composition of cauliflower varies with variation in genotype. The white cauliflower contains glucoiberin and sinigrin whereas green cauliflower contains glucoraphanin and progoitrin [102]. Cauliflower also contains considerable amounts of sulforaphane, indole-3-carbinol and 2-propenyl isothiocyanates, which are formed after the breakdown of glucoraphanin, glucobrassicin, and sinigrin, respectively [103]. Ascorbic acid and its product (dehydroascorbic acid) are involved in many biological activities in plants and animals. Moreover, the antioxidant activity of vitamin C is responsible for the slowdown of cellular aging [104]. Vitamin C helps in various types of health problems like prevention of cancer, cardiovascular, and chronic diseases, as well as reduces hypertension and diabetes [105]. The total phenol content present in the cauliflower is approximately 270 mg per 100 g of gallic acid of the dried edible portion. In cauliflower, carotenoids (lipid-soluble antioxidants) are generally present in relatively low amounts [106]. Purple cauliflowers contain anthocyanins that are present in a complex structure. The compounds like *p*-coumaryl and feruloyl are present in esterified forms of cyanidin-3-sophoroside-5-glucoside which is predominant in cauliflower [107].

1.1.3.6 BULBS

Onion (*Allium cepa*) is the main bulbous crop cultivated all over the world, especially in Asia. Apart from onion, allium species include garlic, leeks, shallot, etc. [108]. In India, different varieties of onions are cultivated, having their distinctive flavor, ranging from sweet to very strong pungency. The pungency of onion arises due to the interaction of S-alk(en)yl-L-cysteine sulfoxide derivatives and allinase enzyme to produce pyruvate and a series of volatile and non-volatile sulfur compounds, which also provide the specific characteristic flavor and odor [109–111]. The important volatile compounds present in the onions are 3,5-Diethyl-1,2,4-trithionale, Propyl 1-propenyl disulfide, n-propanethiol, n-propylthioaldehyde, hydrogen sulfide, and sulfur dioxide [111], whereas non-volatile compounds are Sitosterol, Gitogenin, Oleanolic acid, Amyrin, and Diosgenin. The flavor precursors in the onions, *trans*-(+)-S-(1-propenyl)-L-cysteine sulfoxide is present in the highest concentration, along with a small quantity of (+)-S-methyl-L-cysteine sulfoxide (MCSO) and (+)-S-propyl-L-cysteine sulfoxide. Lachrymatory effect or teary eyes after cutting of onion is due to the presence of thiopropanal S-oxide [112]. Onions are rich source several phytochemicals such as ascorbic acid, flavonoids, and fructooligosaccharides [113, 114]. Phenolic compounds are also present in the onion, which has antioxidant properties as well as beneficial effects against cardiovascular and neurological diseases [112]. The major flavonoids present in the onion are flavones, flavanones, flavonols, isoflavones, flavanonols, flavanols, chalcones, and anthocyanins, which are differentiated based on the degree of unsaturation and the degree of oxidation of the central ring [115]. Quercetin and kaempferol are present in higher concentrations (280–400 mg/kg) as compared to other fruit and vegetable like broccoli (100 mg/kg), and apple (50 mg/kg). Quercetin has health benefits, which inhibits cancer cell growth, as well as also functions as an antioxidant and chelating agent [116, 117]. Anthocyanins are mainly present in red onions (250 mg/kg), whereas flavonols are present in yellow onions [118]. Insulin, kestose, nystose, and fructofuranosylnystose present in the onions have prebiotic effect [119]. Apart from that onion is also good sources of carbohydrate (14.15–14.77%), protein (2.62–1.489%), fat (0.4%), minerals like Ca (46.9–25.7 mg/100 gm), P (50.6–30.3 mg/100 gm) and K (140–129 mg/100 gm) [108].

Onion seeds are also one of the important sources of major and minor nutrients. It contains approximately 15.8% oil, 12.3% protein, 2.4% ash as well as bioactive compounds like phenolic as well as volatile compounds [120]. Onion seeds contain fewer amounts of cysteine derivatives compounds, which make them a functional food. These compounds help to control the obesity problem because they improve the glycemic index [121–123]. Its oil mainly contains 6.4–7.1% palmitic acid (C16:0), 24.8–26% oleic acid (C18:1) and 65.2–64% linoleic acid (C18:2). The total phenolic content (TPC) ranges from 1.84 to 6.87 mg gallic acid equivalents (GAEs) per gram dry extract in the onion seeds. It also contains various volatile compounds, majorly are 2-phenylethanol, 1-hexanol, 1-pentanol, 2-heptanol, 2-octanol, 1-octen-3-ol, 1-heptanol, phenyl methanol, 1-dodecanol, 1-pentanol, 2-heptanol, 1-octen-3-ol, and phenyl methanol. It also contains aromatic hydrocarbons such as *p*-xylene and mesitylene [120].

Garlic is the second-largest bulb crop cultivated in the world, next to the onion. It is recognized by the characteristic flavor and pungent taste, due to complex biochemical reactions [124]. Raw garlic does not contain any strong odor, but when it is crushed or cut, a sulfur odor is immediately produced. The predominant garlic flavor precursor is S-allyl-L-cysteine sulfoxide (ACSO or alliin). When garlic cells are damaged, the allinase (located at vacuole cells) enzyme acts on alliin (located at cytoplasm level) and produce garlic flavoring compounds, such as diallyl thiosulfinate or allyl 2-propenethio-sulfinate (Allicin), diallyl disulfide (DADS) diallyl trisulfide (DATS), allyl sulfide and methanethiosulfonic acid S-methyl ester (MMTC) [125, 126]. Garlic contains organosulfur compounds which are water-soluble (S-allyl-L-cysteine) as well as oil-soluble (allicin, sulfides, ajoenes, and vinyldithiins) [127]. Apart from the organosulfur compounds, it is a good source of bioactive compounds like flavonoids, phenols, antioxidants, polysaccharides, and saponin. It also contains non-sulfur compounds (phosphorous, potassium, calcium, and magnesium), proteins, and fiber. However, variation of organosulfur [126] and phenolic compounds [128] were also observed with garlic genotypes. It helps in the prevention and treatment of cold and cough, cardiovascular diseases, hyperlipidemia, hypertension, and diabetes [65, 129]. The other prospective functions are related to the prevention of cardiovascular and cancer disease, immunity power improvement, detoxification, and hepatitis prevention [127]. In addition, it also has antimicrobial, antioxidant, antifungal, antiparasitic, and antiviral activity [127, 129, 130]. Green garlic contains proteins (20.9–35.7 g/kg), fat (1.6–2.8 g/kg), total carbohydrates (61.3–116.5 g/kg), fiber (33.7–57.0 g/kg), and ash (8.9–14.1 g/kg). It is rich source of antioxidants like ascorbic acid (75.4–459.7 mg/kg), polyphenols (335.3–1895.1 mg/kg), and antioxidant activity (27.0/30.1 mmol TEAC/kg). Garlic leaves contain minerals like calcium (7.55–28.9 g/kg), potassium (15.9–28.0 g/kg), magnesium (0.85–1.32 g/kg), sulfur (2.41–6.22 g/kg), iron (34.4–85.7 mg/kg), and zinc (9.32–13.8 mg/kg) [131].

1.1.3.7 ROOTS AND TUBERS AS VEGETABLES

Roots and tubers are crops that store starch in tubers, rhizomes, roots, corms, and stems. Potatoes and yams are tuber crops, canna, and arrowroots are edible rhizomes, cassava, and sweet potatoes are roots whereas taro and coco-yams are derived from corms, underground stems. Nutritionally (Table 1.1), roots, and tubers have the potential to provide economical sources of energy, in the form of carbohydrates [132]. In general, it contains a low amount of protein ranging from 1–2% on a dry weight basis. Roots and tuber crops like cassava, sweet potatoes, potatoes, and yam contain a small quantity of vitamin C whereas yellow varieties of sweet potatoes, yam, and cassava contain β-carotene.

Roots and tuber crops contain a significant amount of bioactive compounds, such as saponins, phenolic compounds, glycoalkaloids, phytic acids, carotenoids, and ascorbic acid. These compounds play a significant role in antioxidant activity, immunomodulatory, antimicrobial, antidiabetic, antiobesity, and hypocholesterolemic activities (Table 1.1) [132].

Potatoes play a major role in the diet of people from developed countries as compared to the developing country since energy intake from potatoes per person per day in developed and developing countries was 130 and 42 kcal/day, respectively in the diet [133]. It also contains varying amounts of

TABLE 1.1 Major and Minor Nutritional Compositions of Some Selected Roots and Tubers Crops

Nutrients	Potato	Yam	Cassava	Sweet Potato		Giant Taro
				White	Yellow	
Proximate Composition						
Moisture (%)	78.3	71.8	65.5	72.3	70.0	83.0
Protein (g)	2.0	2.0	1.0	1.0	1.2	0.6
Fat (g)	0.1	0.1	0.2	0.3	0.3	–
Total carbohydrate (g)	18.7	24.6	31.4	24.3	26.3	14.8
Fiber (g)	0.4	0.5	1.0	0.8	0.8	–
Ash (g)	0.9	1.0	0.9	0.7	0.7	30
Minerals Content						
Ca (mg)	9	22	26	21	36	50
P (mg)	55	39	32	50	56	–
Fe (mg)	0.7	1.0	0.9	0.9	0.9	1.0
K (mg)	451	294	394	210	304	–
Na (mg)	7	10	2	31	36	–
Vitamins Content						
Carotene (μg)	30	0	0	35	1,680	0
Thiamin (mg)	0.11	0.10	0.05	0.14	0.12	0.05
Riboflavin (mg)	0.04	0.04	0.04	0.05	0.05	–
Niacin (mg)	1.2	0.07	0.6	0.7	0.6	–
Ascorbic acid (mg)	14	–	34	21	30	5
Folic acid (μg)	–	–	24.2	52.0	–	–

Source: Adapted from Ref. [132]

dietary fiber, ascorbic acid, potassium, total carotenoids, and antioxidant phenols such as chlorogenic acid and its polymers, and lesser amounts of protein, other minerals and vitamins [133, 134]. Sweet potatoes are the seventh-largest consumed food crop. It contains a significant amount of carotenoids, thus improves vitamin A content in children, especially in developing countries [135]. The anthocyanins present in the sweet potatoes provide the color to the flesh. Cassava is the fourth important tuber crops and a staple for more than 500 million people in the world due to its high concentration of carbohydrates [136]. A large number of bioactive compounds are associated with it, such as cyanogenic glucosides (linamarin and lotaustralin), noncyanogenic glucosides, hydroxycoumarins (scopoletin), terpenoids, and flavonoids [136, 137]. Genetic modification and crop breeding programs improve the protein and vitamin content [138]. Yam is a member of the monocotyledonous family Dioscoreaceae. It is consumed in different forms, such as raw, cooked, and powder or flour for the preparation of different food products. It contains various types of bioactive components, such as mucin, dioscin, dioscorin, allantoin, choline, polyphenols, diosgenin, vitamins (carotenoids and tocopherols) and minerals (K, P, Ca, Mg, Na, and Fe) [132, 139].

Potato (*Solanum tuberosum*) is an important source of starch and crude fiber. The tuber crops contain diverse groups of phenolic compounds. The phenolic compounds and amino acids have an antioxidant potential against reactive oxygen species, tissue damage, and diseases like atherosclerosis, renal failure, diabetes mellitus, and cancer [140]. Chlorogenic acid is the major phenolic compound that contains up to 80% of the total phenolic in potato [141]. Ascorbic acid improves the immune system as well as plays important roles (enzyme cofactor) in many metabolic functions and improves the bioavailability of iron in humans [142]. Colored potatoes are a rich source of bioactive compounds such as polyphenols, phenolic acids, chlorogenic acid, p-coumaric acid, and caffeic acid. Moreover, it also contains a significantly higher amount of flavonoids, anthocyanins, carotenoids, lutein, and β-carotene [143].

1.1.3.8 BEANS AND PEAS

In India, green beans are consumed as a vegetable, which has been harvested before the senescence and dehiscence of the pods or before reaching the commercial maturity level. The beans are different in color, i.e., yellow, green, reddish, or purple, depending on the variety. These are considered as good source of bioactive compounds such as polyphenols, proteins, and peptides. Apart from that, there are 16 different types of phenolic compounds, which have been identified in the green beans pods, among them 2 are phenolic acids and 14 are flavonoids. The main phenolic acids are hydroxybenzoic and hydroxycinnamic acid, as well as flavonoids are flavonols (quercetin, kaempferol glycosides), flavan-3-ols (catechin and derivatives), and flavanonol (taxifolin glycoside) derivatives [144, 145]. Lignans and other polar compounds such as kutkoside are also present that have potential health benefits [145]. These bioactive compounds help to slow down the growth of cancerous tumors and provide antibacterial, antiviral, antispasmodic, and anti-inflammatory properties [146–148]. Raw green beans contain pigments like Lutein and β-carotence [149]. It has been noticed that the anticancer property of green beans is mainly due to the presence of chlorophyll, resistant starch, soluble, and insoluble dietary fiber, as well as phenolic compounds. Likewise, it has also been noticed that high fiber content can help to control diabetes problems by reducing insulin generation and glucose levels that enter the bloodstream; besides, patients suffering from non-insulin-dependent diabetes can reduce the need for insulin by up to 40% by eating green beans in their regular diet [150].

Green peas acquire one of the special positions among vegetables since it has been recognized as a cheaper source of macronutrients, particularly protein and fiber. The starch and fiber present in the peas make them low glycemic index food and therefore, it is beneficial in the prevention of type-II diabetes mellitus [151]. In addition, peas contain significant amounts of oligosaccharides (raffinose, galactose) that may provide prebiotic effects [152]. It also contains vitamins and minerals, namely ascorbic acid, β-carotene, thiamine, and riboflavin, as well as a rich source of iron, as compared with other vegetables [153]. Peas contain several types of phenolic compounds, which are mostly present in the seed coat, namely, quercetin glycosides, luteolin, apigenin, and proanthocyanadins. Seven phenolic acids and eight flavonoids are identified in the peas by applying LC-ESI-MS analysis. The major compounds identified through this technique are quinic acid, protocatechuic acid, chlorogenic acid, 4-*o*-Caffeoylquinic acid, quercetin, naringenin luteolin, cirsiliol [154]. Pea peel is also one of the important sources of bioactive compounds, especially phenols and flavonoids, having antioxidant and α-amylase inhibitory activity. Ethyl acetate extracted compounds such as phenolic and flavonoids show antibacterial and antifungal properties [155].

1.1.3.9 MUSHROOM

Mushroom comes under the fungi group, which is consumed as a vegetable due to its distinctive nutritional profile. It contains a large amount of protein and carbohydrates. Mushrooms contain both types of carbohydrates, i.e., digestive and non-digestive carbohydrates. The major digestive carbohydrates are trehalose, mannitol, glycogen, and glucose, whereas non-digestive carbohydrates are mannans, chitin, and β-glucan. However, oyster mushroom powder contains approximately 22.4% protein, 56.99% dietary fiber (48.79% insoluble dietary fiber, and 8.20% soluble dietary fiber), 3.32% β-glucan, and 7.79% ash but few amounts of 0.19% sucrose and 2.30% fat [35]. It also contains vitamins (vitamin B and vitamin D), minerals (phosphorus, magnesium, selenium, copper, and potassium), and crude fiber. Mushroom contains bioactive compounds like phenolic compounds, flavonoid (catechins, myricetins, quercetin, kaempferol, naringenin, and resveratrol), ergothioneine, β-carotene, Lycopene, and anthocyanidins [156, 157]. The amount of bioactive compounds present in the mushrooms depends upon the varieties of mushroom, the substrate used for cultivation, cultivation, or fruiting environment (if cultivated), parts of mushroom (stalk or cap), harvesting stage of the mushroom, storage conditions, and processing conditions [158]. It can be used for the preparation of functional as well as nutraceutical foods because it reduces the risk of diseases like Parkinson's, Alzheimer's, hypertension, stroke, and cancer, as well as act as an antibacterial, anti-inflammatory, immunomodulatory, antioxidant, hypoglycemic, and antidiabetic [130, 159–161].

1.2 EXTRACTION OF BIOACTIVE COMPOUNDS FROM VEGETABLES

Extraction is the critical step to identify and characterize the desired natural bioactive compounds present in the vegetables (Table 1.2). All techniques used for extraction of bioactive compounds are mainly based on five key points: (i) to extract defined bioactive compounds from the sample; (ii) to enhance the specificity of analytical methods; (iii) to concentrate the defined compounds; (iv) to change the bioactive compounds into a more specific form for identification and separation; and (v) to supply an effective, efficient, and more reproducible method [162]. Sample preparation before extraction is one of the major factors affecting the yield and retention of bioactive compounds. It includes various steps like drying, homogenization, sieving, etc. [163]. It has been reported in previous studies that freeze-dried fruits and vegetable samples retained more phenolic compounds as compared to heat-drying methods [164]. Various conventional and novel techniques have been used for the extraction of bioactive compounds. Commonly used conventional techniques are Soxhlet extraction, hydro-distillation, and maceration. Novel techniques include pressurized liquid extraction (PLE), SCFE, MAE, ultrasound-assisted extraction (UAE), and enzyme assisted extraction (EAE).

1.2.1 SOLVENT EXTRACTION

Among all the extraction methods, solvent extraction is the most popular and widely used. Nowadays, there are different types of solvents used to extract the bioactive compound from vegetable sources. For extraction of hydrophilic compounds, polar solvents (methanol, ethanol, or ethyl acetate) are used, whereas for lipophilic compounds, dichloromethane or a combination of dichloromethane-methanol in the proportion of 1:1 is generally used. In some cases, hexane can also be used for the extraction of chlorophyll compounds. The key factors considered during the selection of solvents are solubility, toxicity, flammability, cost, and safety. Solvents having the polarity value similar to the polarity of solute, yield better extraction results as compared to other solvents, this is governed by the law of intermiscibility and similarity that means 'like dissolves like' [165]. Several factors affect the yield and composition of the extract. These include temperature, solid to solvent ratio, time of extraction, and particle size of the material. Fine particle size gives a better extraction result due to effective absorption of the solvent by the solutes through diffusion. Higher temperatures are also one of the effective tools for better extraction of bioactive compounds because it increases the solubility of the solute. But increasing the temperature beyond a limit may lead to problems such as (i) faster solvent loss; (ii) extraction of undesirable impurities; and (iii) decomposition of heat-sensitive compounds. The extraction of bioactive compounds increases with an increase in extraction time in the defined period but is not affected after the equilibrium is attained [166]. Attempts have been made to improve the efficiency of the solvents extraction technique by combining it with various other methods like Soxhlet, ultrasound, microwave, etc. [167]. Soxhlet extraction is the most used technique. It is a continuous extraction process based on the principle of reflux and siphoning. The process is carried out repeatedly to achieve complete extraction. However, this process is time-consuming and requires an extensive amount of solvent [168].

1.2.2 HYDRODISTILLATION

Hydrodistillation is also known as steam distillation, where water or steam is used as a solvent. This method is generally used for the extraction of bioactive compounds, particularly those that are volatile or polar. Hydrodistillation can be classified into four types namely – water distillation, water and steam distillation, direct steam distillation and, distillation with cohobation [169, 170]. In this case, a prepared sample and an adequate amount of demineralized water or steam are placed in the closed chamber. After that, the closed chamber is heated for 6–8 hours with the sample. This method is mainly based on three major processes – (i) hydrodiffusion, (ii) hydrolysis, and (iii) decomposition. Higher temperatures can degrade the heat liable compounds during the extraction, which limits its application [171]. It is a complete process since the extracted volatile organic and non-volatile organic compounds (VOCs) can be physically separated by simple steps. The main steps involved in the extraction of VOCs are azeotropic distillation, condensation, collection, and separation from Florentine flask, whereas non- VOCs are extracted by boiling the water. However, this process is time-consuming and requires more energy [172].

1.2.3 PRESSURIZED LIQUID EXTRACTION (PLE)

Pressurized liquid extraction (PLE) is also known as pressurized fluid extraction (PFE), pressurized solvent extraction (PSE), accelerated fluid extraction (AFE), accelerated solvent extraction (ASE), enhanced solvent extraction (ESE), or high-pressure solvent extraction (HSPE), since this extraction technique works at high pressure to maintain solvent as liquid and facilitate effective extraction [173, 174]. High pressure results in reduction in surface tension of the solvent that facilitates the solvent penetration into the pores of matrix. This improves the extraction efficiency [175]. When the water is used as a solvent, then the extraction method is named subcritical water extraction (SWE), hot water extraction (HWE), pressurized hot water extraction (PHWE), high-temperature water extraction (HTWE) superheated water extraction (SHWE), or hot liquid water extraction (HLWE) [176]. In this extraction method generally, five steps are followed: (i) Heating (approximately 25 min) the extraction cell up to the process temperature; (ii) the solvent is driven at a fixed flow rate into the extraction cell up to the predefined pressure (approximately 7.5 MPa); (iii) the outlet valve and micro-meter valve are kept open, and extraction starts; (iv) extraction time of at least 30 min; and (v) extracted material is stored (–18°C) in an amber color vessel, for further analyzes [177]. The solvent and time requirements are less as compared with other extraction techniques since this technique is based on an automation system [178]. The extraction yield of targeted compounds is improved by the selection of accurate solvent, pressure, and temperature. Higher temperature improves the solubility of the compounds as well as mass transfer rate by decreasing the solvent viscosity and surface tension [179].

1.2.4 SUPERCRITICAL FLUID EXTRACTION (SCFE)

Supercritical fluid extraction (SCFE) is based on the technique to convert the gaseous phase into the supercritical fluid stage, where gas and liquid are interchangeable with a slight change in temperature and pressure (Figure 1.3). This technique is performed in two stages – (i) solubilization of chemical compounds and (ii) separation of supercritical solvent. This technique is generally used to isolate nonpolar bioactive compounds like carotenoids and lipids. But there is an option for extraction of polar compounds, namely flavonoids, for these purposes modifiers like ethanol, methanol, water, and acetone are generally used [180]. The extract's solubility also depends on the solvent density. The major supercritical fluids (SFs) used are CO_2, propane, ethane, methanol, nitrous oxide, n-butene, n-pentene, sulfur hexafluoride, and water [181]. SCFE is a rapid extraction process; further does not require cleaning and does not require large amounts of solvents for extraction [182]. One of the most advantages of this technique is the possibility of attachment of chromatographic techniques like gas chromatography (GC) and SF chromatography [181]. SCFE is based on the principle that with slight variations in the pressure, a solvent changes its properties near the critical point [183]. Supercritical carbon dioxide is the most used solvent for extraction using this method. The stage at which carbon dioxide exists in liquid as well as gaseous state, and both the forms are in equilibrium with each other is known as the supercritical state [184]. The major advantage of SCFE is that it is suitable for the extraction of thermolabile compounds as it takes place at low temperature, also there is no solvent residue left after extraction, and is thus considered environment friendly [185].

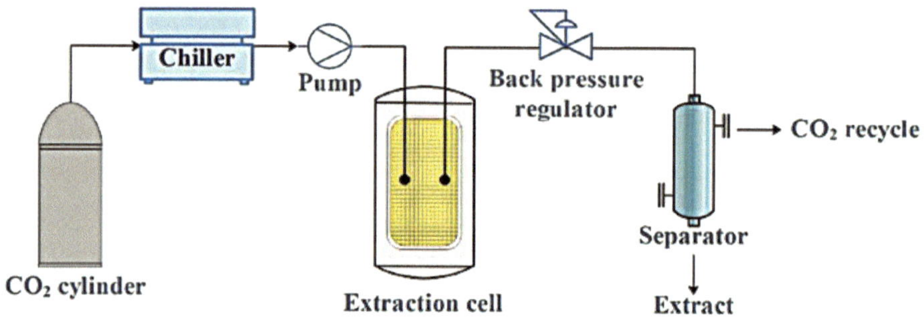

FIGURE 1.3 Diagrammatic representation of supercritical fluid extraction.

1.2.5 MICROWAVE-ASSISTED EXTRACTION (MAE)

Microwave assessed extraction is also known as a green technology since the amount of organic solvent is required for the extraction of bioactive compounds is less as compared to other conventional techniques. Electromagnetic waves having frequency ranges from 300 MHz to 300 GHz comes under the microwaves. In this method, particles are heated in two ways, i.e., ionic conduction (electric field) and dipole rotation (magnetic field) [186]. Due to the heat generated by the microwave-assessed extraction, the solvent penetrates the solid matrix by diffusion, and the solute dissolves until it reaches the concentration limit by the characteristics of a solid [187, 188]. This technique involves three consecutive steps: (i) separation of targeted compounds from the sample matrix by the application of developed heat and pressure; (ii) enhanced diffusion process of the solvent; and (iii) discharge of the targeted compounds from the sample matrix to the solvent [189]. During microwave treatment, heat generated results in the evaporation of residual moisture from the solid matrix. The release of moisture leads to the generation of high vapor pressure, thus resulting in the release of bioactive components due to the rupturing of plant cells [185]. The advantages of this process include its eco-friendly process, requiring less energy due to fast heating, and better recovery in respect to conventional extraction processes [187].

1.2.6 ULTRASOUND-ASSISTED EXTRACTION

Sound waves ranging from 20 kHz to 100 MHz are known as ultrasound. Ultrasound waves create compression and rarefaction phenomena leading to cavitation, which results in the production, growth, and collapse of bubbles. Ultrasonic assisted extraction (UAE) has a better cavitation effect in liquid and liquid-containing solids [162]. UAE is based on the phenomenon of acoustic cavitation. Shockwaves are generated during the collapse of cavitation bubbles leading to fragmentation, increased pore formation thus leading to solubilization of bioactive components into the extraction solvent [190]. Release of bioactive compounds during ultrasonication can also be attributed to 'sonoporation,' that is the formation of pores in cell membrane during cavitation [162]. The most important benefit of UAE has been also observed for extraction of bioactive compounds from the vegetable sample since vegetable contains liquid in solid material [191]. The factors like moisture content, particle size, milling, grinding, and solvent used for extraction important for efficient and effective extraction. The other parameters like temperature, time, pressure, and frequency of sonication are the leading parameters, which affect the action of ultrasound [192]. UAE process of extraction requires relatively low investment, is quite simple to use, versatile, and flexible for its application. It has been used to extract various types of biomolecules like polysaccharides, peptides, pigments, essential oils, and bioactive compounds [193]. Ultrasound can be used in a direct or indirect medium. In a direct medium (without any barrier), ultrasound provides approximately 100 times higher intensity, but in the case of an indirect medium (ultrasonication water bath), the waves should be transported through the water to the sample [194]. The major advantages of this technique include less time required for extraction, less energy requirement, and relatively less amount of solvent required than another extraction method.

1.2.7 ENZYME ASSISTED EXTRACTION (EAE)

Enzymatic extraction is based on the mechanism of breakdown of cell wall and other cell components, thus improving the extraction of target bioactive compounds. Nowadays, enzyme assisted extraction (EAE) technique is gaining more concentration, since focus is more towards the eco-friendly extraction technique. Enzymes, namely amylase, pectinase, cellulase, hemicellulase, hydrolase, ligase, isomerase, and lipoxygenase are derived from microorganisms, animal, and plant sources. Every enzyme has specific catalysis properties and acts on a precise substrate. Enzymes can help to extract, modify, or synthesize the complex bioactive compounds (polysaccharide-lignin network) from natural resources. Subsequently, they enhance the cell wall permeability and finally increase the yield of the bioactive compound [195]. There are mainly two approaches for enzyme oil extraction: (i) enzyme-assisted aqueous extraction (EAAE); and (ii) enzyme-assisted cold pressing (EACP). Mainly, the EAAE method has been used for the improvement of oil recovery by rupturing the cell wall of the seed and hence polysaccharide-protein colloidal bonds breakdown. In the EACP method, enzymes only hydrolyze the seed coats [196]. The extraction depends on intrinsic as well as extrinsic properties like the type of enzyme and its concentration, solid: water ratio, time of hydrolysis, plant material particles size, and moisture content of the materials [197, 198].

Cellulase and pectinase enzyme were found to increase the extraction yield of lycopene and carotenoid from tomato waste [199]. Extraction of phenolic compounds from outer leaves of cauliflower was improved by treatment with polysaccharide degrading enzymes (Viscozyme and Rapidase) increased the extraction yield of kaempferol-glucoside [200].

1.2.8 PULSED ELECTRIC FIELD EXTRACTION

The pulsed electric field (PEF) extraction technique uses moderate electric field strength to high electric field strength, depending upon the nature of the application. In the case of the continuous extraction process, a high electric field (20–80 kV/cm) is used, but in the batch process, a moderate electric field (100–300 V/cm) is used [201]. Its application is based on the principle of electroporation or electro-permeabilization of the cell membrane to the solid matrix. Electroporation depends upon external electrical force, which increases the cell membrane's permeability [192, 202]. The cell membranes lose their structural functionality due to penetration of the membrane by forming the hydrophilic pores, and targeted material is extracted [203]. The electroporation is either reversible or irreversible, but it can be controlled as per the application. Electroporation of irreversible type increases the process of extraction. The electric field strength ranging from 0.1–10 kV/cm is sufficient for delicate plant tissues, whereas tough tissues require slightly high strength ranging from 10–20 kV/cm for effective extraction [204]. The effectiveness of Pulsed electric field treatment is determined by the number of pulses, electrical field strength, treatment temperature, and chemical constituents of the vegetable [205]. The effectiveness of extraction not only depends on the processing parameters but other parameters like the nature of the solvent, cell cytoplasm or vacuoles position, particle size of the sample, and chemical properties of the target compound. PEF extraction is also effective for the extraction of thermolabile compounds as it is a non-thermal technique. Extraction of betalain compounds (betanin and vulgaxanthin) from beetroot was significantly increased on treatment with 20 pulses of an electric field with a strength of 4.38 kV/cm [206].

1.3 ANALYTICAL TECHNIQUES FOR IDENTIFICATION AND CHARACTERIZATION OF BIOACTIVE COMPOUNDS

Various spectrophotometric methods are used for the estimation of different groups of bioactive compounds such as total phenolic content

TABLE 1.2 Extraction of Bioactive Compounds from Various Vegetables Using Different Extraction Techniques

Source	Extraction Technique	Treatment Conditions	Target Compound	References
Onion (skin)	Sonication-assisted extraction	Sonicated at 50% strain and 20 kHz angular frequency for 10 min.	Total flavonoid content.	[207]
Red beetroot	Pulsed electric field treatment	5 pulses, 5 kV/cm (1.2 kJ/kg)	Betanine	[208]
Cauliflower (leaves)	Enzyme-assisted treatment	Polysaccharide-degrading enzymes: Viscozyme L (0.2%) and Rapidase (0.5%) (temperature 35°C, and pH 4.0).	Kaempferol-3-feruloyldiglucoside; kaempferol-3-glucoside.	[200]
Tomato waste	Enzyme aided extraction technique.	Pectinase (optimum temperature: 45°C) and cellulose (optimum temperature: 55°C)	Carotenoid; lycopene	[199]
Tomato waste	High pressure treatment	Pressure range: 100–800 MPa for 1–30 min.	Carotenoid; lycopene	[199]
Spring onion; leaf celery; long bean; mung bean sprout; winged bean.	Solvent extraction	70% Acetone (v/v); 70% ethanol (v/v); 70% methanol (v/v); distilled water.	Total phenol content; total flavonoid content.	[209]
Garlic	Solvent extraction	90% methanol solution with 0.01 N hydrochloric acid.	Alliin; isoalliin; methiin; cycloalliin; ç-L-glutamyl-S-methyl-L-cysteine.	[210]
Garlic	Aqueous extraction	Dulbecco's phosphate buffered saline (DPBS) solution as a buffer.	Phenolic compounds: Gallic acid; syringic acid derivatives; (+)-catechin; p-coumaric acid; p-hydroxybenzoic acid derivatives; epicatechin.	[211]
Bitter gourd (ripe pericarp)	Supercritical fluid extraction	SFE-CO_2 extracts	β-carotene	[212]

TABLE 1.2 *(Continued)*

Source	Extraction Technique	Treatment Conditions	Target Compound	References
Bitter gourd	Ultrasound-assisted extraction	Normal and pulsed mode	Antioxidant activity; total polyphenols content.	[213]
Four-angled bean; French bean; long bean snowpea; snap pea	Solvent extraction	Distilled water; 70% aqueous ethanol.	Antioxidant capacity; Total phenolic content.	[214]
Corchorus olitorius (jute mallow); *Crotalaria ochroleuca* (Slender leaf); *Solanum scabrum* (black nightshade); cleome gynandra (spider plant)	Solvent extraction	Methanol	Alkaloid, flavonoid, saponin, tannin, phenol, anthraquinone, steroids, terpenoids.	[215]

(TPC), total flavonoids, carotenoids, etc. (Table 1.3). But for quantification of individual compounds in these groups, various techniques are used such as GC, gas chromatography-mass spectroscopy (GC-MS), high-pressure liquid chromatography (HPLC), thin-layer chromatography (TLC), column chromatography, fluorescence chromatography and, Nuclear magnetic resonance spectroscopy (NMR) [165]. The most used separation technique is column chromatography as has high capacity, low cost of adsorbents, and is simple to use. Separation is based mainly on the differences in adsorption affinity of different components. The stationary phase (Adsorbent) and composition of the mobile phase are the important factors that affect the separation of components. Polyamide column is used for the separation of polyphenolic compounds namely, phenolic acids, flavonoids, and anthraquinones, where the formation of hydrogen bonds between polyamide and target compounds is the major mechanism of separation [216]. Membrane filtration and gel filtration chromatography are the techniques that carry out the separation of components based on the molecular size of the components. Membrane filtration can also be categorized into ultrafiltration, microfiltration, and nanofiltration depending on the pore size of the membrane. Another type of separation technique is based on the ionic strength of molecules; it is known as ion-exchange chromatography. Alkaloids are separated using the cation-ion exchange resins whereas anion-ion exchange chromatography is used for the separation of phenols and organic acids.

HPLC is the most used technique for the quantification of phenolic components in fruits and vegetable extracts. The most widely used column for polyphenols is C-18 and UV-Vis diode array detector (DAD). Different procedures are used for the detection of different groups of phenolic compounds [217, 218]. Another technique that can be used is superfluid chromatography. It is more versatile as it uses a multi-detector system for separation and identification. The mobile phase used in this method is SF [219]. GC is another widely used technique. In this technique, derivatization of volatile components is done, and mass spectroscopic detection is used [217].

1.4 CONCLUSION

Vegetables, besides being the rich source of micronutrients, are also abundant in various phytochemicals and bioactive compounds. Recent advances

TABLE 1.3 The Techniques Used for the Quantification of Bioactive Components in Various Vegetable Sources

Source	Analytical Technique	Bioactive Compounds	References
Asparagus spears	UHPLC-PDA-MS (Dionex Pump, Accela PDA detector, Column: C8, 2.6-µm particle size, 100-Å pore size, 100 × 2.1 mm)	Quercetin glucosyl rutinoside; Quercetin-3-O-rutinoside (rutin); Kaempferol-3-O-rutinoside (nicotiflorin); Isorhamnetin-O-rutinoside (narcissin); Dicoumaroyl glycerol; Coumaroylferuloyl glycerol; Diferuloyl glycerol	[62]
Garlic	HPLC (aminopropyl-bonded column)	Alliin, isoalliin, methiin, cycloalliin, ç-L-glutamyl-S-methyl-L-cysteine (GSMC), ç-L-Glutamyl-S-(2-propenyl)-L-cysteine (GSAC)) and ç-Lglutamyl-S-(trans-1-propenyl)-L-cysteine (GSPC)	[210]
Green beans	HPLC coupled with time-of-flight – MS (Diode array detector, Zorbax C18 analytical column (4.6 mm × 150 mm, 1.8 mm particle size))	72 phytochemical compounds (10 phenolic acids, 59 flavonoids, 2 lignans and 1 iridoid).	[145]
White cauliflower	HPLC (thermostatically controlled column oven, UV detector)	Protocatechuic acid, quercetin, pyrogallol, vanillic acid, coumaric acid, and kaempferol	[100]
Baby leaf	HPLC–DAD: (Ultraviolet detector, and Diode array detector, YMC, C30 carotenoid column, 250 × 4.6 mm, 5 mm (YMC, Wilmington, NC) and GC–MS: (Slightly polar RXi-5Sil column (Restek; 30 m × 250 µm id × 0.25 µm film))	**Carotenoids:** All-E-violaxanthin, 9′-Z-neoxanthin, All-E-Lutein, All-E-zeaxanthin, All-E-β-cryptoxanthin, α-carotene, and All-E-β-carotene. **Tocopherols:** δ-tocopherol, γ-tocopherol and α-tocopherol. **Fatty Acids:** 16 Fatty acids detected. Major are α-linolenic acid, palmitic acid, and linoleic acid.	[220]
Eryngium caucasicum	HPLC (Eclipse RP-C18 column: 25 cm × 4.6 mm × 5 µm, Supelco, USA)	Gallic acid, *p*-coumaric acid, chlorogenic acid, apigenin, and ferulic acid	[221]
Tomato leaves	HPLC–DAD (reversed-phase ACE 3 C-18-AR column: 150 × 4.6 mm, 3 lm particle size, advanced chromatography technologies, aberdeen, Scotland)	**Phenolic Compounds:** 3-O-caffeoylquinic acid; Caffeoyl-hexoside acid; p-coumaroyl-hexoside acid; Feruloyl-hexoside acid; Sinapoyl-hexoside acid isomer; 5-O-caffeoylquinic acid; p-coumaroylquinic acid; Sinapoyl-hexoside acid; 5-O-feruloylquinic acid; p-coumaroylquinic acid isomer; Feruloylquinic acid isomer; Quercetin-3-O-pentosyl-rutinoside; Quercetin-3-O-rutinoside; Sinapoyl derivate; Kaempferol-3-O-rutinoside. **Alkaloids:** Tomatine, tomatidine.	[222]

TABLE 1.3 *(Continued)*

Source	Analytical Technique	Bioactive Compounds	References
Red chicory	HPLC-PDA/-ESI+-MS (Diode array SPD-M20 UV-Vis detector; Eclipse XDB C18 Column: 4 μm, 4.6 × 150 mm)	**Phenolic Compounds**: Hydroxybenzoic acid; dihydroxybenzoic acid; dihydroxy p-coumaric acid; protocatechuic acid; cyanidin-3-O-glucoside; chlorogenic acid; catechin; cyanidin-3-O-(6″-malonyl-glucoside); quercetin-3-O-rutinoside (Rutin); quercetin-3,4-O-diglucoside; quercetin-3-O-glucoside; quercetin-3-O-(6″-malonyl-glucoside)	[223]
Red onion	HPLC-PDA/-ESI+-MS (Diode array SPD-M20 UV-Vis detector; eclipse XDB C18 column: 4 μm, 4.6 × 150 mm)	**Phenolic Compounds:** Syringic acid; cyanidin-3-O-glucoside and cyanidin-3-olaminaribioside; cyanidin-3-(6″-malonyl-glucoside) and cyanidin-3-(6″-malonyl-laminaribioside); quercetin-3,4-O-diglucoside; quercetin-3-O-glucoside	[223]
Eggplant	HPLC-PDA/-ESI+-MS (Diode array SPD-M20 UV-Vis detector; Eclipse XDB C18 Column: 4 μm, 4.6 × 150 mm)	**Phenolic Compounds:** Hydroxybenzoic acid; delphinidin-3-O-rutinoside; 5-caffeoylquinic acid (chlorogenic acid)	[223]
Sweet potato	HPLC-PDA/-ESI+-MS (Diode array SPD-M20 UV-Vis detector; Eclipse XDB C18 Column: 4 μm, 4.6 × 150 mm)	**Phenolic Compounds:** Hydroxybenzoic acid; dihydroxybenzoic acid; dimethoxy-4-hydroxybenzoic acid (syringic acid); peonidin-3-O-sophoroside-5-O-glucoside; peonidin-3-O-glucoside; cyanidin-3-p-hydroxybenzoyl sophoroside-5-glucoside; chlorogenic acid; peonidin-3-p-hydroxybenzoyl sophoroside-5-glucoside; cyanidin-3-caffeoyl sophoroside-5-glucoside; peonidin-3-caffeoyl sophoroside-5-glucoside; cyanidin-3-caffeoyl-p-hydroxybenzoyl sophoroside-5-glucoside; peonidin-3-dicaffeoyl sophoroside-5-glucoside; peonidin-3-caffeoyl-p-hydroxybenzoyl sophoroside-5-glucoside; peonidin-3-caffeoy-feruloylsophoroside-5-glucoside; ferulic acid	[223]
Black carrot	HPLC-PDA/-ESI+-MS (Diode array SPD-M20 UV-Vis detector; Eclipse XDB C18 Column: 4 μm, 4.6 × 150 mm)	**Phenolic Compounds**: Syringic acid; neochlorogenic acid; chlorogenic acid; cyanidin-3-(p-coumaroyl)-diglucoside-5-glucoside; cyanidin-3-(feruloyl)-glucoside-5-glucoside; caffeic acid; feruloylquinic acid; ferulic acid	[223]
Bitter melon (*Momordica charantia*)	HPLC (ODS C18 column: 250 × 4.6 mm)	Alkaloids; anthraquinone; flavonoids; phenols; sterols; tannins	[224]

TABLE 1.3 *(Continued)*

Source	Analytical Technique	Bioactive Compounds	References
Bitter melon (*Momordica charantia*)	GC-MS (Varian 450GC, 240MS: VF-5 MS Column)	Gentisic acid; Vitamin E; 1-Pentadecyne; cucurbitacin B dihydro; Cis-9-hexadecenal; hexadecanoic acid, methyl ester, pentadecanoic acid 14-methyl-, methyl ester; β-sitoserol; stigmasterol; oleic acid; stigmastan-3-ol; ethyl-4,5-dimethyl-phenol; linoleic acid	[26]
Moringa oleifera leaves	GC-MS (Restek column: 0.25 mm, 60 m, XTI-5)	Methyl undecanoate; undecanoic acid; methyl tetradeca-8,10,12-trienoate; methyl-14-methyl pentadecanoate; n-hexadecanoic acid or palmitic acid; ethyl hexadecanoate or ethyl palmitate; methyl linolelaidate or methyl trans,trans-9,12-octadecadienoate; methyl (11E)-11-octadecenoate; methyl hexadecenoate; cis-octadecenoic acid or cis-oleic acid; octadecanoic acid or stearic acid	[225]
Mushroom	GS-MS (HP 5973 mass detector system; HP-VOC (Agilent Scientific, Palo Alto, CA) capillary column: 60 m length × 0.32 mm i.d. × 1.8 μm film thickness)	1-hexanol; 1-octen-3-ol; benzaldehyde; 3-octanone; benzeneacetaldehyde; 1,3-octadiene; 3-octanol; octanal	[226]

in extraction and purification technologies have increased the opportunities for the utilization of these components in developing functional foods. Nowadays, there has been a shift in trend from synthetic pharmaceuticals and more focus on natural nutraceuticals. Novel extraction techniques can be used to maximize the yield of natural bioactive compounds. A combination of various techniques like microwave-assisted enzymatic extraction, ultrasound-assisted enzymatic extraction, and ultrasonic microwave extraction are also being explored for increasing the extraction yield of bioactive compounds. Advanced analytical techniques have made it easier to determine the complete profiling of different groups of bioactive compounds like phenolics, conjugated phenolics, flavonoids, carotenoids, etc., in various vegetables, thus increasing their applicability in the development of nutraceuticals and drug development.

KEYWORDS

- **bioactive compound**
- **enzyme assisted extraction**
- **extraction method**
- **hydrodistillation**
- **nutraceutical**
- **phytochemicals**
- **supercritical fluid extraction**
- **vegetable**

REFERENCES

1. Witkamp, R. F., & Van, N. K., (2018). *European Journal of Pharmacology, 836*, 102–114.
2. Favela-González, K. M., Hernández-Almanza, A. Y., & De La Fuente-Salcido, N. M., (2020). *Journal of Food Biochemistry, 44*(10), e13414.
3. Nawaz, H., Shad, M. A., & Muzaffar, S., (2018). *Brassica Germplasm-Characterization, Breeding and Utilization*, 7–26.
4. Ramirez, D. A., Altamirano, J. C., & Camargo, A. B., (2021). *Food Chemistry, 337*, 127648.

5. Šamec, D., Urlić, B., & Salopek-Sondi, B., (2019). *Critical Reviews in Food Science and Nutrition, 59*(15), 2411–2422.
6. Mansingh, D. P., Oj, S., Sali, V. K., & Vasanthi, H. R., (2018). *Journal of Biochemical and Molecular Toxicology, 32*(10), e22206.
7. Kodera, Y., Ushijima, M., Amano, H., Suzuki, J. I., & Matsutomo, T., (2017). *Molecules, 22*(4), 570.
8. Yoo, H. G., Lee, B. H., Kim, W., Lee, J. S., Kim, G. H., Chun, O. K., Koo, S. I., & Kim, D. O., (2014). *Journal of Medicinal Food, 17*(11), 1189–1196.
9. Bradley, J. M., Organ, C. L., & Lefer, D. J., (2016). *The Journal of Nutrition, 146*(2), 403S–409S.
10. Ankita, & Prasad, K., (2014). *International Journal of Agriculture and Food Science Technology, 5*(5), 367–374.
11. Ankita, & Prasad, K., (2015). *Biotechnology: An Indian Journal, 11*(11), 426–435.
12. Ankita, & Prasad, K., (2015). *Asian Journal of Chemistry, 27*(10), 3697–3703.
13. Ankita, & Prasad, K., (2015). *Der Pharmacia Lettre, 7*(1), 269–279.
14. Haq, R. U., Kumar, P., & Prasad, K., (2016). *Journal of Food Measurement and Characterization, 10*(3), 701–708.
15. Haq, R. U., & Prasad, K., (2017). *Journal of Food Measurement and Characterization, 11*(4), 1542–1549.
16. Prasad, K., Haq, R. U., Bansal, V., Siddiqui, M. W., & Ilahy, R., (2016). In: Siddiqui, M. W., Bansal, V., & Prasad, K., (eds.), *Plant Secondary Metabolites: Stimulation, Extraction, and Utilization* (Vol. 2, pp. 107–194). Apple Academic Press, CRC Press, Taylor and Francis Group, USA.
17. Prasad, K., Kaushik, V., & Prasad, K. K., (2002). *Beverage and Food World, 29*(12), 24–26.
18. Raees-Ul, H., & Prasad, K., (2015). *South Asian J. Food Technol. Environ., 1*(1), 1–14.
19. Siddiqui, M. W., Lara, I., Ilahy, R., Tlili, I., Ali, A., Homa, F., Prasad, K., et al., (2018). *Comprehensive Reviews in Food Science and Food Safety, 17*(6), 1540–1560.
20. Singh, Y., & Prasad, K., (2012). *Annals of Biology, 28*(2), 140–143.
21. Singh, Y., & Prasad, K., (2013). *International Journal of Agriculture and Food Science Technology, 4*(4), 317–324.
22. Prasad, K., Neeha, V. S., Bansal, V., Siddiqui, M. W., & Brasil, I. M., (2016). In: Siddiqui, M. W., & Prasad, K., (eds.), *Plant Secondary Metabolites: Biological and Therapeutic Significance* (Vol. 1, pp. 101–142). Apple Academic Press, CRC Press, Taylor and Francis Group, USA.
23. Brower, V., (1998). *Nature Biotechnology, 16*(8), 728–731.
24. Zeisel, S. H., (1999). *American Association for the Advancement of Science* (Vol. 285, pp. 1853–1855).
25. Nasri, H., Baradaran, A., Shirzad, H., & Rafieian-Kopaei, M., (2014). *International Journal of Preventive Medicine, 5*(12), 1487.
26. Singh, A., Dubey, R., Paliwal, R. T., Saraogi, G. K., & Singhai, A. K., (2012). *International Journal of Pharmacy and Pharmaceutical Sciences, 4*(4), 39–43.
27. Ahmad, Z., Zamhuri, K. F., Yaacob, A., Siong, C. H., Selvarajah, M., Ismail, A., & Hakim, M. N., (2012). *Molecules, 17*(8), 9631–9640.
28. Ray, A., Joshi, J., & Gulati, K., (2016). In: *Nutraceuticals* (pp. 941–946). Elsevier.
29. Padmavathi, D., (2018). *Int. J. Food Sci. Nut., 3*(2), 214–217.

30. Helal, N. A., Eassa, H. A., Amer, A. M., Eltokhy, M. A., Edafiogho, I., & Nounou, M. I., (2019). *Recent Patents on Drug Delivery & Formulation, 13*(2), 105–156.
31. Swaroop, A., Bagchi, M., Moriyama, H., & Bagchi, D., (2017). In: *Sustained Energy for Enhanced Human Functions and Activity* (pp. 411–419). Elsevier.
32. Guaadaoui, A., Benaicha, S., Elmajdoub, N., Bellaoui, M., & Hamal, A., (2014). *International Journal of Nutrition and Food Sciences, 3*(3), 174–179.
33. Kris-Etherton, P. M., Etherton, T. D., Carlson, J., & Gardner, C., (2002). *Current Opinion in Lipidology, 13*(4), 397–407.
34. Biesalski, H. K., Dragsted, L. O., Elmadfa, I., Grossklaus, R., Müller, M., Schrenk, D., Walter, P., & Weber, P., (2009). *Nutrition, 25*(11, 12), 1202–1205.
35. Han, N. S., Ahmad, W., & Ishak, W. R. W., (2016). *Sains Malaysiana, 45*(11), 1617–1623.
36. Parvathy, K. S., Negi, P. S., & Srinivas, P., (2009). *Food Chemistry, 115*(1), 265–271.
37. Singh, R., Prasad, K. K., Siddiqui, M. W., & Prasad, K., (2016). In: Siddiqui, M. W., & Prasad, K., (eds.), *Plant Secondary Metabolites: Biological and Therapeutic Significance* (Vol. 1, pp. 63–100). Apple Academic Press, CRC Press, Taylor and Francis Group, USA.
38. Bansal, V., Kumar, P., Tuteja, S. K., Siddiqui, M. W., Prasad, K., & Sangwan, R., (2016). In: Siddiqui, M. W., & Bansal, V., (eds.), *Plant Secondary Metabolites: Their Roles in Stress Ecophysiology* (Vol. 3, pp. 215–238). Apple Academic Press, CRC Press, Taylor and Francis Group, USA.
39. Jafari, S. M., Mahdavee, K. K., & Assadpour, E., (2019). *Food Science & Nutrition, 7*(4), 1438–1445.
40. Markom, M., Hasan, M., Daud, W. R. W., Singh, H., & Jahim, J. M., (2007). *Separation and Purification Technology, 52*(3), 487–496.
41. Sarfarazi, M., Jafari, S. M., Rajabzadeh, G., & Feizi, J., (2019). *LWT, 114*, 108428.
42. LaLonde, R. T., (2005). Terpenes and terpenoids. In: *Van Nostrand's Encyclopedia of Chemistry* (pp. 1–6). John Wiley & Sons, Inc., Hoboken, USA.
43. Ziegler, J., & Facchini, P. J., (2008). *Annu. Rev. Plant Biol., 59*, 735–769.
44. Giada, M., (2013). *Oxidative Stress and Chronic Degenerative Diseases – A Role for Antioxidants, 201*(3), 87–112.
45. Bansal, V., Siddiqui, M. W., Singla, M. L., Ghanshyam, C., & Prasad, K., (2016). In: Siddiqui, M. W., Bansal, V., & Prasad, K., (eds.), *Plant Secondary Metabolites: Stimulation, Extraction, and Utilization* (Vol. 2, pp. 233–254). Apple Academic Press, CRC Press, Taylor and Francis Group, USA.
46. Welbaum, G. E., (2015). *Vegetable Production and Practices* (pp. 1–476). CABI, USA.
47. Schreinemachers, P., Simmons, E. B., & Wopereis, M. C. S., (2018). *Global Food Security, 16*, 36–45.
48. Pennington, J. A. T., & Fisher, R. A., (2009). *Journal of Food Composition and Analysis, 22*, S23–S31.
49. Septembre-Malaterre, A., Remize, F., & Poucheret, P., (2018). *Food Research International, 104*, 86–99.
50. Horbowicz, M., Kosson, R., Grzesiuk, A., & Debski, H., (2008). *Vegetable Crops Research Bulletin, 68*, 5.
51. Dhillon, N. P. S., Monforte, A. J., Pitrat, M., Sudhakar, P., Singh, P. K., Reitsma, K. R., Garcia-Mas, J., et al., (2012). *Plant Breeding Reviews, 35*, 85–150.
52. Clinton, S. K., (1998). *Nutrition Reviews, 56*(2), 35–51.

53. Singh, G., Kawatra, A., & Sehgal, S., (2001). *Plant Foods for Human Nutrition, 56*(4), 359–364.
54. Roberts, J. L., & Moreau, R., (2016). *Food & Function, 7*(8), 3337–3353.
55. Santiago-Saenz, Y. O., Hernández-Fuentes, A. D., Monroy-Torres, R., Cariño-Cortés, R., & Jiménez-Alvarado, R., (2018). *Journal of Food Measurement and Characterization, 12*(4), 2855–2864.
56. Noorbakhshnia, M., & Karimi-Zandi, L., (2017). *Physiology & Behavior, 169*, 69–73.
57. Al-Quraishy, S., Dkhil, M. A., & Moneim, A. E. A., (2012). *Pesticide Biochemistry and Physiology, 103*(2), 108–114.
58. Makris, D. P., & Rossiter, J. T., (2001). *Journal of Agricultural and Food Chemistry, 49*(7), 3216–3222.
59. Lee, J. W., Lee, J. H., Yu, I. H., Gorinstein, S., Bae, J. H., & Ku, Y. G., (2014). *Plant Foods for Human Nutrition, 69*(2), 175–181.
60. Di Maro, A., Pacifico, S., Fiorentino, A., Galasso, S., Gallicchio, M., Guida, V., Severino, V., Monaco, P., & Parente, A., (2013). *Food Research International, 53*(1), 180–188.
61. Fuentes-Alventosa, J. M., Jaramillo-Carmona, S., Rodríguez-Gutiérrez, G., Guillén-Bejarano, R., Jiménez-Araujo, A., Fernández-Bolaños, J., & Rodríguez-Arcos, R., (2013). *Food and Bioproducts Processing, 91*(2), 74–82.
62. Ku, Y. G., Bae, J. H., Namieśnik, J., Barasch, D., Nemirovski, A., Katrich, E., & Gorinstein, S., (2018). *Food Analytical Methods, 11*(1), 309–318.
63. Kooti, W., & Daraei, N., (2017). *Journal of Evidence-Based Complementary & Alternative Medicine, 22*(4), 1029–1034.
64. Sowbhagya, H. B., Srinivas, P., & Krishnamurthy, N., (2010). *Food Chemistry, 120*(1), 230–234.
65. Hassanen, N. H. M., Eissa, A. M. F., Hafez, S. A. M., & Mosa, E. A. M., (2015). *Int. J. Curr. Microbiol. App. Sci., 4*(3), 284–296.
66. Mencherini, T., Cau, A., Bianco, G., Loggia, R. D., Aquino, R. P., & Autore, G., (2007). *Journal of Pharmacy and Pharmacology, 59*(6), 891–897.
67. Momin, R. A., & Nair, M. G., (2001). *Journal of Agricultural and Food Chemistry, 49*(1), 142–145.
68. Nickavar, B., Kamalinejad, M., & Izadpanah, H., (2007). *Pak. J. Pharm. Sci., 20*(4), 291–294.
69. Zidorn, C., Jöhrer, K., Ganzera, M., Schubert, B., Sigmund, E. M., Mader, J., Greil, R., et al., (2005). *Journal of Agricultural and Food Chemistry, 53*(7), 2518–2523.
70. Chauhan, E. S., Tiwari, A., & Singh, A., (2016). *International Journal of Home Science, 2*(3), 123–126.
71. Kurilich, A. C., Jeffery, E. H., Juvik, J. A., Wallig, M. A., & Klein, B. P., (2002). *Journal of Agricultural and Food Chemistry, 50*(18), 5053–5057.
72. Rasal, V., Shetty, B., Sinnathambi, A., Yeshmaina, S., & Ashok, P., (2005). *The Internet Journal of Pharmacology, 4*(2), 1–7.
73. Popovich, D. G., Li, L., & Zhang, W., (2010). *Food and Chemical Toxicology, 48*(6), 1619–1626.
74. Puri, R., Sud, R., Khaliq, A., Kumar, M., & Jain, S., (2011). *Indian Journal of Gastroenterology, 30*(5), 233–236.
75. Sabourian, R., Karimpour-Razkenari, E., Saeedi, M., Bagheri, M. S., Khanavi, M., Sadati, N., Akbarzadeh, T., & Rs Ardekani, M., (2016). *Current Pharmaceutical Biotechnology, 17*(11), 974–985.

76. Pandit, S., Kanjilal, S., Awasthi, A., Chaudhary, A., Banerjee, D., Bhatt, B. N., Narwaria, A., et al., (2017). *Journal of Ethnopharmacology, 197*, 165–172.
77. Dandawate, P. R., Subramaniam, D., Padhye, S. B., & Anant, S., (2016). *Chinese Journal of Natural Medicines, 14*(2), 81–100.
78. Güdr, A., (2016). *Iranian Journal of Pharmaceutical Research (IJPR), 15*(1), 301.
79. Kumar, R., Balaji, S., Sripriya, R., Nithya, N., Uma, T. S., & Sehgal, P. K., (2010). *Journal of Agricultural and Food Chemistry, 58*(3), 1518–1522.
80. Horax, R., Hettiarachchy, N., Over, K., Chen, P., & Gbur, E., (2010). *Journal of Agricultural and Food Chemistry, 58*(3), 1892–1897.
81. Ibrahim, T. A., El-Hefnawy, H. M., & El-Hela, A. A., (2010). *Natural Product Research, 24*(16), 1537–1545.
82. Yadav, M., Lavania, A., Tomar, R., Prasad, G., Jain, S., & Yadav, H., (2010). *Applied Biochemistry and Biotechnology, 160*(8), 2388–2400.
83. Grossmann, M. E., Mizuno, N. K., Dammen, M. L., Schuster, T., Ray, A., & Cleary, M. P., (2009). *Cancer Prevention Research, 2*(10), 879–886.
84. Liu, X. R., Deng, Z. Y., Fan, Y. W., Li, J., & Liu, Z. H., (2010). *Spectroscopy and Spectral Analysis, 30*(8), 2265–2268.
85. Anilakumar, K. R., Kumar, G. P., & Ilaiyaraja, N., (2015). *International Journal of Nutrition and Food Sciences, 4*(1), 75–83.
86. Shih, C. C., Lin, C. H., Lin, W. L., & Wu, J. B., (2009). *Journal of Ethnopharmacology, 123*(1), 82–90.
87. Tsai, C. H., Chen, E. C. F., Tsay, H. S., & Huang, C. J., (2012). *Nutrition Journal, 11*(1), 1–9.
88. Giovannini, P., Howes, M. J. R., & Edwards, S. E., (2016). *Journal of Ethnopharmacology, 184*, 58–71.
89. Nhiem, N. X., Yen, P. H., Ngan, N. T. T., Quang, T. H., Kiem, P. V., Minh, C. V., Tai, B. H., et al., (2012). *Journal of Medicinal Food, 15*(4), 369–377.
90. Alam, S., Asad, M., Asdaq, S. M. B., & Prasad, V. S., (2009). *Journal of Ethnopharmacology, 123*(3), 464–469.
91. Bakare, R. I., Magbagbeola, O. A., & Okunowo, O. W., (2010). *Journal of Medicinal Plants Research, 4*(21), 2189–2193.
92. Mesia, G. K., Tona, G. L., Nanga, T. H., Cimanga, R. K., Apers, S., Cos, P., Maes, L., et al., (2008). *Journal of Ethnopharmacology, 115*(3), 409–415.
93. Khandaker, M., Akter, S., & Imam, M. Z., (2018). *Food Science and Human Wellness, 7*(1), 34–48.
94. Gemede, H. F., Ratta, N., Haki, G. D., Woldegiorgis, A. Z., & Beyene, F., (2015). *J. Food Process Technol., 6*(458), 2.
95. Durazzo, A., Lucarini, M., Novellino, E., Souto, E. B., Daliu, P., & Santini, A., (2019). *Molecules, 24*(1), 38.
96. Elkhalifa, A. E. O., Alshammari, E., Adnan, M., Alcantara, J. C., Awadelkareem, A. M., Eltoum, N. E., Mehmood, K., et al., (2021). *Molecules, 26*(3), 696.
97. Avallone, S., Tiemtore, T. W. E., Mouquet-Rivier, C., & Trèche, S., (2008). *Journal of Food Composition and Analysis, 21*(7), 553–558.
98. Cartea, M. E., Francisco, M., Soengas, P., & Velasco, P., (2011). *Molecules, 16*(1), 251–280.
99. Picchi, V., Migliori, C., Scalzo, R. L., Campanelli, G., Ferrari, V., & Di Cesare, L. F., (2012). *Food Chemistry, 130*(3), 501–509.

100. Ahmed, F. A., & Ali, R. F. M., (2013). *BioMed Research International*, pp. 1–9.
101. Verkerk, R., Van, D. G. M. S., Dekker, M., & Jongen, W. M. F., (1997). *Cancer Letters, 114*(1, 2), 193–194.
102. Schreiner, M., (2005). *European Journal of Nutrition, 44*(2), 85–94.
103. Agerbirk, N., De Vos, M., Kim, J. H., & Jander, G., (2009). *Phytochemistry Reviews, 8*(1), 101–120.
104. Wada, L., & Ou, B., (2002). *Journal of Agricultural and Food Chemistry, 50*(12), 3495–3500.
105. Bremus, C., Herrmann, U., Bringer-Meyer, S., & Sahm, H., (2006). *Journal of Biotechnology, 124*(1), 196–205.
106. Krinsky, N. I., (1998). *Annals of the New York Academy of Sciences, 854*(1), 443–447.
107. Scalzo, R. L., Genna, A., Branca, F., Chedin, M., & Chassaigne, H., (2008). *Food Chemistry, 107*(1), 136–144.
108. Arshad, M. S., Sohaib, M., Nadeem, M., Saeed, F., Imran, A., Javed, A., Amjad, Z., & Batool, S. M., (2017). *Cogent Food & Agriculture, 3*(1), 1280254.
109. Schwimmer, S., & Weston, W. J., (1961). *Journal of Agricultural and Food Chemistry, 9*(4), 301–304.
110. Bacon, J. R., Moates, G. K., Ng, A., Rhodes, M. J. C., Smith, A. C., & Waldron, K. W., (1999). *Food Chemistry, 64*(2), 257–261.
111. Liguori, L., Califano, R., Albanese, D., Raimo, F., Crescitelli, A., & Di Matteo, M., (2017). *Journal of Food Quality, 201*, 7.
112. Griffiths, G., Trueman, L., Crowther, T., Thomas, B., & Smith, B., (2002). *Phytotherapy Research, 16*(7), 603–615.
113. Slimestad, R., Fossen, T., & Vågen, I. M., (2007). *Journal of Agricultural and Food Chemistry, 55*(25), 10067–10080.
114. Mollavali, M., Bolandnazar, S. A., Schwarz, D., Rohn, S., Riehle, P., & Zaare, N. F., (2016). *Journal of Agricultural and Food Chemistry, 64*(1), 71–77.
115. Pérez-Gregorio, R. M., García-Falcón, M. S., Simal-Gándara, J., Rodrigues, A. S., & Almeida, D. P. F., (2010). *Journal of Food Composition and Analysis, 23*(6), 592–598.
116. Chu, Y. H., Chang, C. L., & Hsu, H. F., (2000). *Journal of the Science of Food and Agriculture, 80*(5), 561–566.
117. Torel, J., Cillard, J., & Cillard, P., (1986). *Phytochemistry, 25*(2), 383–385.
118. De Ancos, B., Colina-Coca, C., González-Peña, D., & Sánchez-Moreno, C., (2015). *Biotechnology of Bioactive Compounds: Sources and Applications – Section I*, 3–36.
119. Benítez, V., Mollá, E., Martín-Cabrejas, M. A., Aguilera, Y., López-Andréu, F. J., Cools, K., Terry, L. A., & Esteban, R. M., (2011). *Plant Foods for Human Nutrition, 66*(1), 48–57.
120. Yalcin, H., & Kavuncuoglu, H., (2014). *Journal of Applied Botany and Food Quality*, 87.
121. Lu, X., Wang, J., Al-Qadiri, H. M., Ross, C. F., Powers, J. R., Tang, J., & Rasco, B. A., (2011). *Food Chemistry, 129*(2), 637–644.
122. Roldán, E., Sanchez-Moreno, C., De Ancos, B., & Cano, M. P., (2008). *Food Chemistry, 108*(3), 907–916.
123. Parry, J., Hao, Z., Luther, M., Su, L., Zhou, K., & Yu, L. L., (2006). *Journal of the American Oil Chemists' Society, 83*(10), 847–854.
124. Randle, W. M., & Lancaster, J. E., (2002). *Allium Crop Sci.*, 329.

125. Yamaguchi, Y., & Kumagai, H., (2020). *Experimental and Therapeutic Medicine, 19*(2), 1528–1535.
126. Bajac, J., Nikolovski, B., Kocić-Tanackov, S., Tomšik, A., Mandić, A., Gvozdanović-Varga, J., Vlajić, S., Vujanović, M., & Radojković, M., (2018). In: *Proceedings of the IV International Congress Food Technology, Quality and Safety, Novi Sad, Serbia* (p. 82).
127. Santhosha, S. G., Jamuna, P., & Prabhavathi, S. N., (2013). *Food Bioscience, 3*, 59–74.
128. Volk, G. M., & Stern, D., (2009). *HortScience, 44*(5), 1238–1247.
129. Majewski, M., (2014). *Roczniki Państwowego Zakładu Higieny, 65*(1).
130. El-Sayed, H. S., Chizzola, R., Ramadan, A. A., & Edris, A. E., (2017). *Food Chemistry, 221*, 196–204.
131. Piatkowska, E., Kopec, A., & Leszczynska, T., (2015). *Żywność Nauka Technologia Jakość, 1*(98), 181–192.
132. Chandrasekara, A., & Josheph, K. T., (2016). *International Journal of Food Science*, 1–15.
133. Burlingame, B., Mouillé, B., & Charrondiere, R., (2009). *Journal of Food Composition and Analysis, 22*(6), 494–502.
134. Hale, A. L., Reddivari, L., Nzaramba, M. N., Bamberg, J. B., & Miller, J. C., (2008). *American Journal of Potato Research, 85*(5), 332–341.
135. Van, J. P. J., Marais, D. W., Harmse, E., Nestel, P., & Rodriguez-Amaya, D. B., (2006). *Journal of Food Composition and Analysis, 19*(4), 321–329.
136. Blagbrough, I. S., Bayoumi, S. A. L., Rowan, M. G., & Beeching, J. R., (2010). *Phytochemistry, 71*(17, 18), 1940–1951.
137. Prawat, H., Mahidol, C., Ruchirawat, S., Prawat, U., Tuntiwachwuttikul, P., Tooptakong, U., Taylor, W. C., et al., (1995). *Phytochemistry, 40*(4), 1167–1173.
138. Montagnac, J. A., Davis, C. R., & Tanumihardjo, S. A., (2009). *Comprehensive Reviews in Food Science and Food Safety, 8*(3), 181–194.
139. Bhandari, M. R., Kasai, T., & Kawabata, J., (2003). *Food Chemistry, 82*(4), 619–623.
140. Wegener, C. B., Jansen, G., & Jürgens, H. U., (2015). *Functional Foods in Health and Disease, 5*(3), 108–116.
141. Brown, C. R., (2005). *American Journal of Potato Research, 82*(2), 163–172.
142. Hacisevkİ, A., (2009). *Journal of Faculty of Pharmacy of Ankara University, 38*(3), 233–255.
143. Vaitkevičienė, N., Kulaitienė, J., Jarienė, E., Levickienė, D., Danillčenko, H., Średnicka-Tober, D., et al., (2020). *Sustainability, 12*(7), 2701.
144. Petropoulos, S. A., Taofiq, O., Fernandes, Â., Tzortzakis, N., Ciric, A., Sokovic, M., Barros, L., & Ferreira, I. C. F. R., (2019). *Journal of the Science of Food and Agriculture, 99*(13), 6049–6059.
145. Abu-Reidah, I. M., Arráez-Román, D., Lozano-Sánchez, J., Segura-Carretero, A., & A. Fernández-Gutiérrez, A., (2013). *Phytochemical Analysis, 24*(2), 105–116.
146. Dicko, M. H., Gruppen, H., Traoré, A. S., Voragen, A. G. J., & Van, B. W. J. H., (2006). *Biotechnology and Molecular Biology Reviews, 1*(1), 21–38.
147. Reynoso-Camacho, R., Ramos-Gomez, M., & Loarca-Pina, G., (2006). *Advances in Agricultural and Food Biotechnology*, 217–236.
148. Johnson, I. T., (2004). *Mutation Research/Fundamental and Molecular Mechanisms of Mutagenesis, 551*(1, 2), 9–28.
149. De Sá, M. C., & Rodriguez-Amaya, D. B., (2004). *Journal of Food Composition and Analysis, 17*(1), 37–51.

150. Osowski, A., Pietrzak, M., Wieczorek, Z., & Wieczorek, J., (2010). *Journal of Toxicology and Environmental Health, Part A, 73*(17, 18), 1141–1149.
151. Trinidad, T. P., Mallillin, A. C., Loyola, A. S., Sagum, R. S., & Encabo, R. R., (2010). *British Journal of Nutrition, 103*(4), 569–574.
152. Fernando, W., Hill, J., Zello, G., Tyler, R., Dahl, W., & Van, K. A., (2010). *Beneficial Microbes, 1*(2), 197–207.
153. Nilsson, J., Stegmark, R., & Åkesson, B., (2004). *Food Chemistry, 86*(4), 501–507.
154. Chahbani, A., Fakhfakh, N., Balti, M. A., Mabrouk, M., El-Hatmi, H., Zouari, N., & Kechaou, N., (2018). *Food Bioscience, 25*, 32–38.
155. Hadrich, F., El Arbi, M., Boukhris, M., Sayadi, S., & Cherif, S., (2014). *Journal of Oleo Science, 63*(11), 1177–1183.
156. Patel, S., & Goyal, A., (2012). *3 Biotech, 2*(1), 1–15.
157. Blumfield, M., Abbott, K., Duve, E., Cassettari, T., Marshall, S., & Fayet-Moore, F., (2020). *The Journal of Nutritional Biochemistry, 84*, 108453.
158. Guillamón, E., García-Lafuente, A., Lozano, M., Rostagno, M. A., Villares, A., & Martínez, J. A., (2010). *Fitoterapia, 81*(7), 715–723.
159. Rathore, H., Prasad, S., & Sharma, S., (2017). *Pharma Nutrition, 5*(2), 35–46.
160. Valverde, M. E., Hernández-Pérez, T., & Paredes-López, O., (2015). *International Journal of Microbiology*, 1–14.
161. Xu, T., & Beelman, R. B., (2015). *Adv. Food Technol. Nutr. Sci. Open J., 1*(2), 62–66.
162. Azmir, J., Zaidul, I. S. M., Rahman, M. M., Sharif, K. M., Mohamed, A., Sahena, F., Jahurul, M. H. A., et al., (2013). *Journal of Food Engineering, 117*(4), 426–436.
163. Luthria, D. L., Mukhopadhyay, S., & Krizek, D. T., (2006). *Journal of Food Composition and Analysis, 19*(8), 771–777.
164. Hung, P. V., & Duy, T. L., (2012). *International Food Research Journal, 19*(1), 327–332.
165. Zhang, Q. W., Lin, L. G., & Ye, W. C., (2018). *Chinese Medicine, 13*(1), 1–26.
166. Hayouni, E. A., Abedrabba, M., Bouix, M., & Hamdi, M., (2007). *Food Chemistry, 105*(3), 1126–1134.
167. Szentmihályi, K., Vinkler, P., Lakatos, B., Illés, V., & Then, M., (2002). *Bioresource Technology, 82*(2), 195–201.
168. Heleno, S. A., Diz, P., Prieto, M. A., Barros, L., Rodrigues, A., Barreiro, M. F., & Ferreira, I. C. F. R., (2016). *Food Chemistry, 197*, 1054–1063.
169. Vankar, P. S., (2004). *Resonance, 9*(4), 30–41.
170. Manousi, N., Sarakatsianos, I., & Samanidou, V., (2019). In: *Engineering Tools in the Beverage Industry* (pp. 283–314). Elsevier.
171. Wu, C., Wang, F., Liu, J., Zou, Y., & Chen, X., (2015). *Integrative Medicine Research, 4*(3), 171–177.
172. Petigny, L., Périno, S., Minuti, M., Visinoni, F., Wajsman, J., & Chemat, F., (2014). *International Journal of Molecular Sciences, 15*(5), 7183–7198.
173. Marchese, S., Perret, D., Bafile, E., Gentili, A., Caretti, F., & Berardino, M., (2009). *Chromatographia, 70*(5), 761–767.
174. Mustafa, A., & Turner, C., (2011). *Analytica Chimica Acta, 703*(1), 8–18.
175. Garcia-Mendoza, M. P., Paula, J. T., Paviani, L. C., Cabral, F. A., & Martinez-Correa, H. A., (2015). *LWT-Food Science and Technology, 62*(1), 131–137.
176. Carabias-Martínez, R., Rodríguez-Gonzalo, E., Revilla-Ruiz, P., & Hernández-Méndez, J., (2005). *Journal of Chromatography A, 1089*(1, 2), 1–17.

177. Machado, A. P. D. F., Pasquel-Reátegui, J. L., Barbero, G. F., & Martínez, J., (2015). *Food Research International, 77*, 675–683.
178. Luthria, D. L., (2008). *Food Chemistry, 107*(2), 745–752.
179. Ibañez, E., Herrero, M., Mendiola, J. A., & Castro-Puyana, M., (2012). In: *Marine Bioactive Compounds* (pp. 55–98). Springer.
180. Herrero, M., Castro-Puyana, M., Mendiola, J. A., & Ibañez, E., (2013). *TrAC Trends in Analytical Chemistry, 43*, 67–83.
181. Da Silva, R. P. F. F., Rocha-Santos, T. A. P., & Duarte, A. C., (2016). *TrAC Trends in Analytical Chemistry, 76*, 40–51.
182. Oroian, M., & Escriche, I., (2015). *Food Research International, 74*, 10–36.
183. Palenzuela, B., Arce, L., Macho, A., Munoz, E., Rios, A., & Valcarcel, M., (2004). *Analytical and Bioanalytical Chemistry, 378*(8), 2021–2027.
184. Werner, B. G., & Hotchkiss, J. H., (2006). *Journal of Dairy Science, 89*(3), 872–881.
185. Handa, S. S., (2008). *Extraction Technologies for Medicinal and Aromatic Plants, 1*, 21–40.
186. Ekezie, F. G. C., Sun, D. W., & Cheng, J. H., (2017). *Trends in Food Science & Technology, 67*, 160–172.
187. Angiolillo, L., Del Nobile, M. A., & Conte, A., (2015). *Current Opinion in Food Science, 5*, 93–98.
188. Letellier, M., & Budzinski, H., (1999). *Analusis, 27*(3), 259–270.
189. Alupului, A., Calinescu, I., & Lavric, V., (2012). *UPB Science Bulletin, Series B, 74*(2), 1454–2331.
190. Roselló-Soto, E., Koubaa, M., Moubarik, A., Lopes, R. P., Saraiva, J. A., Boussetta, N., Grimi, N., & Barba, F. J., (2015). *Trends in Food Science & Technology, 45*(2), 296–310.
191. Herrera, M. C., & De Castro, M. D. L., (2005). *Journal of Chromatography A, 1100*(1), 1–7.
192. Rajha, H. N., Boussetta, N., Louka, N., Maroun, R. G., & Vorobiev, E., (2015). *Separation and Purification Technology, 146*, 243–251.
193. Tiwari, B. K., (2015). *TrAC Trends in Analytical Chemistry, 71*, 100–109.
194. Kek, S. P., Chin, N. L., & Yusof, Y. A., (2013). *Food and Bioproducts Processing, 91*(4), 495–506.
195. Puri, M., Sharma, D., & Barrow, C. J., (2012). *Trends in Biotechnology, 30*(1), 37–44.
196. Latif, S., & Anwar, F., (2009). *European Journal of Lipid Science and Technology, 111*(10), 1042–1048.
197. Niranjan, K., & Hanmoungjai, P., (2004). *Nutritionally Enhanced Edible Oil Processing.* (pp. 1–314). AOCS Publishing, USA.
198. Dominguez, H., Nunez, M. J., & Lema, J. M., (1995). *Food Chemistry, 54*(2), 223–231.
199. Strati, I. F., Gogou, E., & Oreopoulou, V., (2015). *Food and Bioproducts Processing, 94*, 668–674.
200. Huynh, N. T., Smagghe, G., Gonzales, G. B., Van, C. J., & Raes, K., (2014). *Journal of Agricultural and Food Chemistry, 62*(30), 7468–7476.
201. Xi, J., Li, Z., & Fan, Y., (2021). *Critical Reviews in Food Science and Nutrition, 61*(10), 1738–1750.
202. Panja, P., (2018). *Current Opinion in Food Science, 23*, 173–182.
203. Redondo, D., Venturini, M. E., Luengo, E., Raso, J., & Arias, E., (2018). *Innovative Food Science & Emerging Technologies, 45*, 335–343.

204. Ranjha, M. M. A. N., Kanwal, R., Shafique, B., Arshad, R. N., Irfan, S., Kieliszek, M., Kowalczewski, P. Ł., et al., (2021). *Molecules, 26*(16), 4893.
205. Heinz, V., Toepfl, S., & Knorr, D., (2003). *Innovative Food Science & Emerging Technologies, 4*(2), 167–175.
206. Nowacka, M., Tappi, S., Wiktor, A., Rybak, K., Miszczykowska, A., Czyzewski, J., Drozdzal, K., Witrowa-Rajchert, D., & Tylewicz, U., (2019). *Foods, 8*(7), 244.
207. Sagar, N. A., Pareek, S., & Gonzalez-Aguilar, G. A., (2020). *Journal of Food Science and Technology, 57*(7), 2423–2432.
208. López, N., Puértolas, E., Condón, S., Raso, J., & Alvarez, I., (2009). *Journal of Food Engineering, 90*(1), 60–66.
209. Sulaiman, S. F., Sajak, A. A. B., Ooi, K. L., & Seow, E. M., (2011). *Journal of Food Composition and Analysis, 24*(4, 5), 506–515.
210. Ichikawa, M., Ide, N., Yoshida, J., Yamaguchi, H., & Ono, K., (2006). *Journal of Agricultural and Food Chemistry, 54*(5), 1535–1540.
211. Jakuczyk, A., Matysiak, M., Binduga, U. E., & Gminski, J., (2018). *Polish Journal of Food and Nutrition Sciences, 68*(1), 73–81.
212. Patel, A. S., Kar, A., Dash, S., & Dash, S. K., (2019). *Scientific Reports, 9*(1), 1–10.
213. Chakraborty, S., Uppaluri, R., & Das, C., (2020). *Food and Bioproducts Processing, 120*, 114–122.
214. Wen, T. N., Prasad, K. N., Yang, B., & Ismail, A., (2010). *Innovative Food Science & Emerging Technologies, 11*(3), 464–469.
215. Mibei, E. K., Ojijo, N. K. O., Karanja, S. M., & Kinyua, J. K., (2012). *Annals Food Science and Technology, 13*(1), 37–42.
216. Gao, M., Wang, X., Gu, M., Su, Z., Wang, Y., & Janson, J. C., (2011). *Journal of Separation Science, 34*(15), 1853–1858.
217. Naczk, M., & Shahidi, F., (2006). *Journal of Pharmaceutical and Biomedical Analysis, 41*(5), 1523–1542.
218. Sakakibara, H., Honda, Y., Nakagawa, S., Ashida, H., & Kanazawa, K., (2003). *Journal of Agricultural and Food Chemistry, 51*(3), 571–581.
219. Kamangerpour, A., Ashraf-Khorassani, M., Taylor, L. T., McNair, H. M., & Chorida, L., (2002). *Chromatographia, 55*(7), 417–421.
220. Saini, R. K., Shang, X. M., Ko, E. Y., Choi, J. H., Kim, D., & Keum, Y. S., (2016). *Journal of Food Measurement and Characterization, 10*(2), 341–349.
221. Bimakr, M., Ganjloo, A., & Noroozi, A., (2019). *Journal of Food Measurement and Characterization, 13*(3), 1839–1851.
222. Taveira, M., Ferreres, F., Gil-Izquierdo, A., Oliveira, L., Valentão, P., & Andrade, P. B., (2012). *Food Chemistry, 135*(2), 748–755.
223. Frond, A. D., Iuhas, C. I., Stirbu, I., Leopold, L., Socaci, S., Andreea, S., Ayvaz, H., et al., (2019). *Molecules, 24*(8), 1536.
224. Supe, U., & Daniel, P., (2015). *Research Journal of Pharmacy and Technology, 8*(12), 1649–1652.
225. Igwe, K. K., Nwankwo, P. O., Otuokere, I. E., Ijioma, S., & Amaku, F., (2015). *International Journal of Research in Pharmaceutical Sciences, 2*(11), 1–6.
226. Zhang, Z. M., Wu, W. W., & Li, G. K., (2008). *Journal of Chromatographic Science, 46*(8), 690–696.

CHAPTER 2

Technological and Analytical Aspects of Bioactive Compounds and Nutraceuticals from Plant (Plantation Crops) Sources

RAHUL THAKUR, VIVEK GUPTA, POOJA SINGHA, and AMIT BARAN DAS

Department of Food Engineering and Technology, Tezpur University, Tezpur, Assam, India

ABSTRACT

The plantation crops such as tea, coffee, cocoa, coconut, palm oil (PO), sugarcane, etc., are abundant sources of bioactive compounds. Epidemiological studies have found that consumption of plantation crops have effect on anti-cancer, anti-diabetic, antioxidant, and lipid profile regulator for human health. It contained various types of bioactive compounds like terpenes, terpenoids, alkaloids, and phenolic compounds. A systematic and integrated technique for discovering these compounds with good health benefits is essential, given the wide spectrum of bioactive chemicals and the large number of plantation crops. As a result, choosing the right extraction process for quantitative and qualitative studies on bioactive compounds from plantation crops is critical. Extraction is the primary step in any bioactive chemical study, and it has a significant impact on the ultimate outcome. Bioactive substances are typically discovered as a mixture of several molecules with varied degrees of polarity, making separation and identification difficult. As a result, choosing an analytical approach for bioactive component identification and quantification is a critical stage in the research process. In this chapter, bioactive compounds of plantation crops are discussed and their extraction technique. In addition, the analytical technique and functional activity of the bioactive compounds are discussed.

Bioactive Compounds and Nutraceuticals from Plant Sources: Extraction Technology, Analytical Techniques, and Potential Health Prospects. Charanjit Singh Riar, Parmjit S. Panesar (Eds.)

2.1 INTRODUCTION

Natural materials, such as plant extracts, whether as pure biochemicals or standardized extracts, provide limitless possibilities for new medicine development. According to data provided by the World Health Organization (WHO), there are over 20,000 different medicinal plants in 91 nations around the world, including 12 mega-biodiversity countries [1, 2]. Conventional medicine plants offer a variety of biochemicals that can be utilized to treat both chronic and infectious ailments [3]. As a result of the advent of unfavorable effects and microbial resistance to chemically created treatments, men turned to ethnopharmacognosy. Thousands of phytochemicals derived from plants have been revealed to be safe and effective substitutes with fewer adverse effects. Clinical trials aimed at determining the bioavailability, pharmacokinetics, safety, effectiveness, and drug interactions of freshly produced bioactive substances and their formulations or extracts must be thoroughly assessed. Clinical trials are meticulously prepared to protect participants' health as well as to answer particular research queries by assessing both instant and long-term adverse effects, as well as their effects, before the treatment is extensively used on patients [1]. Extraction of these bioactive compounds from plantation crops is one of the major challenges.

Pharmacological screening, extraction, isolation, and characterization of bioactive substance, clinical examination and toxicity evaluation are the first steps in using biologically active compounds from plant materials. This chapter covers extraction, separation, health benefits, chemical composition, and characterization of bioactive compounds extracted from plantation crops using conventional phytochemical screening assays, chromatographic methods including Fourier transform mass spectrometry (FTMS) and HPLC and HPLC/MS. Also offers researchers with information that might be useful in developing upcoming strategies, like selecting potential bioactive components to build functional meals for a variety of non-communicable diseases.

2.2 CLASSIFICATION AND SYNTHESIS OF BIOACTIVE COMPOUNDS

The categorization of bioactive substances into multiple groups is currently inconclusive, and it relies on the classification's purpose. Biosynthetic classifications, for example, will not equal the extent of pharmacological categorization due to their simplicity in describing biosynthetic processes. Plant bioactive chemicals are grouped into three groups, according to Croteau et al. [4]: (i) terpenes and terpenoids (about 25,000 varieties); (ii) alkaloids (roughly

12,000 types); and (iii) phenolic compounds (approximately 8,000 types). Figure 2.1 depicts the general architectures of many bioactive chemical types.

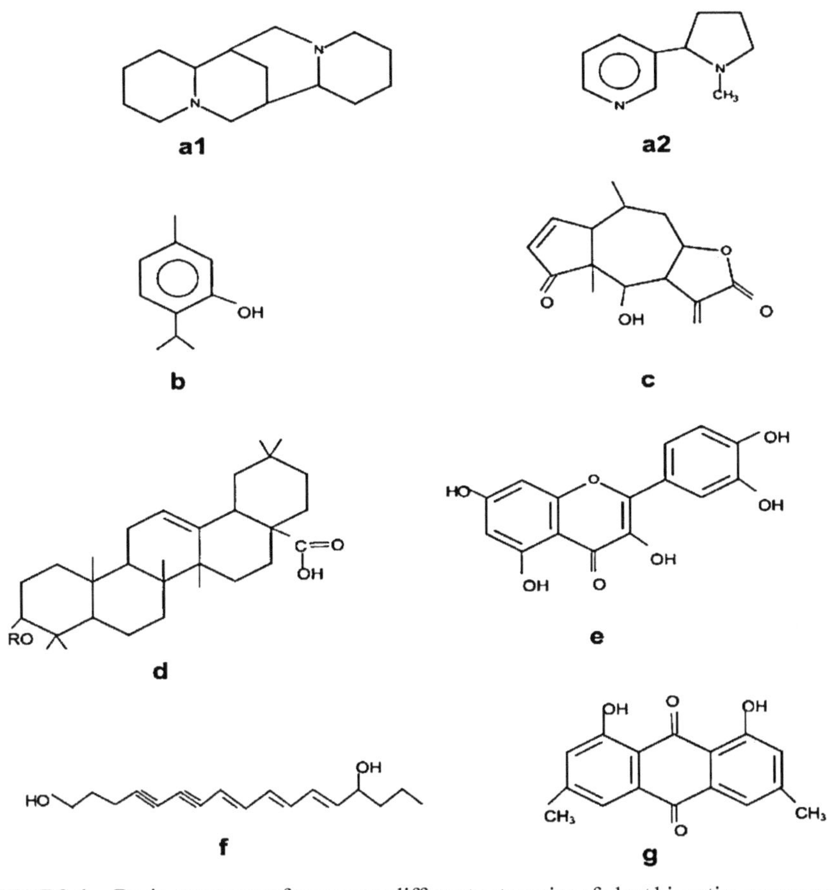

FIGURE 2.1 Basic structures of numerous different categories of plant bioactive compounds: alkaloids (a1 and a2), monoterpenes (b); sesqueterpenes (c); triterpenes, saponins, steroid (d); flavonoids (e); polyacetylenes (f); and polyketides (g).

Most of the bioactive chemicals are classified into one of many families, all with distinct structural properties resulting from the mode they are formed in nature (biosynthesis). Malonic acid, shikimic acid, mevalonic acid, and non-mevalonate (MEP) are the four primary pathways to produce secondary metabolites or bioactive compounds [5]. Aromatic amino acids (from the shikimic acid route) and aliphatic amino acids (from the tricarboxylic acid cycle) both create alkaloids. To make phenolic chemicals, malonic acid and shikimic acid pathways are used. Terpenes are generated via pathways; mevalonic acid and MEP [6] (Figure 2.2).

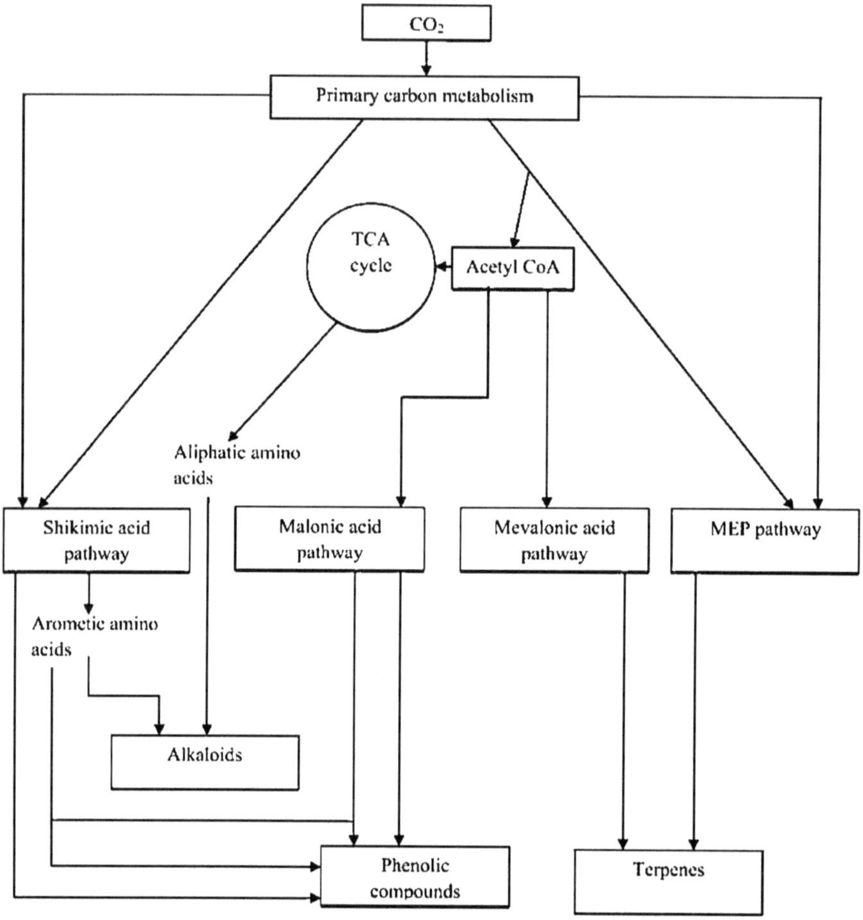

FIGURE 2.2 A simplified diagram of the three primary classes of plant bioactive chemicals' manufacturing routes.

Source: Adapted from Ref. [6]

2.3 CHEMICAL COMPOSITIONS

Phytochemicals are plant compounds that protect plant cells from environmental threats such as pollution, stress, dehydration, UV exposure, and pathogenic assault [7, 8]. Phytochemicals can be found in a broad variety of foods, including tea, coffee, cocoa, oil palm, cashew nut, coconut, areca nut, rubber, sugarcane, legumes, whole grains, nuts, seeds, fungus, herbs, and spices [8, 9]. For example, the most frequent chemicals in cocoa shells are flavanols: epicatechin, catechin, and procyanidins, which are stored in cocoa seed cotyledons [10]. Similarly, in teas alkaloids and flavonoids are present [11, 12] and catechin and epicatechin are the most abundant polyphenolics in cashew nut Testa. Plants create these substances to defend themselves, but recent study has shown that many phytochemicals can also protect humans against illnesses [13]. Plant phytochemicals are physiologically active, naturally existing chemical compounds that provide therapeutic effects in addition to those provided by micronutrients and macronutrients. These chemicals are minor metabolites in plants that have biological features such antioxidant activity, antimicrobial activity, immune system modulation, enzyme detoxification control, platelet aggregation reduction, anticancer action and hormone metabolism [14]. Aromatic plants, commonly called herbs and spices, have been utilized in traditional medicine and food preservation since antiquity. Oregano, rosemary, sage, anise, basil, and other well-known fragrant herbs come from the Mediterranean region. They include a variety of physiologically active chemicals, mostly polyphenolics that have been discovered to have antioxidant, antimicrobial, antifungal, antiprotozoal, antiparasitic, and anti-inflammatory activities [15]. Classification of different phytochemicals/bioactive compounds can be seen in Figure 2.3. The lists of bioactive and disease-preventing phytochemicals found in plants, and bioactive compounds present in plantation crops can be understood from Tables 2.1 and 2.2, respectively.

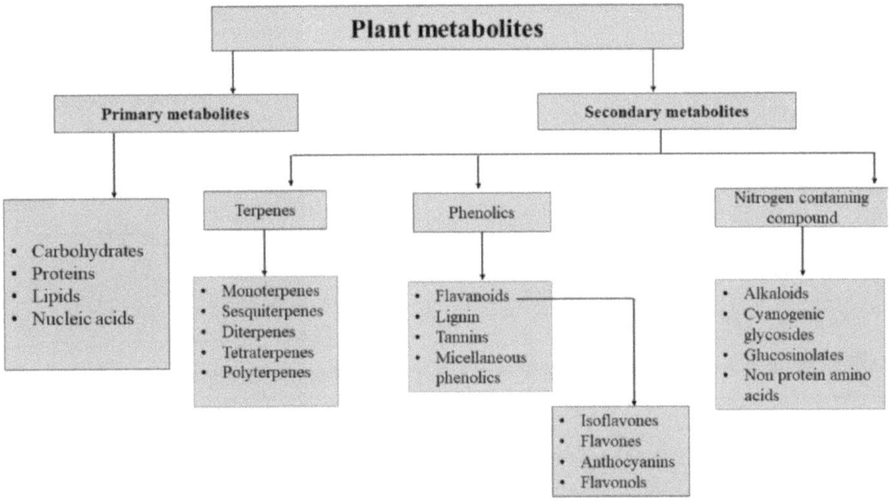

FIGURE 2.3 Classification of different phytochemicals/bioactive compounds.

Source: Adapted from: Ref. [113].

TABLE 2.1 Lists the Bioactive and Disease-Preventing Phytochemicals Found in Plants

Classification	Main Groups of Compounds	Biological Function
NSA (non-starch poly-saccharides)	Cellulose, hemicellulose, gums, mucilages, pectins, lignins	Water holding capacity, delay in nutrient absorption, binding toxins and bile acids.
Antibacterial and antifungal	Terpenoids, alkaloids, phenolics.	Inhibitors of microorganisms reduce the risk of fungal infection.
Antioxidants	Polyphenolic compounds, flavonoids, carotenoids, tocopherols, ascorbic acid.	Oxygen free radical quenching, inhibition of lipid peroxidation.
Anticancer	Carotenoids, polyphenols, curcumine, flavonoids.	Inhibitors of the tumor inhibited the development of lung cancer, anti-metastatic activity.
Detoxifying agents	Reductive acids, tocopherols, phenols, indoles, aromatic isothiocyanates, coumarins, flavones, carotenoids, retinoids, cyanates, phytosterols.	Inhibitors of procarcinogen activation, inducers of drug binding of carcinogens, inhibitors of tumorogenesis.
Other	Alkaloids, terpenoids, volatile flavor compounds, biogenic amines.	Neuropharmacological agents, antioxidants, cancer chemoprevention.

TABLE 2.2 Bioactive Compounds Present in Plantation Crops

Plant	Bioactive Compound	References
Rosmarinus Officinalis Leaves	Flavonoids, arnosol, rosmaridiphenol, rosmadial, Rosmarinic acid, and carnosic acid.	[114]
Arecanut (*Areca catechu* L.)	Gallic acid, (–)-gallocatechin, (–)-epigallocatechin, D-catechin, isorhamnetin-3-*O*-galactoside, arecaine, arecoline, and guvacoline.	[115]
Cocoa (*Theobroma cacao* L.)	Flavanols: epicatechin, catechin, and procyanidins.	[10]
Oolong tea	Alkaloids: Theobromine, caffiene, ophylline; flavonoids: catechins, tannins, flavonols, and flavonol glycosides.	[12]
Green tea	Alkaloids: Caffeine and flavonoids: tannins, catechins, gallic acid.	[11]
Sugarcane	Apigenin, luteolin, tricin, caffeic acid, hydroxycinnamic acid and sinapic acid.	[116]
Black tea	Catechin (C), (–)-epicatechin (EC), (–)-epicatechin-3-gallate (ECG), (–)-epigallocatechin (EGC), (–)-epigallocatechin-3-gallate (EGCG), theaflavin (TF1), heaflavin-3-gallate (TF2a), theaflavin-3′-gallate (TF2b), theaflavin-3 and 3′-digallate (TF3)	[117]

TABLE 2.2 *(Continued)*

Plant	Bioactive Compound	References
Rubber	Trolox equivalent antioxidant capacity by 2,2′-azinobis (3-ethylbenzothiazoline-6-sulfonic acid) diammonium salt and 1,1-diphenyl-2-picrylhydrazyl (DPPH) methods, total phenolics (TotP), non-tannic phenolics, tannic phenolics and total flavonoids.	[118]
Coconut	Phenolic acid, favonoids, phytosterols, D5-Avenasterol, cycloartenol, campesterol, and β-sitosterol, β-stigmasterol	[119]
Cashew nut	Beta-carotene, lutein, zeaxanthin, alpha-tocopherol, gamma-tocopherol, thiamin, stearic acid, oleic acid, and linoleic acid.	[120]
Oil palm	Tocopherols, tocotrienols, carotenoids, phytosterols, phenolic compounds and other phytonutrients.	[121]
Coffee	Caffeine, chlorogenic acids (CGAs), trigonelline, tryptophan alkaloids, diterpenes.	[122]

2.4 EXTRACTION TECHNOLOGY OF PLANTATION CROP BIOACTIVE COMPOUNDS

Given the wide range of bioactive chemicals and the large number of plant species, a standardized and integrated strategy to identifying these compounds with beneficial health effects is required. The quality and quantity of bioactive substances affect their efficacy. The selection of a suitable extraction technique is crucial for qualitative and quantitative research of bioactive chemicals from plant materials [6]. Extraction is the initial stage in every bioactive molecule investigation, and it has a big impact on the result.

It is a crucial step in the isolation and recovery of bioactive molecules from plants. It transforms an actual matrix into a sample appropriate for further analysis. Sample preparations have been carried out by numerous ethnic groups across the world since ancient times, employing decoction, infusion, maceration, digesting, and percolation processes to create extract and thereby use plants for their medicinal benefit. During the modernization and industrialization era, an enhanced extraction process called "Soxhlet Extraction" was established in the 18th century, which is a more sophisticated version of digesting and decoction procedures. However, traditional extraction procedures, such as Soxhlet, have the drawbacks of requiring a higher amount of solvent and taking longer to extract chemicals with poorer yields. The demands of the rising herbal/nutraceutical medicine-based businesses include the exponentially growing demand for herbal produces and/or extracts for wider and safer uses, as well as the timely convenience of high-quality goods with low processing costs and better output. Alternate and non-conventional extraction practices are in high demand to tackle these issues [1, 16, 17]. So basically, there are two types of extraction techniques: conventional and non-conventional. Organic solvents, heating, and agitation are all required in traditional procedures. Soxhlet, maceration, and hydro-distillation are examples of this sort of procedure. Modern techniques or non-traditional techniques, such as enzyme-assisted extraction (EAE), supercritical fluid extraction (SFE), microwave-assisted extraction (MAE), pulse electric field extraction (PEF), subcritical water extraction (SCWE) and ultrasound assisted extraction (UAE), are green or clean techniques because they use less energy and use organic solvents, which are environmentally friendly.

2.4.1 CONVENTIONAL EXTRACTION

To extract bioactive compounds from plant sources, a variety of classic extraction techniques can be applied. The bulk of these techniques rely on the solvents' extraction capabilities, as well as the application of heat and/or mixing. Soxhlet extraction, maceration, and hydro-distillation are the three most used methods for extracting bioactive compounds from plants.

2.4.1.1 SOXHLET EXTRACTION

Soxhlet is a standard method and the fundamental reference for assessing the performance of other solid–liquid extraction (or leaching) procedures. It has been in use for quite some time. Apart from the extraction of thermolabile compounds, which has a limited range of applications, Soxhlet extraction is a general and well-established method that outperforms other classic extraction procedures [18]. As the liquid reaches the overflow level, a siphon aspirates the thimble-holder solution and reloads it into the distillation flask, transferring the extracted solutes back into the bulk liquid. Distillation separates the solute from the solvent in the solvent flask. The solute is left in the flask while fresh solvent is introduced into the plant's solid bed. The technique is carried out again and again until the extraction is complete [17]. While utilizing the Soxhlet extraction method to extract specified nutraceuticals, an appropriate extracting solvent should be used. Extracts and extract compositions will vary depending on the solvent used. The most common solvent for extracting edible oils from plant sources is hexane. Hexane has a relatively limited boiling point range of 63–69°C, making it an effective oil solvent in case of oil solubility and recovery. Whereas, the major component of industrial hexane, n-hexane, is ranked No. 1 harmful air pollutants [19]. On the other hand, Due to health, environmental, and safety considerations, the practice of alternate solvents like isopropanol, ethanol, hydrocarbons, and even water has grown. Also, under every given set of circumstances, it was discovered that d-limonene extracted much more oil than hexane [20].

Alternative solvents frequently result in lower recovery because of a lower molecular affinity between the solute and the solvent. Alternative solvents may have greater prices. To boost the polarity of the liquid phase, one co-solvent is occasionally added. It has been found that combining solvents like isopropanol and hexane improves extraction yield and kinetics.

2.4.1.2 MACERATION

The simplest extraction method is maceration, which is immersing the plant sample in a solvent appropriate for extracting the bioactive component for an extended period of time(s). For a long period of time, maceration has been exploited in the making of tonics at home. It grew in popularity as a low-cost source of essential oils and bioactive chemicals. Maceration is usually divided into many phases for small-scale extraction. To begin, the surface area of plant materials is increased by grinding them into fine particles. This allows for optimum solvent mixing. Second, a suitable solvent called menstruum is added to a closed vessel during the maceration process. Finally, the liquid is strained out, but the marc, which is the solid residue left over from the extraction process, is forced to recover a substantial amount of occluded solutions. Filtration is used to extract contaminants from the strained and pressurized liquid. In maceration, occasional shaking helps extraction in two ways: (i) it increases diffusion; and (ii) it removes concentrated solution from the sample surface, allowing more solvent to reach the menstruum, resulting in a better extraction yield [7].

2.4.1.3 HYDRO-DISTILLATION

Hydro-distillation is a centuries-old technique for extracting bioactive compounds and essential oils from plants. It may be done before the dehydration of plant materials and does not require the use of organic solvents. The three types of hydro-distillation are water distillation, water, and steam distillation, and direct steam distillation. The plant components are first packed in a still chamber; after that, enough water is added, and the procedure is brought to a boil for 6 to 8 hours. Due to indirect cooling with water, the vapor mixture of water and oil condenses. The condensed mixture is passed through a separator, which easily separates the oil and bioactive compounds from the water. Hydro-distillation is made up of three physical and chemical processes: diffusion, hydrolysis, and heat decomposition. Some volatile compounds may be lost at high extraction temperatures. Because of this limitation, it can only be used to extract thermally labile compounds [6, 21, 22].

2.4.2 NON-CONVENTIONAL EXTRACTION

Lengthy extraction times, the need for costly and high-quality solvent, limited extraction selectivity, evaporation of a large volume of solvent, and heat breakdown of thermolabile chemicals are all important drawbacks of traditional extraction. New and promising extraction strategies are developed to address the limitations of traditional extraction methods. Non-conventional extraction techniques are the name given to these methods. Some of the most promising methods are pulsed electric field-assisted extraction (PEFAE), enzyme-assisted extraction, ultrasound-assisted extraction, SCFE, MAE, and SCWE. Because they comply with the Environmental Protection Agency's (EPA) rules, these methods are referred to as "green techniques." Just a few of these include less hazardous chemical synthesis, relatively safe chemicals, safe solvent auxiliaries, configuration for energy efficiency, use of renewable feedstock, reduction of derivatives, catalysis, design to avoid degradation, atom economy, and time analysis for pollution control, and essentially safer chemistry for safety practices [17, 18, 23].

2.4.2.1 ENZYME ASSISTED EXTRACTION (EAE)

Plant tissues are one of the most common sources for antioxidant extraction. Polysaccharides such as cellulose, hemicellulose, and pectins found in plant cell walls function as barriers to the discharge of intracellular chemicals. Cellulase, gluconase, xylanase, glucosidase, and pectinase are the enzymes that help disintegrate structure of cell wall and depolymerize the polysaccharides of plant cell wall, allowing associated chemicals to be released. As a result, these enzymes have been proposed as instruments for optimizing chemical extraction from plant matrix [24–28]. The majority of bioactive compounds, such as flavonoids, interact with the cell wall components (cellulose, hemicellulose, and pectin) in various ways. Cell wall-hydrolyzing enzymes must break these interactions in order for these chemicals to be released. In glucosides, Glycosidase breaks the 1,4-glucosidic bonds. While phenolic molecules are frequently related to cell wall polysaccharides cell wall breakdown is an important step in releasing phenolic compounds from the cell wall. Because they are able to hydrolyze the ester-linked phenolic acids, xylanases, gluconases, and cellulases have been demonstrated to be useful in this regard. However, the compositional and structural properties of compounds have a role in their release [26, 29, 30].

Enzyme-assisted extraction could be used to gain useful compounds from waste materials and agricultural by-products, like gallic acid, which also can be used to make food additives like pyrogallol and propyl gallate, as well as assist as an intermediate in the pharmaceutical chemistry synthesis of the antibacterial drug trimethroprim. The ability of enzymes to catalyze a wide range of processes for the creation of bioactive chemicals is a promising strategy that should be explored further in terms of activity, robustness, and efficiency [31–33].

2.4.2.2 MICROWAVE-ASSISTED EXTRACTION (MAE)

Microwave-assisted extraction (MAE) is a unique method that uses microwave radiation to extract soluble molecules from a range of materials into a fluid. Microwaves are electromagnetic fields with frequencies ranging from 300 MHz to 300 GHz [34]. The electric and magnetic fields, for example, are made up of two perpendicular oscillating fields. Microwave heating theory is based on the notion that it has a direct effect on polar materials. Ionic conduction and dipole rotation processes convert electromagnetic energy to heat. The resistance of the medium to the movement of ions generates heat during the ionic conduction mechanism [35]. Ions, on the other hand, maintain their orientation along constantly changing field indications. Because of the rapid changes in direction, molecules collide, causing heat to be generated. Three phases are assumed to be involved in MAE: first, isolation of solutes from active sites of sample matrix at increased temperature and pressure; second, solvent diffusion through sample matrix; and third, solute transfer from sample matrix to solvent [36, 37]. MAE can extract bioactive chemicals faster and with higher recovery than traditional extraction methods. Cravotto et al. [38] mention several advantages of MAE, including faster heating for the extraction of bioactive compounds from plant materials, lower temperature gradients, smaller equipment, and higher extract yield. MAE can extract bioactive chemicals faster and with a higher recovery rate than traditional extraction methods. It is a selective approach for extracting more intact organic and organometallic compounds. MAE is also considered a green technique because it uses less organic solvents [36].

2.4.2.3 SUPERCRITICAL FLUID EXTRACTION (SFE)

Supercritical fluids (SFs) provide several benefits over traditional extraction methods, including high dissolving power, a higher diffusion coefficient, and ease of separation. They may quickly diffuse through solid materials and hence provide quicker extraction rates. Other advantages include the use of solvents that are generally known as safe (GRAS), higher extraction efficiency in terms of increased yields and shorter extraction times, and the ability to directly couple with analytical chromatographic methods like supercritical fluid chromatography (SFC) or gas chromatography (GC) [23]. del Pilar Garcia-Mendoza et al. [39]; and Paes et al. [40] extracted anthocyanin using a SFs extraction method. They found that extracting anthocyanin using SFs was more efficient than extracting anthocyanin with pressurized liquids or a normal extraction procedure. Researchers used supercritical CO_2, ethanol, and water as solvents to extract anthocyanins and luteolin, finding that pure supercritical CO_2 had the lowest extraction yield while ethanol had the greatest. However, there are a few drawbacks to this method, including the need for a co-solvent, high operating pressure, and greater capital and operational expenses. In the SFE process, raw material is employed in an extraction vessel with pressure and temperature controls to ensure that the extraction process runs smoothly. The fluid is then pumped into the extraction container by a pump. The choice of SFs is important to the process's performance, and this approach can use a wide variety of compounds as solvents [41, 42].

2.4.2.4 ULTRASOUND-ASSISTED EXTRACTION

Ultrasound is a form of sound wave that is audible to humans but not to other animals. In chemistry, the frequency range is usually 20 kHz to 100

MHz. It travels across a medium by compressing and expanding, much like other waves. This results in a phenomenon known as cavitation, which refers to the formation, development, and collapse of bubbles. The transfer of kinetic energy of motion into heating the contents of the bubble can yield a significant quantity of energy. Bubbles have a temperature of roughly 5,000 K, a pressure of 1,000 atm, and a heating and cooling rate of more than 1,010 K/s, according to Suslick [43]. Cavitation occurs only in liquids and liquids containing solids. The major advantage of UAE may be seen in solid plant samples because ultrasonic energy aids in the leaching of organic and inorganic chemicals from the plant matrix [44]. Ultrasound amplification of mass transfer and faster solvent access to plant cell materials are two possible mechanisms. The ultrasonic extraction technique comprises two primary physical phenomena: (i) diffusion over the cell wall; and (ii) rinsing the contents of the cell once the walls have been broken [45]. For attaining efficient and successful extraction, moisture content of the sample, milling degree, particle size, and solvent are all critical components. Furthermore, the effect of ultrasound is governed by temperature, pressure, frequency, and period of sonication. Ultrasound assisted extraction (UAE) have also been integrated with different traditional ways since they are said to improve the effectiveness of a traditional system. An ultrasonic device is positioned in an optimal location in a solvent extraction machine to improve extraction efficiency [46].

UAE has several benefits, including reduced energy consumption, extraction time, and solvent usage. For extraction, ultrasound energy also allows for better mixing, lower thermal gradients and extraction temperatures, faster energy transfer, selective extraction, faster response to process extraction control, smaller equipment, faster start-up, higher production, and the elimination of process steps [6, 47].

2.4.2.5 PULSE ELECTRIC FIELD (PEF) EXTRACTION

Pulse electric field extraction (PEF) works by destroying the cell membrane structure to increase extraction. An electric potential travels through the membrane of a live cell when it is suspended in an electric field. Electric potential divides molecules in the cell membrane based on their charge because of the dipole nature of membrane molecules. Repulsion between charged molecules that produce holes in weak sections of the membrane occurs when the trans membrane potential exceeds a threshold value of roughly 1 V, causing a significant increase in permeability [48]. Plant materials are commonly treated with PEF using a circuit with exponential decay pulses. Plant materials are placed in a chamber with two electrodes for treatment. Depending on the treatment chamber design, the PEF process can run in continuous or batch mode [6, 49]. The efficiency of PEF treatment is greatly influenced by process variables such as treatment temperature, specific energy input, field strength, pulse number, and material properties [50].

PEF can improve mass transfer throughout extraction by disrupting the membrane structure of plant materials, resulting in faster extraction and less waste. PEF has been used to increase cell membrane permeability and thereby improve intracellular chemical release from plant tissue [51]. Plant tissue cell membranes are damaged by PEF treatment at a modest electric field (500 and 1,000 V/cm; for 10^{-4}–10^{-2} s) with no temperature increase [52, 53]. As a result, PEF can help to prevent heat-sensitive chemicals from degrading [54]. PEF can also be used as a pre-treatment technique on plant materials prior to traditional extraction to reduce extraction effort [55].

2.4.2.6 SUBCRITICAL WATER EXTRACTION (SCWE)

The four phases of the SCWE technology are listed below. At high temperature and pressure, the solute is desorbing at different active sites in the sample matrix in the first step. The diffusion of the isolates into the matrix is the focus of the second step. Depending on the sample matrix, the solutes may split themselves from the sample matrix into the extraction fluid in the third step. The solution from the extraction cell is drained and obtained using chromatography in the final stage [56, 57]. Previous study has proven that the extraction of subcritical water follows the thermodynamic model. To extract a constituent from a matrix, this approach requires two steps – (i) the chemicals must be removed from the surface from their original binding location in the matrix of the sample; and (ii) the chemicals must be eluted from the sample in the same way as front elution chromatography is done.

The following two features may be used to explain why SCWE efficiency should be improved: (i) increased surface destabilization; and (ii) better solubility and mass transfer effects. The features of subcritical water change as the temperature rises, as does its ability to dissolve analytes. This is attributed to a reduction in viscosity and an increase in diffusibility, allowing the matrix particles to penetrate the matrix more effectively. Fresh water is continuously injected into the subcritical water dynamic extraction process, improving extraction yield and mass transfer efficiency. Moreover, high temperatures and pressure can harm a material's surface. The active areas in the matrix, van der Waals forces, hydrogen bonds, and dipole attraction of solute molecules that cause solute-matrix interactions can be addressed by increasing the temperature [58, 59]. When water's temperature goes over the boiling point, sufficient pressure must be supplied to keep it liquid. The dissolution of analytes from matrix pores can be aided by applying pressure. SCW of phenolic compounds from a variety of foods is becoming more common. Subcritical water is described as water with a temperature between 100 and 374°C and a pressure high enough to keep it liquid (below the critical pressure of 22 MPa). SCW provides several advantages over standard extraction approaches, with a better extraction quality, lower solvent costs, shorter extraction time, and environmental friendliness. SCW is the most potential engineering technology for ecologically friendly extraction of different chemicals from plants and algae [42, 60, 61].

To analyze extracts generated from plant resources, the researchers utilized ethanol, subcritical water, and boiling water. The highest extraction yield of polyphenolic concentrations is done by subcritical water, although acidic hydrolysis boosted polyphenolic and flavonoid levels. As measured by DPPH, acidification increased the antioxidant activities of -carotene and ferric thiocyanate [62, 63]. Another study looked at how modifying subcritical-water extraction conditions influenced the total polyphenolic content and antioxidant activity (such as duration and temperature). The greatest antioxidant activity and total polyphenolic content were obtained at T = 200°C, P = 1.4 MPa, and t = 60 minutes (as assessed by DPPH). Under these conditions, SCW was shown to be the most suitable and successful method for extracting antioxidants and nutraceuticals from plantation crops, yielding the highest quantities of polymethoxylated flavones (sinensetin, nobiletin, and tangeretin) [63].

2.5 ANALYTICAL TECHNIQUES FOR QUALITATIVE AND QUANTITATIVE ANALYSIS

Bioactive compounds are generally found as a mixture of various molecules with varying degrees of polarity, making it difficult to separate and identify them. Separation techniques such as flash chromatography, Sephadex chromatography, and HPLC are often used in the separation of bioactive compounds. The pure compounds' structure and biological activity are then determined. Bioactive compounds can also be obtained and identified using non-chromatographic techniques such as mass spectrometry (MS) and Fourier-transform infrared spectroscopy (FTIR) [1]. These techniques can be divided into two parts: chromatographic techniques and non-chromatographic techniques.

2.5.1 CHROMATOGRAPHIC TECHNIQUES

2.5.1.1 THIN LAYER CHROMATOGRAPHY

Thin layer chromatography (TLC) is inexpensive and quick method for analyzing plant extracts and natural biomolecules that is utilized all over the world. It's simple to execute and doesn't cost a lot of money because it doesn't require any large machinery. TLC is also adaptable to extracts of varied polarity, and the ability to analyze a large number of samples at once makes it a quick screening technique. A thin glass plate covered with aluminum oxide or silica gel serves as the solid phase in thin layer chromatography (TLC). The mobile phase is a solvent that is chosen depending on the properties of the components in the combination. The "Like Dissolves Like" solubility rule is observed. The more time a compound spends in the mobile phase, the more its physical characteristics resemble those of the mobile phase [64]. Through TLC, compound in the mixture can be identified by comparing its Rf value from the Rf value of a known compound. Other tests include spraying phytochemical screening reagents, which create color changes based on the phytochemicals present in a plant extract, or observing the plate under UV light. Purity and identification of isolated chemicals have also been confirmed using this method [1].

2.5.1.2 HIGH PERFORMANCE THIN LIQUID CHROMATOGRAPHY (HPTLC)

This is a type of planer chromatography in which the sample components are separated on high-performance layers, and data is collected using an advanced workstation. This is a reliable, fast, and efficient instrument for quantitative chemical analysis. Despite the fact that it is based on TLC, It has been modified in a number of ways in order to increase the separation resolution of the chemicals to be separated and to allow quantitative analysis of the compounds. This method employs high-quality TLC plates with smaller particle sizes in the stationary phase, resulting in a superior result. The separation of compounds can be increased by utilizing a multiple development device to develop the plate many times. In addition, high-performance thin liquid chromatography (HPTLC) appears to be more sensitive and reliable than traditional TLC. Considering that the distance of migration in HPTLC can be half that of standard TLC, the expense of the plate for both ways to do the same analysis is about the same. HPTLC, on the other hand, is more expensive in terms of the initial financial commitment required to purchase the main and auxiliary equipment as well as software [65]. Sorbent particles of size 5–7 microns and a layer thickness of 150–200 microns are pre-coated over the surface of the plate. Table 2.3 shows the primary differences between TLC and HPTLC in terms of sorbent particle and pore size [66].

TABLE 2.3 Differences between HPTLC and TLC

Criteria	HPTLC	TLC
Layer of sorbent	100 μm.	250 μm
Efficiency	High due to smaller particle size generated.	Less
Separations	3–5 cm.	10–15 cm
Analysis time	Shorter migration distance and the analysis time is greatly reduced.	Less
Solid support	Silica gel for normal phase and C8, C18 for reverse phase.	Silica gel, alumina
Sample spotting	Autosampler	Manual spotting
Scanning	UV/Visible/fluorescence scanner scans the entire chromatogram qualitatively and quantitatively and the scanner is an advanced type of densitometer.	Not possible

Source: Adapted from: Ref. [66].

2.5.1.3 HIGH PERFORMANCE LIQUID CHROMATOGRAPHY (HPLC)

HPLC is a chromatographic process for separating a mixture of substances that is used in phytochemical and analytical chemistry to identify, measure, and purify the individual components of a mixture. For the isolation of natural compounds, HPLC is a flexible, efficient, and widely utilized technology [67, 68]. This approach is now widely used among many analytical techniques, and it is the method of choice for fingerprinting studies for quality control of chemicals from diverse herbal plants [69]. Identification of compounds through HPLC is a significant aspect of any HPLC technique. UV detectors

(UVD) are the most popular of all detectors due to its high sensitivity [70]. In addition, the majority of naturally occurring substances have some UV absorption at low wavelengths (190–210 nm) [67]. In contemporary HPLC, a non-polar solid phase, such as C18, and a polar liquid phase, often a mixture of water and another solvent are utilized. Before flowing through a diode array detector (DAD), the analyte must be eluted down a column at a high pressure of up to 400 bars (DAD). A DAD studies the analytes' absorption spectra to help in analyte identification. HPLC is an excellent supplement to GC for identifying chemicals that cannot be vaporized, or that break down at high temperatures. UV detection's great sensitivity comes in handy when a component of interest is only present in trace concentrations in a sample. Phytochemicals are also detected using other detection methods, such as a DAD paired with a mass spectrometer (MS) [71]. For the examination of complicated botanical extracts, liquid chromatography with mass spectrometry (LC/MS) is also a useful tool [72]. LC gives a lot of information for structure determination of substances when tandem MS is applied. As a result, when a pure standard is not available, the combination of HPLC and MS allows for the quick and accurate identification of chemical constituents in medicinal plants [73].

2.5.2 NON-CHROMATOGRAPHIC TECHNIQUES

UV-visible spectroscopy is used to measure and characterize biomolecules in both their pure and crude forms. Aromatic bioactive compounds are potent chromophores in the UV range; therefore, UV-Vis spectroscopy is the most commonly utilized approach for its quantitative analysis [74]. UV-Vis spectroscopy can easily identify phenolic chemicals such as tannins, polymer dyes, anthocyanidins, and phenols [75]. UV-Vis techniques were proven to be less discriminating and to provide more detailed information on total quantity of polyphenol. UV-VIS spectra are used to measure total flavones (320 nm), phenolic acids (360 nm), phenolic extract (280 nm), and anthocyanidins (520 nm). When compared to other procedures, this methodology is not only time-saving but also cost-effective [76].

2.5.2.1 FOURIER-TRANSFORM INFRARED SPECTROSCOPY

Fourier-transform infrared spectroscopy (FTIR) is a useful technique for identifying functional groups in plant extracts. It aids in molecular identification and structure determination [77, 78]. Additionally, a pure compound spectrum is generally so unique that they resemble as a "fingerprint" of similar molecules. The spectrum of an unknown molecule can be recognized by comparing it to the spectrum of a known compound for most typical plant compounds. In this technique, samples are prepared in a various way. For liquid samples, the easiest method is to sandwich one drop of the sample between two sodium chloride plates. The drop produces a thin layer between the plates. Solid samples can be crushed and compressed into a thin pellet for examination using potassium bromide (KBr). Solid samples are dissolved in the solvent system like methylene chloride and then pour the solution onto a single salt plate. The solvent is subsequently evaporated from the plate, leaving a thin coating of the original substance. The FTIR is a high-resolution analytical technique used to identify chemical components and illuminate structural details. To fingerprint herbal extracts or powders, FTIR provides a rapid and non-destructive analysis [78, 79].

2.5.2.2 PHYTOCHEMICAL SCREENING ASSAY

Phytochemicals, often known as secondary metabolic products, are substances derived from plants. The phytochemical screening test is a simple, quick, and inexpensive process that allows researchers to quickly identify the many types of phytochemicals present in plant extracts. It's also beneficial for analyzing bioactive compounds. After obtaining the crude extractor active fraction from plant material, phytochemical screening may be carried out using the proper assays [80].

2.5.2.3 IMMUNOASSAY TECHNIQUE

Immunoassays, which utilize monoclonal antibodies to detect pharmaceuticals and low molecular weight natural bioactive chemicals, are becoming increasingly relevant in bioactive compound research. For receptor binding investigations, enzyme assays, and qualitative and quantitative analytical procedures, they have great specificity and sensitivity. In many circumstances, MAb-based enzyme-linked immunosorbent assays (ELISA) are more sensitive than traditional HPLC procedures. Hybridoma technology is a technique for producing monoclonal antibodies in specialized cells [81]. For the synthesis of monoclonal antibodies against plant medicines using hybridoma technology, the following stages are followed:

- A rabbit is immunized by administering plant medicines repeatedly to produce drug-specific antibodies.
- Spleen cells from two different species of animals are cultivated independently and have the potential to develop specific antibodies against plant medicines.
- The hybridoma approach combines spleen cells with myeloma (cancerous) cells in the presence of polyethylene glycol to produce hybrid cells (PEG). The hybrid cells are chosen for their capacity to make antibodies on hypoxanthine aminopterin thymidine (HAT) media.
- Selected hybridoma cells are cultivated in high quantities to produce monoclonal antibodies against certain plant medicines.
- Using ELISA, monoclonal antibodies are utilized to check for comparable medicines in the plant extract combination [80].

2.5.2.4 NUCLEAR MAGNETIC RESONANCE (NMR) SPECTROSCOPY

The magnetic characteristics of atomic nuclei, particularly the nucleus of the hydrogen atom, the proton, carbon, and a carbon isotope, are studied using nuclear magnetic resonance spectroscopy (NMR). Many researchers have used NMR spectroscopy to analyze molecules by gathering differences between a range of magnetic nuclei and generating an intelligible representation of where these nuclei are in the molecule. It also displays which atoms are present in neighboring groupings. Finally, it may be argued that each of these settings has many atoms [75]. C-NMR with radiolabeled carbon is used to determine the kinds of carbon contained in a chemical [82]. H-NMR is used to determine which forms of hydrogen are present in a chemical and how the hydrogen atoms are linked together.

2.5.2.5 MASS SPECTROMETRY (MS)

In mass spectrometry (MS), molecules are blasted with a mixture of electrons or lasers and then transformed into very energetic charged ions. A mass spectrum is a graph showing the relative big number of fragmented ions vs the mass/charge ratio of these ions. The relative molecular mass (molecular weight) may be determined with great precision using MS, as can precise molecular formula using information about where the molecule has been broken [83]. When tandem MS is employed, it offers a lot of information about the structural compounds. As a result, both HPLC and tandem MS methods may be used to identify chemical compounds in plants quickly and precisely, especially when a pure form of bioactive chemicals is not available [79]. By using MALDI-TOF-MS and UPLCPDA techniques, the phenolic compounds like quinic acid, kaempferol, crysin, and chlorogenic acid were identified in a hydrom [84].

2.6 POTENTIAL HEALTH PERSPECTIVES

Because of propensity of plantation crops to donate hydrogen atoms or electrons, phenolic compounds are known to be main antioxidative or free radical inhibitory agents. Antioxidants can help to reduce heart disease, arthritis, neurodegenerative illnesses, cancer, and brain dysfunction. Anti-hypersensitive activity has been discovered in several bioactive peptides isolated from different crops.

Plantation crops including tea, coffee, cocoa, coconut, palm oil (PO), sugarcane, etc., can be very beneficial as anti-cancer, anti-diabetic, anti-oxidant, and lipid profile regulator for human health. Thus according to data in the literature, plant-derived bioactive chemicals exhibit biological features such as anti-diabetic action and anti-cancer, oxidoreductive balance and brain health, blood lipid profile improvement, cardiovascular disease amelioration, and gut health management.

2.6.1 TEA (OOLONG TEA, GREEN TEA, AND BLACK TEA)

Tea has long been utilized as a medicinal in China. Tea was a good medicine that could treat hundreds of ailments, according to the records in "A Supplement to Materia Medica." Although tea is not a magical treatment, many scientific studies reveal that most of the chemical components in tea are linked to a variety of useful medicines. The chemical content of different types of teas is what distinguishes them the most. Various modern medical studies have described the association between chemicals and health advantages [85, 86]. According to early research, tea includes caffeine that increases lipolysis by acting on lipid droplets but not hormone-sensitive lipase (HSL). This suggests that tea could be a useful basic treatment for fatty liver and obesity caused by a high-fat diet. Tea polymerized polyphenols from tea extract suppress pancreatic lipase, according to another study [87]. According to one animal study, tea extract and catechins can lower body weight and body fat content through lipid metabolism. In addition, some researchers discovered that the polysaccharides in tea had a good hypoglycemic impact, which reduces blood sugar by preventing the a-amylase enzyme from working. Other research evaluated the use of tea to lower the risk of diabetes and to control diabetes, including both animal tests and clinical trials [88]. Gamma-aminobutyric acid (GABA), an amino acid found naturally in Camellia sinensis and tea, can help to reduce stress and anxiety. As a result, GABA in Oolong tea (OT) may be used as a natural sleep aid. It's also a potential treatment for pancreatic ductal adenocarcinoma prevention (PDAC). GABA may also have anti-cancer properties in non-PDAC tumors, according to the researchers. The majority of oolong GABA tea is processed in Japan and Taiwan [89, 90].

In Taiwan, 1,507 people were studied over several decades to see how tea consumption affected their risk of newly diagnosed hypertension (711 men and 796 women). The participants in the study were all over the age of 20 and did not have a history of hypertension. The use of moderate-strength tea on a regular basis can lower the risk of hypertension, according to this study. As a result, the researcher believes that tea extractions can greatly lower the chance of acquiring hypertension [91]. Tea extracts may have a hypocholesterolemic impact due to epicatechin gallate and epigallocatechin gallate. Furthermore, specific OT fractions may suppress erythrocyte hemolysis. Meanwhile, several studies have shown that OT-polymerized polyphenols (OTPP) may help to reduce postprandial triglyceride rises. This suggests that OT (beverage) enhanced with OTPP has significant inhibitory effects on postprandial serum triglyceride and hyperlipidemia increases, and that OT (beverage) may be effective in reducing the risk of lifestyle-related illnesses. In addition, long-term OT intake (drinking) was linked to a lower risk of ischemic stroke and dyslipidemia in another research. Long-term OT use, according to researchers, may slow the progression of atherosclerosis in people with coronary artery disease [92]. Variable tea kinds have different *in vitro* antioxidant functional power, according to research, and this power is largely influenced by the total phenolics concentration in the tea. Furthermore, epigallocatechin, and epigallocatechin gallate make up around 70% of total catechins in tea. This explains why tea has a strong antioxidant impact. Green and black tea has received more investigation in this area than OT. Tea has also been shown to protect DNA from oxidation in several studies [93, 92]. Tea has been shown to protect teeth in studies. Early study suggests that polymerized polyphenols in tea extract can dramatically reduce dental plaque deposition and caries development in people. Tea also has cariostatic action, even after *Streptococcus sobrinus* has established itself in the oral cavity and is more effective in preventing caries than drinking water. There is some evidence that tea, in addition to polyphenols, includes additional cariostatic compounds that inhibit the virulence of *Streptococcus sobrinus* [94, 95].

2.6.2 COFFEE

Coffee is a stimulating and soothing beverage. According to the International Coffee Organization, about 136.5 million bags (60 kg) of coffee were drunk in 2011. According to the National Coffee Association, 64% of adults in the United States over the age of 18 consume coffee. Aside from providing a surge of energy, a morning cup of coffee has other health advantages. According to a study published in the New England Journal of Medicine, coffee use may help to prevent dementia and Alzheimer's disease. The Journal of Medicine is a publication dedicated to the study of medicine. Coffee consumption throughout adulthood may help to postpone or perhaps prevent the above-mentioned condition. Drinking caffeinated coffee can lessen the chance of getting basal cell carcinoma, according to a study done by Jiali Han, a professor at Harvard Medical School who led a study involving over 1,10,000 individuals. Coffee has been shown to protect against cancer and other cardiovascular disorders due to its antioxidant action. Low Density Lipoprotein is also protected against oxidation. Several polyphenol components work together to provide this protective effect. Individual phenolics' physical and chemical characteristics have a significant impact on their antioxidant activity. Coffee consumption of one or two cups per day may lower the risk of heart failure. Drinking 4 cups or more per day reduces the risk of diabetes [96–98].

2.6.3 CASHEW NUTS

Several bioactive and health-promoting compounds (phenolics and flavonoids) have been discovered in tree nuts and their oils. The lipid component of tree nuts, which comprises unsaturated fatty acids, phytosterols, and tocols, is thought to have multiple cardioprotective benefits, according to epidemiological studies. Studies have also revealed that consuming tree nut oils has even more health benefits than eating entire tree nuts, probably owing to the substitution of unsaturated lipids and/or other components in the oil extracts for dietary carbs [99]. The need of a well-balanced diet for a long and healthy life is well acknowledged. Diet-related illnesses occur in all nations because of nutritional deficits or over intake of unhealthy foods. Overconsumption of animal-based meals causes illnesses such as atherosclerotic heart disease and some forms of cancer. Animal-based foods are high in saturated fat. Excess saturated fat is also harmful to one's health. Dieticians all around the globe advocate cutting back on meat and animal products while increasing fruit and nut consumption. Proteins, the body-building ingredient required for tissue growth and maintenance, are abundant in nuts. They're also high in unsaturated fats and provide a concentrated source of energy. The cashew kernel, with its unique combination of lipids, carbs, and proteins, is a high-energy source and delicious snack food, especially for persons on the go [97].

Cashews, like other plant-based foods, are cholesterol-free. Cashew kernels are sodium-free and comprise 7% of the daily required dietary fiber in each serving. They're high in iron and potassium, and they're low in micronutrients like thiamin, riboflavin, niacin, folic acid, and fiber. A well-balanced diet should include a high proportion of monounsaturated fat and a low proportion of saturated and polyunsaturated fats, with the latter two being roughly equal in quantity. In this regard, the lipid composition of cashew kernels is well-balanced. The monounsaturated fat to saturated fat ratio is roughly 4:1. The saturated fat to polyunsaturated fat ratio is likewise well-balanced, at about 1:1.

2.6.4 COCOA

Several studies have shown that polyphenol-rich cocoa shell extracts can be utilized in dental care because these polyphenols have anti-glucosyltransferase action. As a result, the impact of cocoa shells on mutans streptococci caries-inducing activities *in vitro* and experimental dental caries in particular pathogen-free rats infected with mutans streptococci was investigated. According to the findings, cocoa shell extract has significant anti-cariogenic capabilities. Various studies have extracted cariostatic chemicals, higher-molecular-weight polyphenolic compounds, and unsaturated free fatty acids including oleic and linoleic acids from cocoa shells. Anti-glucosyltransferase (GTF) activity was found in higher-molecular-weight polyphenolic compounds, and unsaturated free fatty acids had antibacterial action against *S. mutans* [100, 101]. The antioxidant content in hot chocolate is nearly double that of red wine. The concentration of cocoa was two to three times higher than green tea and four to five times higher than black tea. When chocolate is cooked, more antioxidants are released. About 611 milligrams of the phenolic component gallic acid equivalents (GAE) and 564 milligrams of the flavonoid epicatechin equivalents are found in a cup of hot chocolate (ECE). Internal hemorrhages, albuminuria (the amount of albumin in the urine, which might suggest kidney disease), and diabetes are all treated with the antioxidant gallic acid. Although a conventional bar of chocolate contains significant antioxidant activity, the health advantages may be overshadowed by the saturated fats present – cocoa has significantly less fat per serving than a standard chocolate bar, which has 8 grams of fat. Hot chocolate helps enhance blood flow, reduce blood pressure, and promote heart health because flavonoids assist your body metabolize nitric oxide. Hot chocolate's antioxidants also prevent platelets in your blood from clumping together and creating clots. Hot chocolate can also help you think more clearly. The flavonoids improve brain blood flow and oxygenation. Researchers believe that because dementia is caused by a reduction in blood supply to the brain, cocoa might be used to cure it [97].

Also, the researchers wanted to see if cocoa shell flavonoids might prevent the human immunodeficiency virus (HIV) from causing cytopathic effects in cell culture. They found that cocoa shell pigment suppressed HIV reproduction in two separate testing setups, but that the action of cocoa shell pigment was greatest when it was administered simultaneously with viral adsorption. Cocoa shell pigment is resistant to pH and heat, and it has no toxicity after oral administration, all of which indicate that it has medical promise, particularly as an antiviral agent [10, 102].

2.6.5 PALM OIL (PO)

Several studies have demonstrated the importance of PO intake in the prevention or onset of various diseases. Meta-analytical investigations revealed evidence in favor of the good benefits of PO intake, as well as data in favor of individuals for whom PO use would induce pathologies or worsen disorders. We learned about the involvement of PO in pathologies including obesity, cardiovascular disease, diabetes, and cancer through the literature [103]. PO is a herbal supplement liquid derived from the same-named tree that can be good to your health and used to treat some medical conditions. PO is widely available in supermarkets, medicine shops, and online pharmacies, and is deemed safe for everyday usage due to the absence of recorded severe side effects. However, before beginning any PO treatment program, talk to your doctor about how this supplement will help your unique medical problem [104]. PO supplements are potent antioxidants that can help your body cleanse itself of dangerous free radicals in the circulation. These free radicals, which are basically toxic substances, can raise the risk of heart disease and other cardiac problems. Cardiac disorders such as atherosclerosis (hardening of the arteries) and other cardiac issues can be prevented and treated using herbal supplements such as PO. Moreover, this herbal supplement can reduce the sticky quality of blood platelets, reducing the danger of stroke or internal blood clots, which can harm the body's important organs [97]. Other advantages of PO include the supplement's active components and nutritional worth. Tropical oils, such as PO, include both saturated and trans-fatty acids, which are essential for optimal biological function and balanced meals.

2.6.6 SUGARCANE

In several regions of the world, sugarcane juice is used to treat a variety of human ailments. Since time immemorial, it has been utilized in India's Ayurveda and Unani medical systems as a solo medicament or in combination with other plant products. Sugarcane extracts have been found to have a wide range of biological effects, including anti-inflammatory, acetylcholine release regulation, anti-thrombosis, immunostimulant, vaccine adjuvant, and anti-stress properties. Sugarcane juice has a wide range of biological impacts on infection resistance [105]. In ancient medicine, individuals with jaundice and those with liver ailments were recommended to ingest sugarcane extract to treat their illnesses. Aphrodisiac, laxative, demulcent, antibacterial, and tonic are all uses for sugarcane juice. Sugarcane juice is said to be good for the liver since it regulates bilirubin levels, and it is suggested that drinking a lot of it would help you get rid of your jaundice quickly. Modern pharmacological research has shown that sugarcane has a variety of bioactivities. Although apigenin, tricin, and luteoline glycosides such as schaftoside, vitexin, orientin, and swertisin were detected in abundance in sugarcane juice, several policosanols and steroids were also discovered in various locations of *S. officinarum*. Researchers have concentrated their efforts on developing some of the cheapest crop's lead compounds for various diseases based on these bioactivities and chemical elements of sugarcane [106, 107].

Sugarcane juice affects the numbers of macrophages, neutrophils, and natural killer cells, as well as the natural immunity of host cells against various microbial illnesses such as viral, bacterial, and protozoan diseases. Sugarcane juice by-products have been found to have a wide range of biological activities, including antioxidant, preventive, and other physiological properties. Antioxidants are abundant in sugarcane juice. Many antioxidants are being studied as possible therapeutic agents because free radicals have been implicated in the genesis of various human diseases. Lipid peroxidation is a major cause of membrane damage in many human disorders. Antioxidants are chemicals that can stop the chain reaction of free radicals before they harm important molecules. Antioxidant supplementation became a popular treatment technique for lowering the risk of illnesses caused by free radicals [105, 108]. Sugarcane juice's polyphenols boost metabolism and help prevent weight gain during pregnancy, while its low glycemic index keeps energy levels up. Sugarcane juice is high in calcium, magnesium, and iron, so drinking it regularly can help strengthen immunity and prevent mineral deficit during pregnancy. Due to the presence of potassium, the juice can also be used to aid digestion and alleviate constipation. It aids in the correct functioning of the digestive system and helps to keep stomach illnesses at bay. In addition, diabetic individuals should consume sugarcane juice. It contains natural sugar with a low glycemic index, which protects diabetics from experiencing a rapid spike in blood glucose levels. Also, in an alkaline condition, cancer cannot survive. Sugarcane juice is high in calcium, magnesium, potassium, iron, and manganese, and it has been shown that frequent use of sugarcane juice can help prevent cancer, particularly prostate and breast cancer [105, 109].

2.6.7 COCONUT OIL (CO)

In numerous studies, saturated fatty acids in the oil were discovered to be responsible for some of the nutritional and therapeutic advantages of CO, resulting in features such as oxidation resistance, extended shelf life, and exceptional taste and flavor in cooking. CO was later discovered to be made up of medium-chain fatty acids (MCFAs), which account for 64% of total fat and make it unique and noteworthy. CO includes phospholipids, tocopherol, and other minor ingredients in addition to MCFAs. CO is also known as lauric oil since it contains between 40% and 50% lauric acid in its total fatty acid content. The substantial content of lauric acid contributes to CO's chemical distinction from other oils [110, 111]. Several research investigations have proved the positive effects of CNO, dispelling long-held preconceptions about it as a saturated fatty acid source. This wonderful meal, according to health experts, has exceptional functional features such as antimicrobial, HIV prevention activity, and cardio-protective action. CO also has anti-diabetic properties since it helps to keep blood sugar levels in check. CO contains medium-chain triglycerides (MCTs), which increase insulin secretion and sensitivity, according to research. CO is also getting traction as a possible cancer-fighting and chemotherapy-protective drug. Apart from its main health benefits, CO's high vitamin E content helps to soothe hair and skin [112].

2.7 CONCLUSION

The plantation crops are good source of bioactive compounds and nutraceuticals. All the bioactive compounds and nutraceuticals have significant health benefit. The activity of the bioactive compounds depends on the type of extraction process, as they are sensitive to the processing factors. Non-conventional extraction methods mitigate the negative effects of traditional extraction methods. Because bioactive chemicals are commonly present in complex or combination form, advanced analytical and quantitative techniques are required. Furthermore, each bioactive molecule has a unique function in the human body.

KEYWORDS

- **chromatographic techniques**
- **Fourier-transform infrared spectroscopy**
- **high performance thin liquid chromatography**
- **immunoassay technique**
- **mass spectrometry**
- **nuclear magnetic resonance spectroscopy**
- **phytochemical screening assay**
- **quantitative analysis**

REFERENCES

1. Sasidharan, S., Chen, Y., Saravanan, D., Sundram, K. M., & Latha, L. Y., (2011). Extraction, isolation, and characterization of bioactive compounds from plants' extracts. *African Journal of Traditional, Complementary and Alternative Medicines, 8*(1).
2. Cos, P., Vlietinck, A. J., Berghe, D. V., & Maes, L., (2006). Anti-infective potential of natural products: How to develop a stronger *in vitro* 'proof-of-concept.' *Journal of Ethnopharmacology, 106*(3), 290–302.
3. Duraipandiyan, V., Ayyanar, M., & Ignacimuthu, S., (2006). Antimicrobial activity of some ethnomedicinal plants used by Paliyar tribe from Tamil Nadu, India. *BMC Complementary and Alternative Medicine, 6*(1), 1–7.
4. Croteau, R., Kutchan, T. M., & Lewis, N. G., (2000). Natural products (secondary metabolites). *Biochemistry and Molecular Biology of Plants, 24*, 1250–1319.
5. Tiaz, L., & Zeiger, E., (2006). Secondary metabolites and plant defense. *Plant Physiology* (4th edn., pp. 283–308). Sinauer Associates, Inc., Sunderland, Massachusetts.
6. Azmir, J., Zaidul, I. S. M., Rahman, M. M., Sharif, K. M., Mohamed, A., Sahena, F., & Omar, A. K. M., (2013). Techniques for extraction of bioactive compounds from plant materials: A review. *Journal of Food Engineering, 117*(4), 426–436.
7. Gibson, E. L., Wardle, J., & Watts, C. J., (1998). Fruit and vegetable consumption, nutritional knowledge and beliefs in mothers and children. *Appetite, 31*(2), 205–228.
8. Mathai, K., (2000). Nutrition in the adult years. In: Mahan, L. K., & Escott-Stump, S., (eds.), *Krause's Food, Nutrition, and Diet Therapy* (10th edn., Vol 271, pp. 274–275).
9. Patel, A., Temgire, S., & Borah, A., (2021). Agro-industrial waste as source of bioactive compounds and their utilization: A review. *The Pharma Innovation, 10*(5), 192–196.
10. Panak, B. J., Ačkar, Đ., Jokić, S., Jozinović, A., Babić, J., Miličević, B., & Pavlović, N., (2018). Cocoa shell: A by-product with great potential for wide application. *Molecules, 23*(6), 1404.
11. Chu, D. C., & Juneja, L. R., (1997). General chemical composition of green tea and its infusion. *Chemistry and Applications of Green Tea*, 13–22.
12. Ng, K. W., Cao, Z. J., Chen, H. B., Zhao, Z. Z., Zhu, L., & Yi, T., (2018). Oolong tea: A critical review of processing methods, chemical composition, health effects, and risk. *Critical Reviews in Food Science and Nutrition, 58*(17), 2957–2980.
13. Rao, B. N., (2003). Bioactive phytochemicals in Indian foods and their potential in health promotion and disease prevention. *Asia Pacific Journal of Clinical Nutrition, 12*(1).
14. Saxena, M., Saxena, J., Nema, R., Singh, D., & Gupta, A., (2013). Phytochemistry of medicinal plants. *Journal of Pharmacognosy and Phytochemistry, 1*(6).
15. Christaki, E., Bonos, E., Giannenas, I., & Florou-Paneri, P., (2012). Aromatic plants as a source of bioactive compounds. *Agriculture, 2*(3), 228–243.
16. Belwal, T., Ezzat, S. M., Rastrelli, L., Bhatt, I. D., Daglia, M., Baldi, A., & Atanasov, A. G., (2018). A critical analysis of extraction techniques used for botanicals: Trends, priorities, industrial uses and optimization strategies. *TrAC Trends in Analytical Chemistry, 100*, 82–102.
17. Wang, L., & Weller, C. L., (2006). Recent advances in extraction of nutraceuticals from plants. *Trends in Food Science & Technology, 17*(6), 300–312.

18. De Castro, M. L., & Garcıa-Ayuso, L. E., (1998). Soxhlet extraction of solid materials: An outdated technique with a promising innovative future. *Analytica Chimica Acta, 369*(1, 2), 1–10.
19. Zarnowski, R., & Suzuki, Y., (2004). Expedient Soxhlet extraction of resorcinolic lipids from wheat grains. *Journal of Food Composition and Analysis, 17*(5), 649–663.
20. Hanmoungjai, P., Pyle, L., & Niranjan, K., (2000). Extraction of rice bran oil using aqueous media. *Journal of Chemical Technology & Biotechnology: International Research in Process, Environmental & Clean Technology, 75*(5), 348–352.
21. Vankar, P. S., (2004). Essential oils and fragrances from natural sources. *Resonance, 9*(4), 30–41.
22. Silva, L. V., Nelson, D. L., Drummond, M. F. B., Dufossé, L., & Glória, M. B. A., (2005). Comparison of hydrodistillation methods for the deodorization of turmeric. *Food Research International, 38*(8, 9), 1087–1096.
23. Da Silva, R. P., Rocha-Santos, T. A., & Duarte, A. C., (2016). Supercritical fluid extraction of bioactive compounds. *TrAC Trends in Analytical Chemistry, 76*, 40–51.
24. Moore, J., Cheng, Z., Su, L., & Yu, L., (2006). Effects of solid-state enzymatic treatments on the antioxidant properties of wheat bran. *Journal of Agricultural and Food Chemistry, 54*(24), 9032–9045.
25. Chen, S., Xing, X. H., Huang, J. J., & Xu, M. S., (2011). Enzyme-assisted extraction of flavonoids from Ginkgo biloba leaves: Improvement effect of flavonol transglycosylation catalyzed by *Penicillium decumbens* cellulase. *Enzyme and Microbial Technology, 48*(1), 100–105.
26. Kim, Y. J., Kim, D. O., Chun, O. K., Shin, D. H., Jung, H., Lee, C. Y., & Wilson, D. B., (2005). Phenolic extraction from apple peel by cellulases from *Thermobifida fusca. Journal of Agricultural and Food Chemistry, 53*(24), 9560–9565.
27. Wilkins, M. R., Widmer, W. W., Grohmann, K., & Cameron, R. G., (2007). Hydrolysis of grapefruit peel waste with cellulase and pectinase enzymes. *Bioresource Technology, 98*(8), 1596–1601.
28. Gil-Chávez, G. J., Villa, J. A., Ayala-Zavala, J. F., Heredia, J. B., Sepulveda, D., Yahia, E. M., & González-Aguilar, G. A., (2013). Technologies for extraction and production of bioactive compounds to be used as nutraceuticals and food ingredients: An overview. *Comprehensive Reviews in Food Science and Food Safety, 12*(1), 5–23.
29. Fu, Y. J., Liu, W., Zu, Y. G., Tong, M. H., Li, S. M., Yan, M. M., & Luo, H., (2008). Enzyme assisted extraction of luteolin and apigenin from pigeonpea [*Cajanuscajan* (L.) Millsp.] leaves. *Food Chemistry, 111*(2), 508–512.
30. Yang, Y. C., Li, J., Zu, Y. G., Fu, Y. J., Luo, M., Wu, N., & Liu, X. L., (2010). Optimization of microwave-assisted enzymatic extraction of corilagin and geraniin from *Geranium sibiricum* Linne and evaluation of antioxidant activity. *Food Chemistry, 122*(1), 373–380.
31. Puri, M., Sharma, D., & Barrow, C. J., (2012). Enzyme-assisted extraction of bioactives from plants. *Trends in Biotechnology, 30*(1), 37–44.
32. Wang, T., Jónsdóttir, R., Kristinsson, H. G., Hreggvidsson, G. O., Jónsson, J. Ó., Thorkelsson, G., & Ólafsdóttir, G., (2010). Enzyme-enhanced extraction of antioxidant ingredients from red algae *Palmaria palmata. LWT-Food Science and Technology, 43*(9), 1387–1393.
33. Curiel, J. A., Betancor, L., De Las, R. B., Munoz, R., Guisan, J. M., & Fernandez-Lorente, G., (2010). Hydrolysis of tannic acid catalyzed by immobilized-stabilized

derivatives of tannase from *Lactobacillus plantarum*. *Journal of Agricultural and Food Chemistry*, *58*(10), 6403–6409.
34. Paré, J. J., Bélanger, J. M., & Stafford, S. S., (1994). Microwave-assisted process (MAP™): A new tool for the analytical laboratory. *TrAC Trends in Analytical Chemistry*, *13*(4), 176–184.
35. Letellier, M., & Budzinski, H., (1999). Microwave-assisted extraction of organic compounds. *Analusis*, *27*(3), 259–270.
36. Alupului, A., Calinescu, I., & Lavric, V., (2012). Microwave extraction of active principles from medicinal plants. *UPB Science Bulletin, Series B*, *74*(2), 1454–2331.
37. Jain, T., Jain, V., Pandey, R., Vyas, A., & Shukla, S. S., (2009). Microwave assisted extraction for phytoconstituents – An overview. *Asian Journal of Research in Chemistry*, *2*(1), 19–25.
38. Cravotto, G., Boffa, L., Mantegna, S., Perego, P., Avogadro, M., & Cintas, P., (2008). Improved extraction of vegetable oils under high-intensity ultrasound and/or microwaves. *Ultrasonics Sonochemistry*, *15*(5), 898–902.
39. Del Pilar Garcia-Mendoza, M., Espinosa-Pardo, F. A., Baseggio, A. M., Barbero, G. F., Junior, M. R. M., Rostagno, M. A., & Martínez, J., (2017). Extraction of phenolic compounds and anthocyanins from Juçara (*Euterpe edulis* Mart.) residues using pressurized liquids and supercritical fluids. *The Journal of Supercritical Fluids*, *119*, 9–16.
40. Paes, J., Dotta, R., Barbero, G. F., & Martínez, J., (2014). Extraction of phenolic compounds and anthocyanins from blueberry (*Vaccinium myrtillus* L.) residues using supercritical CO_2 and pressurized liquids. *The Journal of Supercritical Fluids*, *95*, 8–16.
41. Sihvonen, M., Järvenpää, E., Hietaniemi, V., & Huopalahti, R., (1999). Advances in supercritical carbon dioxide technologies. *Trends in Food Science & Technology*, *10*(6, 7), 217–222.
42. Kumar, K., Yadav, A. N., Kumar, V., Vyas, P., & Dhaliwal, H. S., (2017). Food waste: A potential bioresource for extraction of nutraceuticals and bioactive compounds. *Bioresources and Bioprocessing*, *4*(1), 1–14.
43. Suslick, K. S., (1990). Effects of ultrasound on surfaces and solids. *Advances in Sonochemistry*, *1*, 197–230.
44. Herrera, M. C., & De Castro, M. L., (2005). Ultrasound-assisted extraction of phenolic compounds from strawberries prior to liquid chromatographic separation and photodiode array ultraviolet detection. *Journal of Chromatography A*, *1100*(1), 1–7.
45. Mason, T. J., Paniwnyk, L., & Lorimer, J. P., (1996). The uses of ultrasound in food technology. *Ultrasonics Sonochemistry*, *3*(3), S253–S260.
46. Vinatoru, M., Toma, M., Filip, P., Achim, T., Stan, N., Mason, T. J., & Lazurca, D., (1998). Ultrasonic reactor dedicated to the extraction of active principles from plants. *Romanian Patent*, (98–01014).
47. Chemat, F., Tomao, V., & Virot, M., (2008). Ultrasound-assisted extraction in food analysis. *Handbook of Food Analysis Instruments*, 85–103.
48. Bryant, G., & Wolfe, J., (1987). Electromechanical stresses produced in the plasma membranes of suspended cells by applied electric fields. *The Journal of Membrane Biology*, *96*(2), 129–139.
49. Puértolas, E., López, N., Saldaña, G., Álvarez, I., & Raso, J., (2010). Evaluation of phenolic extraction during fermentation of red grapes treated by a continuous pulsed electric fields process at pilot-plant scale. *Journal of Food Engineering*, *98*(1), 120–125.

50. Heinz, V., Toepfl, S., & Knorr, D., (2003). Impact of temperature on lethality and energy efficiency of apple juice pasteurization by pulsed electric fields treatment. *Innovative Food Science & Emerging Technologies*, *4*(2), 167–175.
51. Toepfl, S., Mathys, A., Heinz, V., & Knorr, D., (2006). Potential of high hydrostatic pressure and pulsed electric fields for energy efficient and environmentally friendly food processing. *Food Reviews International*, *22*(4), 405–423.
52. Fincan, M., & Dejmek, P., (2002). In situ visualization of the effect of a pulsed electric field on plant tissue. *Journal of Food Engineering*, *55*(3), 223–230.
53. Lebovka, N. I., Bazhal, M. I., & Vorobiev, E., (2002). Estimation of characteristic damage time of food materials in pulsed-electric fields. *Journal of Food Engineering*, *54*(4), 337–346.
54. Ade-Omowaye, B. I. O., Angersbach, A., Taiwo, K. A., & Knorr, D., (2001). Use of pulsed electric field pre-treatment to improve dehydration characteristics of plant-based foods. *Trends in Food Science & Technology*, *12*(8), 285–295.
55. López, N., Puértolas, E., Condón, S., Raso, J., & Alvarez, I., (2009). Enhancement of the extraction of betanine from red beetroot by pulsed electric fields. *Journal of Food Engineering*, *90*(1), 60–66.
56. Zhang, J., Wen, C., Zhang, H., Duan, Y., & Ma, H., (2020). Recent advances in the extraction of bioactive compounds with subcritical water: A review. *Trends in Food Science & Technology*, *95*, 183–195.
57. Hawthorne, S. B., Yang, Y., & Miller, D. J., (1994). Extraction of organic pollutants from environmental solids with sub-and supercritical water. *Analytical Chemistry*, *66*(18), 2912–2920.
58. Ong, E. S., Cheong, J. S. H., & Goh, D., (2006). Pressurized hot water extraction of bioactive or marker compounds in botanicals and medicinal plant materials. *Journal of Chromatography A*, *1112*(1, 2), 92–102.
59. Cvjetko, B. M., Vidović, S., Radojčić, R. I., & Jokić, S., (2015). Green solvents for green technologies. *Journal of Chemical Technology & Biotechnology*, *90*(9), 1631–1639.
60. Herrero, M., Cifuentes, A., & Ibañez, E., (2006). Sub-and supercritical fluid extraction of functional ingredients from different natural sources: Plants, food-by-products, algae, and microalgae: A review. *Food Chemistry*, *98*(1), 136–148.
61. Zakaria, S. M., & Kamal, S. M. M., (2016). Subcritical water extraction of bioactive compounds from plants and algae: Applications in pharmaceutical and food ingredients. *Food Engineering Reviews*, *8*(1), 23–34.
62. Min, K. Y., Lee, K. A., Kim, H. J., Kim, K. T., Chung, M. S., Chang, P. S., & Paik, H. D., (2014). Antioxidative and anti-inflammatory activities of *Citrus unshiu* peel extracts using a combined process of subcritical water extraction and acid hydrolysis. *Food Science and Biotechnology*, *23*(5), 1441–1446.
63. Putnik, P., Bursać, K. D., Režek, J. A., Barba, F. J., Cravotto, G., Binello, A., & Shpigelman, A., (2017). Innovative "green" and novel strategies for the extraction of bioactive added value compounds from citrus wastes: A review. *Molecules*, *22*(5), 680.
64. Singhal, S., Singhal, N., & Agarwal, S., (2009). *Pharmaceutical Analysis-II, Thin Layer Chromatography, Pragati Prakashan* (1st edn., pp. 98–111).
65. Yan, Y. Z., Xie, P. S., Lam, W. K., Chui, E., & Yu, Q. X., (2010). Study on triterpenoic acids distribution in *Ganoderma* mushrooms by automatic multiple development high performance thin layer chromatographic fingerprint analysis. *Journal of AOAC International*, *93*(5), 1384–1389.

66. Ingle, K. P., Deshmukh, A. G., Padole, D. A., Dudhare, M. S., Moharil, M. P., & Khelurkar, V. C., (2017). Phytochemicals: Extraction methods, identification, and detection of bioactive compounds from plant extracts. *Journal of Pharmacognosy and Phytochemistry*, *6*(1), 32–36.
67. Cannell, J. P., (1998). *Natural Products Isolation* (p. 560). Human Press. Totawa, New Jersy.
68. Piana, M., Zadra, M., De Brum, T. F., Boligon, A. A., Gonçalves, A. F. K., Da Cruz, R. C., & Athayde, M. L., (2013). Analysis of rutin in the extract and gel of viola tricolor. *Journal of Chromatographic Science*, *51*(5), 406–411.
69. Fan, X. H., Cheng, Y. Y., Ye, Z. L., Lin, R. C., & Qian, Z. Z., (2006). Multiple chromatographic fingerprinting and its application to the quality control of herbal medicines. *Analytica Chimica Acta*, *555*(2), 217–224.
70. Li, H. B., Jiang, Y., & Chen, F., (2004). Separation methods used for *Scutellaria baicalensis* active components. *J. Chromatogr. B.*, *812*, 277–290.
71. Tsao, R., & Deng, Z., (2004). Separation procedures for naturally occurring antioxidant phytochemicals. *J. Chromatogr. B.*, *812*, 85–99.
72. Cai, Z., Lee, F. S. C., Wang, X. R., & Yu, W. J., (2002). A capsule review of recent studies on the application of mass spectrometry in the analysis of Chinese medicinal herbs. *J. Mass Spectrom.*, *37*, 1013–1024.
73. Ye, M., Han, J., Chen, H., Zheng, J., & Guo, D., (2007). Analysis of phenolic compounds in rhubarbs using liquid chromatography coupled with electrospray ionization mass spectrometry. *J. Am. Soc. Mass Spectrom.*, *18*, 82–91.
74. Kemp, W., (1991). *Organic Spectroscopy*. Macmillan Education Ltd. Sound Mills Basing Stole, Hampshire.
75. Urbano, M., De Castro, M. D. L., Pe´rez, P. M., Garcı´a-Olmo, J., & Gomez-Nieto, M., (2006). Ultraviolet-visible spectroscopy and pattern recognition methods for differentiation and classification of wines. *Food Chemistry*, *97*(1), 166–175.
76. Shrinet, K., Singh, R. K., Chaurasia, A. K., Tripathi, A., & Kumar, A., (2021). *Bioactive Compounds and Their Future Therapeutic Applications*, 337–362.
77. Eberhardt, T. L., Li, X., Shupe, T. F., & Hse, C. Y., (2007). Chinese tallow tree (*Sapium sebiferum*) utilization: Characterization of extractives and cell-wall chemistry. *Wood and Fiber Science*, *39*(2), 319–324.
78. Hazra, K. M., Roy, R. N., Sen, S. K., & Laskar, S., (2007). Isolation of antibacterial pentahydroxy flavones from the seeds of *Mimusops elengi* Linn. *African Journal of Biotechnology*, *6*(12).
79. Kemp, W., (1991). *Energy and the Electromagnetic Spectrum* (pp. 1–17). Organic spectroscopy. Springer.
80. Farnsworth, N., (1984). The role of medicinal plants in drug development. *Natural Products and Drug Development*, *17*, 30–34.
81. Shoyama, Y., Tanaka, H., & Fukuda, N., (2003). Monoclonal antibodies against naturally occurring bioactive compounds. *Cytotechnology*, *31*, 9–27.
82. Zweig, G., & Sherma, J., (1972). Chromatography. *Analytical Chemistry*, *44*(5), 42–79.
83. Christophoridou, S., Dais, P., Tseng, L. H., & Spraul, M., (2005). Separation and identification of phenolic compounds in olive oil by coupling high-performance liquid chromatography with post-column solid-phase extraction to nuclear magnetic resonance spectroscopy (LC-SPE-NMR). *Journal of Agricultural and Food Chemistry*, *53*(12), 4667–4679.

84. Narayan, C., & Kumar, A., (2013). Identification and characterization of phenolic compounds in hydro methanolic extract of *Achyranthes aspera* (HMEA) by UPLC and MALDI-TOF-MS and *in vivo* antioxidant activity. *Oriental Pharmacy and Experimental Medicine, 13*(1), 51–59.
85. Liang, H. R., (2013). *Tea Therapy Prescription.* Wan Li Book Co. Ltd., Hong Kong.
86. Kong, D., Wu, J., Sun, S., Wang, Y., & Xu, P., (2014). A comparative study on antioxidant activity and inhibitory potential against key enzymes related to type 2 diabetes of four typical teas. *J. Food Nutr. Res., 2*(9), 652–658.
87. Nakai, M., Fukui, Y., Asami, S., Toyoda-Ono, Y., Iwashita, T., Shibata, H., Mitsunaga, T., et al., (2005). Inhibitory effects of oolong tea polyphenols on pancreatic lipase *in vitro. J. Agric. Food Chem., 53*, 4593–4598.
88. Wang, K., Liu, F., Liu, Z., Huang, J., Xu, Z., Li, Y., Chen, J., et al., (2010). Analysis of chemical components in oolong tea in relation to perceived quality. *Int. J. Food Sci. Technol., 45*(5), 913–920.
89. Anju, P., Moothedath, I., & Shree, A. B. R., (2014). Gamma amino butyric acid accumulation in medicinal plants without stress. *Ancient Sci. Life, 34*(2), 68–72.
90. Boonstra, E., De Kleijn, R., Colzato, L. S., Alkemade, A., Forstmann, B. U., & Nieuwenhuis, S., (2015). Neurotransmitters as food supplements: The effects of GABA on brain and behavior. *Front. Psychol., 6*, 1–6.
91. Yang, L., Shu, L., Yao, D. D., Jia, X. B., & Yu, S. M., (2014). Study on the glucose-lowering effect of puerarin in STZ-induced diabetic mice. *Chin. J. Hosp. Pharm, 34*, 1338–1342.
92. Zhu, Q. Y., Hackman, R. M., Ensunsa, J., Holt, R. R., & Keen, C. L., (2002). Antioxidative activities of oolong tea. *J. Agric. Food Chem., 50*, 6929–6934.
93. Yang, Z. B., Zhao, J. S., Zhuang, L. L., & Hu, Y. X., (1998). Human aging causes an oolong tea anti-aging mechanism. *Fujian Tea, 4*, 45–47.
94. Ooshima, T., Minami, T., Aono, W., Tamura, Y., & Hamada, S., (1994). Reduction of dental plaque deposition in humans by oolong tea extract. *Basic Sci. Caries Res., 28*(3), 146–149.
95. Hamilton-Miller, J. M. T., (2001). Anti-cariogenic properties of tea (*Camellia sinensis*). *J. Med. Microbiol., 50*(4), 299–302.
96. Benoit, S., Christophe, C., Angelika, T., & Anne, C., (2001). Health effects and safety considerations. In: Clarke, R. J., & Vitzthum, O. G., (eds.), *Coffee Recent Developments, Agricultural Series* (pp. 165–183).
97. Abraham, V. K., (2015). 16 role of plantation crops in. *Agriculture and Food Technology in Human Life*, 235.
98. George, S. E., Ramalakshmi, K., & Mohan, R. L. J., (2008). A perception of health benefits of coffee. *Critical Reviews in Food Science and Nutrition, 48*(5), 464–486.
99. Singh, N., & Kaur, A., (2013). Plantation crops and tree nuts: Composition, phytochemicals and health benefits. *Handbook of Plant Food Phytochemicals: Sources, Stability, and Extraction*, 163–179.
100. Ooshima, T., Osaka, Y., Sasaki, H., Osawa, K., Yasuda, H., Matsumura, M., & Matsumoto, M., (2000). Caries inhibitory activity of cacao bean husk extract in *in-vitro* and animal experiments. *Archives of Oral Biology, 45*(8), 639–645.
101. Osawa, K., Miyazaki, K., Shimura, S., Okuda, J., Matsumoto, M., & Ooshima, T., (2001). Identification of cariostatic substances in the cacao bean husk: Their

anti-glucosyltransferase and antibacterial activities. *Journal of Dental Research, 80*(11), 2000–2004.
102. Unten, S., Ushijima, H., Shimizu, H., Tsuchie, H., Kitamura, T., Moritome, N., & Sakagami, H., (1991). Effect of cacao husk extract on human immunodeficiency virus infection. *Letters in Applied Microbiology, 13*(6), 251–254.
103. Absalome, M. A., Massara, C. C., Alexandre, A. A., Gervais, K., Chantal, G. G. A., Ferdinand, D., & Jean-Paul, C., (2020). Biochemical properties, nutritional values, health benefits and sustainability of palm oil. *Biochimie, 178*, 81–95.
104. Mancini, A., Imperlini, E., Nigro, E., Montagnese, C., Daniele, A., Orrù, S., & Buono, P., (2015). Biological and nutritional properties of palm oil and palmitic acid: Effects on health. *Molecules, 20*(9), 17339–17361.
105. Chinnadurai, C., (2017). Potential health benefits of sugarcane. In: *Sugarcane Biotechnology: Challenges and Prospects* (pp. 1–12). Springer, Cham.
106. El-Abasy, M., Motobu, M., Shimura, K., Na, K. J., Kang, C. B., Koge, K., & Hirota, Y., (2002). Immunostimulating and growth-promoting effects of sugar cane extract (SCE) in chickens. *Journal of Veterinary Medical Science, 64*(11), 1061–1063.
107. Xu, F., Sun, R. C., Sun, J. X., Liu, C. F., He, B. H., & Fan, J. S., (2005). Determination of cell wall ferulic and p-coumaric acids in sugarcane bagasse. *Analytica Chimica Acta, 552*(1, 2), 207–217.
108. Brash, D. E., & Harve, P., (2002). New careers for antioxidants. *Proc. Natl. Acad. Sci. USA, 99*(22), 13969–13971.
109. McClatchey, W., (2002). From polynesian healers to health food stores: Changing perspectives of *Morinda citrifolia* (Rubiaceae). *Integrative Cancer Therapies, 1*(2), 110–120.
110. Deen, A., Visvanathan, R., Wickramarachchi, D., Marikkar, N., Nammi, S., Jayawardana, B. C., & Liyanage, R., (2021). Chemical composition and health benefits of coconut oil: An overview. *Journal of the Science of Food and Agriculture, 101*(6), 2182–2193.
111. Gans, W. M., & Kauwell, G. P., (2017). *Coconut Oil: A Heart-Healthy Fat?* (Vol. 2017, No. 1, p. 5). FSHN-17-1/FS289, 2/2017. EDIS.
112. Krishna, A. G., Gaurav, R., Singh, B. A., Kumar, P. P., & Preeti, C., (2010). Coconut oil: Chemistry, production, and its applications: A review. *Indian Coconut Journal, 53*(3), 15–27.
113. Mondal, S., Soumya, N. P. P., Mini, S., & Sivan, S. K., (2021). Bioactive compounds in functional food and their role as therapeutics. *Bioactive Compounds in Health and Disease, 4*(3), 24–39.
114. Borrás-Linares, I., Stojanović, Z., Quirantes-Piné, R., Arráez-Román, D., Švarc-Gajić, J., Fernández-Gutiérrez, A., & Segura-Carretero, A., (2014). Rosmarinus officinalis leaves as a natural source of bioactive compounds. *International Journal of Molecular Sciences, 15*(11), 20585–20606.
115. Wang, R., Pan, F., He, R., Kuang, F., Wang, L., & Lin, X., (2021). Arecanut (*Areca catechu* L.) seed extracts extracted by conventional and eco-friendly solvents: Relation between phytochemical compositions and biological activities by multivariate analysis. *Journal of Applied Research on Medicinal and Aromatic Plants, 25*, 25.
116. Thao, N. T., & Tuan, H. Q., (2017). Recovering bioactive compounds from cane sugar wastes. In: Thao, N. T., & Tuan, H. Q., (eds.), *Recovering Bioactive Compounds from Agricultural Wastes*. Wiley.

117. Skotnicka, M., Wynimko, J. C., Jankun, J., & Jankun, E. S., (2011). The black tea bioactivity: An overview. *Central-European Journal of Immunology, 36*(4), 284–292.
118. Piluzza, G., Campesi, G., Molinu, M. G., Re, G. A., & Sulas, L., (2020). Bioactive compounds from leaves and twigs of guayule grown in a Mediterranean environment. *MDPI Plants, 9*(4), 442–448.
119. Ngampeerapong, C., Chavasit, V., & Durst, R., (2018). Bioactive and nutritional compounds in virgin coconut oils. *Malaysian Journal of Nutrition, 24*(2), 257–267.
120. Trox, J., V, V., Vetter, W., Stuetz, W., Scherbaum, V., Gola, U., & Biesalski, H., (2010). Bioactive compounds in cashew nut (*Anacardium occidentale* L.) kernels: Effect of different shelling methods. *J. Agric. Food Chem., 58*(9), 5341–5346.
121. Teixeira, B. C., Macedo, G. A., Macedo, J. A., Da Silva, L. H., & Rodrigues, A. M., (2013). Simultaneous extraction of oil and antioxidant compounds from oil palm fruit (*Elaeis guineensis*) by an aqueous enzymatic process. *Bioresource Technology, 129*, 575–581.
122. Hu, G. L., Wang, X., Zhang, L., & Qiu, M., (2019). The sources and mechanisms of bioactive ingredients in coffee. *Journal: Food & Function, 10*(6), 3113–3126.
123. Weggler, B. A., Gruber, B., Teehan, P., Jaramillo, R., & Dorman, F. L., (2020). Inlets and sampling. In: *Separation Science and Technology* (Vol. 12, pp. 141–203). Academic Press.
124. Hussain, M. K., Saquib, M., & Khan, M. F., (2019). Techniques for extraction, isolation, and standardization of bio-active compounds from medicinal plants. In: *Natural Bio-Active Compounds* (pp. 179–200). Springer, Singapore.
125. Khaw, K. Y., Parat, M. O., Shaw, P. N., & Falconer, J. R., (2017). Solvent supercritical fluid technologies to extract bioactive compounds from natural sources: A review. *Molecules, 22*(7), 1186.
126. Panzella, L., Moccia, F., Nasti, R., Marzorati, S., Verotta, L., & Napolitano, A., (2020). Bioactive phenolic compounds from agri-food wastes: An update on green and sustainable extraction methodologies. *Frontiers in Nutrition, 7*, 60.
127. Samtiya, M., Aluko, R. E., Dhewa, T., & Moreno-Rojas, J. M., (2021). Potential health benefits of plant food-derived bioactive components: An overview. *Foods, 10*(4), 839.

CHAPTER 3

Technological and Analytical Aspects of Bioactive Compounds and Nutraceuticals from Fruit Sources

JOSÉ DE JESÚS LIRA-RICÁRDEZ,[1] MARÍA EUGENIA MENDOZA JASSO,[1,2] and LUCÍA ORTEGA CABELLO[1]

[1]*Biological Systems Department, Metropolitan Autonomous University, Xochimilco Campus, Mexico City, Mexico*

[2]*Pharmacy Department, Chemistry Faculty, National Autonomous University of Mexico, Mexico City, Mexico*

ABSTRACT

The consumption of fruits has been of importance because they have been related to an improvement of the quality of life of humans by preventing chronic-degenerative disease or increasing their lifespan because of their high antioxidant activity, mainly linked to the presence of polyphenols and other compounds such as terpenoids and essential nutrients such as vitamins and minerals. Due to the latter, it is important to explore classical as well modern extraction techniques that allow the extraction of large phytochemical content with minimum social and environmental impact as it is with supercritical fluids (SFs) or microwave-assisted extraction (MAE). The bioactive compounds that explain some of the therapeutical activities of fruits, even if some of them are already known such as resveratrol, quercetin, and gallic acid, among others; it is needed to develop more precise analytical techniques that allow a fast and reliable identification even with low phytochemical content, as it is in the case of high-performance liquid chromatography (LC) or gas chromatography (GC) coupled to mass spectrometry (MS); because to the best of our knowledge these phytochemical

Bioactive Compounds and Nutraceuticals from Plant Sources: Extraction Technology, Analytical Techniques, and Potential Health Prospects. Charanjit Singh Riar, Parmjit S. Panesar (Eds.)

are not in a free form, but they are found bonded to either a sugar, a protein or a fatty acid moiety; or even these techniques can be useful in determining their relationship between their activities and their presence or absence in certain fruits through metabolomic analysis as it is in the case of superfruits or the renovated interest of lactic acid fermentation.

3.1 INTRODUCTION

Many of the diseases or disorders known come from the consumption of fast food or junk food due to our lifestyle, to optimize times to fulfill a saturated daily agenda. To overcome these disorders physicians and nutritionists prescribe the use of supplementary food, mainly composed of nutraceuticals, as a combination between nutrition and pharmaceutics, which concept arose in 1989 by Stephen De Felice as "a food (or part of a food) that provides medicinal or health benefits, including in the prevention and/or treatment of a disease," later on, the concept has evolved by adding the characteristics that nutraceuticals are found within pharmaceutical forms such as pills, tablets, capsules, and powders, among others [1]. Due to this concept evolution, Dutta et al. [2] examined the possibility to consider nutraceuticals either as a food or as a drug, in which they answer that may be considered as both because they can promote health by helping in disease prevention and because it can be used to complement other treatments.

The mechanism by which nutraceuticals exert their promoting health activities are diverse which could be with an antioxidant activity which eventually can promote the increase of life expectancy, as it is in the case of citric fruits and grapes; other prevention mechanisms can be to improve gastrointestinal health mainly attributed to flavonoids. These mechanisms lead to disease prevention such as cardiovascular diseases attributed to polyphenols found on berries and apples, diabetes with the aid of bitter melon and pomegranate, osteoarthritis because of banana consumption and pomegranate, oral diseases due to polyphenols present on grapes, among other diseases [2]. Some of these nutraceuticals have been found to have an important activity against SARS-CoV-2 like polyphenols, vitamins, fatty acids, and some carotenoids to avoid or minimize cardiovascular damage by maintaining redox homeostasis which also results in the enhancement of the immunological response [3]. The latter allows focusing on the importance of phytochemicals present on food present on the dairy intake [4]. Fruits have a special importance because of the variety of bioactive compounds other than polyphenols such as essential oils, carotenoids, vitamins, and minerals [5–8]; where the polyphenol content provides fruits a high antioxidant activity [9].

Ghaffari & Roshanravan [10] wrote a table with common nutraceuticals along with their common chemical structure and food source in which it can be seen that lycopene can be found in watermelon, pink grapefruits, apricots; vitamin C can be found on strawberries and mango; vitamin E in nuts; flavonoids can be found on apple, berries, red grapes, citrus, apricot, in the case of banana inulin can be found, among other examples. Similarly, Prakash et al. [5] reported also a table on the main sources of different classes of polyphenols in which it can be seen that fruits are the main sources of these phytochemicals, as well as some terpenoids reported by Keservani et al. like limonene on cherries and lutein in plum, prunes, and oranges [11].

For this reason, the development of extraction and analytical techniques for structural elucidation and quantification is of importance in the research of nutraceuticals. In the case of classical extraction techniques the choice of solvent is a determinant factor for effective extraction, for example, dimethylformamide can extract efficiently antioxidants, while for the extraction of phenolic compounds methanol can be used to increase their amount from dried powder, the solvents to be used according to their polarity are hexane, chloroform, ethylacetate, acetone, methanol, and water as reported by Thavamoney et al. with different parts of *Dacryodes rostrata* where in the seed with ethanol polyphenols and flavonoids were extracted with ethanol, while anthocyanins were found on the peel using water and carotenoids were found on the peel with hexane followed by ethanol. However, the use of solvents becomes limited due to the generation of hazardous residues, therefore ecofriendly alternatives are currently developing such as microwave and ultrasound [12, 13].

The complexity of the matrix must be analyzed by several techniques such as mass spectrometry (MS), nuclear magnetic resonance (NMR), high-performance liquid chromatography (HPLC), gas chromatography (GC), capillary electrophoresis (CEP); in which a careful consideration of their physicochemical properties will influence the sample preparation, separation, and analytical technique as it is in the case of lipids (fatty acids, sterols, terpenes, and polyphenol derivatives) the methods of choice are GC and GC-MS, and with water-soluble vitamins, peptides, saccharides, flavonoids, and stilbenes among others HPLC and HPLC-MS are preferred techniques [14–16].

3.2 FRUITS

3.2.1 IMPORTANCE OF FRUITS ON AN EVERYDAY DIET

Since ancient times, fruits have been considered as valuable sources of nutrients such as minerals (calcium, magnesium, iron, and potassium), vitamins (vitamin C and E), and fibers and in later times, as sources of valuable phytochemicals which have become useful whether as food additives or bioactive compounds on their own. The bioactivity of these phytochemicals has been associated inversely with the prevention of some chronic degenerative diseases such as cancer, Alzheimer's disease, cardiovascular diseases, and aging, among other diseases [17–19].

Nevertheless, there is controversy in this association with specific diseases such as rectal cancer, bladder cancer, non-Hodgkin lymphoma, and prostate cancer, among others showed null association with fruit intake and disease prevention; the latter has allowed establishing a level of evidence to associate the health outcomes with fruit intake into convicting, probable, possible, limited, insufficient, and no evidence [20]. Being the reason why there is a need to explore the phytochemicals present in fruits to explain the preventive actions of fruits. Among the phytochemicals found on fruits are carotenoids, phenolic acids, flavonoids, coumarins, stilbenes, and lignans [19]. The importance of the before mentioned phytochemicals lies in their antioxidant activity, in which the final result is chemoprevention, either by blocking or suppressing free radicals. Because of the latter, WHO recommends a diary intake of 400 g of fruit per person, in which it has to be each of the seven color classes for the intake of a variety of phytochemicals [21, 22].

However, fruit intake is limited due to availability, economic, and cultural status, as well as doubts on their preservation and pesticide content [22], in which as alternatives to freezing and canned fruit, dried fruits have also been a preservation alternative in which depending on the fruit sugars can be added, where drying without sugars present a similar nutritional content compared to fresh fruit [23]. From all these factors cost is the outstanding barrier of fruit intake where Suchánková et al. report that a high fresh fruit intake is associated with people of high economic income [24].

The PBH Foundation of the United States of America (USA) have implemented strategies to increase the fruit intake by extending the influence to the elderly population, establishing partnerships with food industries and restaurants, making points of view of homemade food being cheaper and better than outdoor food, review storing techniques such as freezing and canning, use of fruits as an easier way to ingest different fruit nutrients as well as the use of special days to promote the intake of fruits, for example on February 27th the national strawberry day is celebrated, in Table 3.1 some national days can be seen for food intake in the USA [25].

TABLE 3.1 Some National Days for Promoting Fruit Intake [25]

January to June	July to December
National Apricot Day (January 23rd)	National Berry Month (July)
National Strawberry Day (February 27th)	National Watermelon Day (August 3rd)
National Oranges and Lemons Day (March 31st)	National Acorn Squash Day (September 7th)
National Raisin Day (April 30th)	National Cranberry Month (October)
Pick Strawberries Day (May 20th)	National Fig Week (first week of November)
National Papaya Month (June)	National Kiwifruit Day (first day of winter)

3.2.2 ANATOMY AND CLASSIFICATION OF FRUITS

The fruit is the recipient of the seeds of plants when they mature mainly constituted by the pericarp (exocarp, mesocarp, and endocarp) and seeds [26]; fruits can also be considered as the organ derived from the gynoecium [27]. The different layers of the pericarp play different roles (Figure 3.1) which could be either for defense or fruit dissemination or sometimes layers can work together to contribute to the lignifying process of fruits of the endocarp.

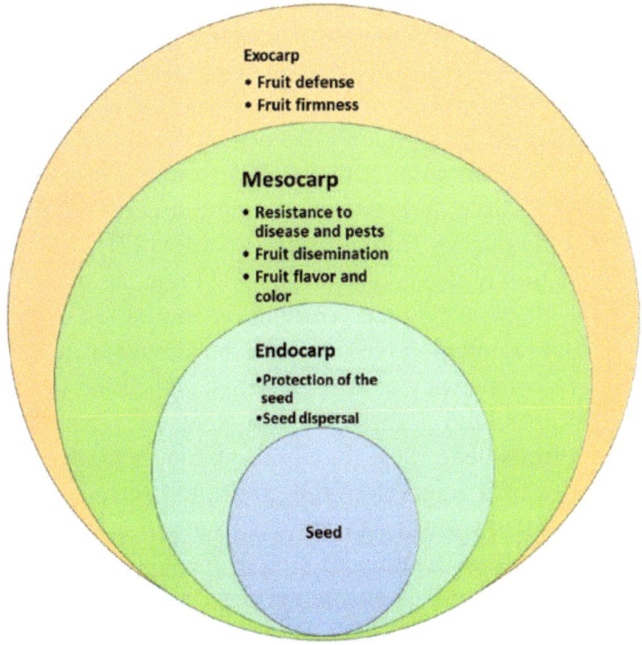

FIGURE 3.1 Layers of the pericarp found on fruits.

Source: Image of own making.

In the case of the exocarp, its function of defense can be as a physical barrier, synthesis of different compounds such as flavonoids for chemical defense from oxidative stress and execute some molecular pathways. In the case of mesocarp and endocarp, metabolite accumulation contributes to fruit characteristics such as color and flavor. In the endocarp, it has been detected in some fruits the presence of calcium as well as cell wall degrading enzymes as well as lignins that aid in the germination process once the seed is dispersed when there are fruits that possess two endocarps, as it is in the case of dried fruits [27, 28]. Depending on the layer of the pericarp different compounds can be found [26]. An example of this could be the results reported by Oranusi et al. [29] where they examined the antimicrobial activity of the exocarp, mesocarp, and the endocarp of *Cola lepidota* fruits and found in the exocarp the presence of tannins, saponins, flavonoids, terpenoids, and phenols; in the mesocarp flavonoids, anthocyanins, terpenoids, phenols coumarins, and betacyanins; and in the endocarp tannins, saponins, flavonoids, and phenols. Other examples are the study reported by Cervantes-Hernández et al. [30] in *Capsicum frutescens fruits* in the pericarp glycosides and terpenoids were detected, the placenta carried capsaicinoids, alkaloids, and tocopherols, while the seeds contained fatty acids and saponins; and that reported by Lachowicz et al. [31] in saskatoon berry where phenolic compounds were present in a higher amount on peel compared to other parts of the fruit, while the pulp was rich in flavonols and seeds were rich in polymeric procyanidins. Fruits are classified by Candolle in 1813 according to the number of ovaries that fruits come from in: simple, they come from a single ovary; multiple, they come from multiple ovaries from a single flower; and aggregate, they come from multiple ovaries from multiple flowers [32].

This classification is also taken by the PBH Foundation [25] and by Jiang and Song [33] with more detailed information about their classification. For example, the PBH Foundation [25] deepens the classification of simple fruits according to their consistency or how protected is the seed in berries, hesperidium, pepos, drupes, and pomes. Jiang & Song [33] prefer extending the classification of fruits better in eight classes: berry, pepo, hesperidium, false berries, compound fruit, aggregate fruit, multiple fruits, and other accessory fruit; looking at Figure 3.2 it can be seen that some of these classes are enclosed within the classification of simple fruits as reported by the PBH Foundation [25]. Another classification criteria are based on their geographical distribution (Figure 3.3). Where tropical fruits require warm weather to grow; sub-tropical fruits are those which grow in the Mediterranean region and unlike the tropical fruits they may endure cold weather, and temperate fruits grow in cold climates [34].

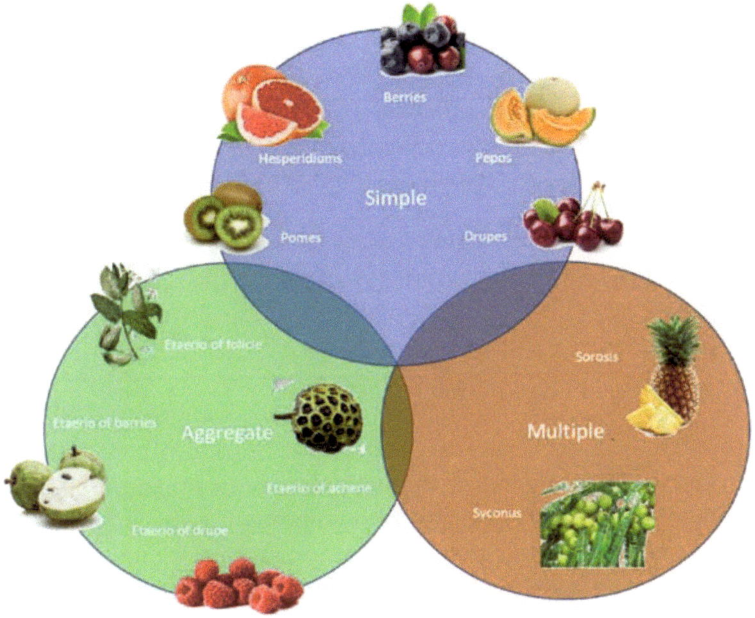

FIGURE 3.2 Classification of fruits according to Candolle in 1813.

Source: Image of own making.

FIGURE 3.3 Classification of fruits based on geographical distribution.

Source: Image of own making.

Other classification criteria can be carried out based on anatomical features of different fruits, reported by Cerri & Reale [27] in which they compared the histology of drupes, berries, and hesperidium in the category of fleshy fruits; legumes, caryopsis, and achenes in the category of dry fruits, where the authors conclude that improving the taxonomy of fruits is of importance for the understanding of plant physiology as well as metabolomics.

3.3 CHEMICAL COMPOSITION OF FRUITS AND BIOLOGICAL ACTIVITIES

3.3.1 METABOLITES FOUND ON FRUITS

As previously seen, the chemical composition of fruits is becoming important for nutritional value as well as biological activities that aids in the control of chronic diseases. In this regard, it is important to consider the secondary metabolism because it provides aroma, color, and nutritional value, this phase of metabolism comes from the interaction of the plants with the environment, assuring diversity of compounds that could be useful in health apart of micronutrient supplement in which it can be considered three main groups of secondary metabolites according to Pott et al. which are terpenoids or isoprenoids, phenolic, and nitrogen or sulfur-containing compounds, being terpenoids and phenolic compounds the principal compounds produced by fruits [35].

Phenolic compounds because of the presence of hydroxyl groups present very good antioxidant activity since these groups can work as electron donors, thus presenting influence over chronic degenerative diseases as well as in infectious diseases caused by *Bacillus subtilis, Escherichia coli, Klebsiella pneumoniae,* and *Staphylococcus aureus,* among others. It has been inferred that phenolic acids exhibit a stronger antibacterial activity than other phenolic compounds because of their small size, thus improving their interaction with the cell membrane, being proposed as adjuvants for resistant bacteria by blocking the resistance mechanisms, making synergism with antibiotics or by modifying the virulence factors of bacteria [36]. Many efforts have been implemented in extracting phenolic compounds from residues such as fruit peels, in this regard, Suleira et al. [37] report phenolic acid, tannin, and flavonoid content of 20 fruit peels where citric fruits presented the highest amount of phenolic compounds, however, only grapefruit presented the highest antioxidant activity in which the authors also attempted to relate the composition with the climate reaching to the conclusion of the importance of studying the production of compounds according to the cultivation conditions.

In the case of terpenoids, Khan et al. [38] reviewed the gastroprotective effects of terpenoids, in this review the authors propose different mechanisms under which terpenoids such as β-pinene and limonene can modulate antibiotic resistance, lupeol with ursolic acid (UA) and nomilin help in the control of inflammation by immunomodulation processes by activating the macrophages, as reported by Wahdaningsih et al. on dragon fruit [39], α-pinene and citral can reduce gastric injury and lupeol along with limonene can reduce the risk of gastric cancer [38] as well as other types of cancer either by inducing apoptosis, activity on reactive oxygen species or inhibiting enzymes such as topoisomerase II [4]. These terpenoids can be found on lemon, apple, berries, and mango. Other uses of terpenoids are to prevent cardiovascular diseases by triterpenoids from grapes [40]. Among the fruits that present higher terpenoid content are citric fruits while other fruits like apple, peach, and carrot have a low content. Among the terpenoids found on citric fruits are myrcene, β-pinene, limonene, linalool, and borneol; while on other fruits terpenoids like geranyl acetone, valencene, bisabolene, farnesene, and ionone among others [41]. There are other types of terpenoids known as hidden terpenoids which are synthesized via de isoprenoid pathway but do not have the known terpenoid structure such as psoralen, angelicin, 8-geranyloxypsoralen, and bergamottin, which are found on the oil glands of the citric peels [42].

3.4 EXTRACTION TECHNIQUES OF BIOACTIVE COMPOUNDS FROM FRUITS

The post-harvest processing (includes cutting, ripening, drying, and storage) of fruits for extraction of bioactive compounds can directly affect their compounds content and inherent stability, associated with physical, chemical, and physiological characteristics of the peel (exocarp), flesh (mesocarp) and endocarp [43, 44]. Bioactive compound content fruit-related studies delve into the application of traditional post-harvest procedures and propose novel and alternative techniques for enhanced production of potential nutraceuticals, such as potential increasing of bioactive compounds production after UV-B irradiation post-harvest treatment, such as lycopene and ascorbic acid production from "Moneymaker" cultivar tomato fruits (*Solanum lycopersicum* L.) [45], or flavan-3-ols (catechin and procyanidins), flavonols (quercetin and their glycosides) and dihydrochalcones (phloridzin and its glycoside) from "Red Delicious" cultivar apple peel (*Malus domestica* L.) [46].

3.4.1 DRYING OF FRUITS

Drying is the most ancient and important post-harvest procedure for preservation of bioactive compounds, relating to the loss of moisture from pericarp tissues and related products, such as peel, pomace, pulp, or juice, to prevent the microbial growing, reduce chemical deterioration and prolong shelf-time [47, 48]. This process can be released around 3 different methodologies: Air or gas contact under atmospheric pressure (representing 85% of the convective industrial dryers), heat transfer from heated surfaces, and vacuum-related dehydration [48, 49]. Figure 3.4 shows the classification of some drying techniques used on fruits.

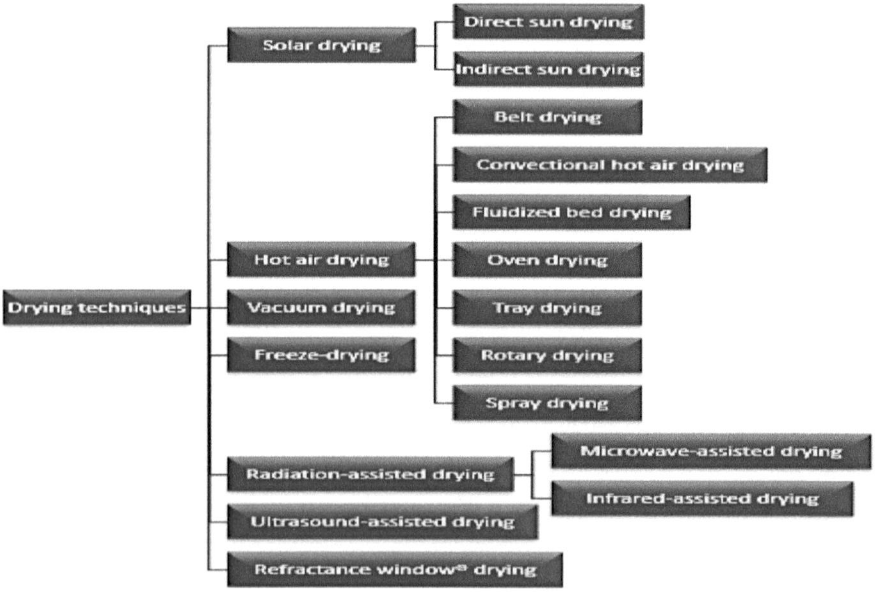

FIGURE 3.4 Classification of drying techniques.

Source: Image of own making, information obtained from Refs. [48, 50].

Solar drying is one of the most common drying techniques for fruits, for their cheapest functionating and easy building of solar dryers, along that can operate into direct or indirect irradiation, without any kind of adverse mishaps (such as cloudy weather, pollution, and/or insect infestation); however, solar irradiation exposure may affect the food nutraceutical properties, by activation of photolytic reactions with antioxidants, vitamins, and another photosensitive compound [47, 51]. Recent studies relate the kinetics and the content of the bioactive compounds of some fruits that have been submitted to direct or indirect solar drying like blackberry fruits (*Rubus fruticosus* L.) where indirect solar irradiation can enhance the total flavonoid content (TFC) of dried fruits, but decrease the total phenolics and total anthocyanidins content, comprising with direct solar irradiation and freeze-drying [52], and different fruit tissues from pomegranate fruits (*Punica granatum* L.) like aril, flavedo (exocarp), albedo (mesocarp) and pomace, where the direct solar drying decreases their total phenolics and flavonoids content comprising with solar tunnel and mechanical cabinet drying, which reflects in differences into color tissue associated with lack preservation of the anthocyanidin content into arils and carotenoid content into flavedo and decreasing of the antioxidant potential [53].

Convectional hot air is another technique for drying fruits tissues that provide evaporation of moisture from fruit tissues to the atmospheric environment through diffusional and capillary phenomenon and can be realized on tray, tunnel, continuous belt, kiln, and cabinet method, where a heated air flux cross around the complete fruit or sliced material [47, 49, 54]. However, hot air drying is related to heat-sensitive degradation and oxidation-related browning, which is associated with poor preservation and quality of the bioactive compounds, as happened on berries like blueberry (*Vaccinium corymbosum* L.) where anthocyanidin content decreases, along with their antioxidant properties [44]. For that reason, novel hot air drying techniques like spray drying can be applied to fruit juice or fruit wastes (pulp or pomace) passing through a high-pressure atomizer to generate specific size droplets that are quickly evaporated by hot air, where the temperature of droplets maintain a low temperature and perform better preservation of bioactive compounds, with low moisture and possible micro-and nanoencapsulation of nutraceuticals when they are associated to encapsulating materials such as maltodextrin and polysaccharides (chitosan, xanthan gum), such as the case of some polyphenolic-rich fruits like blueberries, maqui berry (*Aristotelia chilensis* (Molina) Stuntz), and jaboticaba (*Plinia cauliflora* (Mart.) Kausel) [55–58].

For these considerations, other drying techniques have been evaluated versus traditional methodologies for better preservation of bioactive compounds and good nutraceutical potentials, such as lyophilization (also known as freeze-drying) where ice sublimation can improve dehydration of fleshy fruits like figs (*Ficus carica* L.), blueberries (genus *Vaccinium*) and strawberries (genus *Fragaria*), potimarron pumpkin (*Cucurbita maxima* Dushesne) and cherry laurel (*Prunus laurocereasus* L.) with good retention of physicochemical properties, especially about easy rehydration and conservation of volatile compounds associated with odor and flavor [47, 54, 59, 60]. Recently, novel drying techniques have been developed for a quick and potent dehydrating of fruits and related material, like microwave-assisted and infrared-assisted methodology, where interactions between water molecules and electromagnetic radiation allow uniform dehydration, and also can be coupled with conventional techniques like convectional air as an enhancer step; like dehydration of pomegranate arils and myrtle fruits (*Myrtus communis* L.) where phenolic content was preserved with excellent antioxidant potential [48, 61, 62]. Refractance window technique (RW®) supposes to be a novel nutraceutical drying methodology that involves an indirect dehydrating interaction between thermal energy from pumped hot water and the fruit material passing through along an adequate plastic thin layer or convoyed belt; this technique has been evaluated for some fruit material such as cornelian cherry pulp (*Cornus mas* L.) where ascorbic acid content was higher than conventional hot air-dried material; and guava pulp (*Psidium guajava* L.) where carotenoid content is highly retained comprised than freeze-drying technique [63, 64]. Table 3.2 summarizes the examples mentioned earlier as well as other spices, where classical and novel drying techniques have been used on fruits depending on the material.

3.4.2 EXTRACTION OF BIOACTIVE COMPOUNDS FROM FRUITS

Extraction of bioactive compounds with nutraceutical potential has to be adequate about post-harvest treatments, the complexity of the raw material or total compounds, polarity, and solubility. Traditional extraction procedures like hydro distillation, maceration, and Soxhlet extraction are the most commonly used techniques for extraction but exhibits several disadvantages like time-consuming realization, utilization of hazardous organic solvents, and presence of solvent residual into extracts. Cold and/or green extraction procedures such as microwave-assisted, ultrasound-assisted, supercritical fluid (SF), and pressurized liquid extraction (PLE) have been released as a suitable way to extract compounds of interest with a considerable minimal

TABLE 3.2 Different Drying Techniques Performed on Fruits and Their Derivatives

Fruit Organ	Plant Material	Drying Methods	Observations	References
Peel or albedo (exocarp)	*Punica granatum*	Oven-drying with different temperatures (40°, 50°, and 60°C) and freeze-drying.	The freeze-dried product presented the highest rutin, hesperidin, and catechins content, while 60°C presented the highest punicalin and *p*-coumaric content.	[65]
	Citrus grandis	Freeze-drying and oven 50° and 60°C drying.	60°C oven-dried product exhibit the highest antioxidant potential and total phenolic content.	[66]
	Citrus reticulata	Sun drying, freeze-drying, vacuum drying, and short-wave infrared drying.	Freeze-dried and infrared dried products presented the highest total phenolic content and antioxidant potential.	[67]
	Musa paradisiaca	Sun drying, hot air-drying, dehumidified air drying, freeze-drying, vacuum drying, and microwave drying.	The freeze-dried product presented the highest total flavonoid content, versus the microwave-dried product presented the highest total phenolic content.	[68]
Aril (endocarp)	*Punica granatum*	Natural-sun drying, solar tunnel drying, mechanical cabinet drying, and oven drying.	Mechanical cabinet dried product presented the highest total phenolic content and enhanced antioxidant potential.	[53]
	Cucurbita maxima	Air-flow drying, high pressure followed by flash pressure-drop toward a vacuum (VMFD), swell drying followed by flash pressure-drop toward a vacuum (SD), and freeze-drying.	VMFD-aqueous extract presents the highest total phenolic content versus VMFD-ethanol extract present the highest total flavonoid content.	[59]
	Lycium barbarum × *L. chinense*	Air-thermal drying and freeze-drying.	Freeze-dried products exhibit the highest antioxidant potential and total phenolic content.	[69]
	Plinia cauliflora (*Myrciaria jaboticaba*)	Chlorine-pretreatment with sliced jaboticaba dried by hot-air dehydration chamber.	60°C–240 minutes of drying-product present the highest anthocyanidin content, and 60°C–240 minutes of drying-product present the highest total phenolics and ascorbic acid content.	[70]

TABLE 3.2 *(Continued)*

Fruit Organ	Plant Material	Drying Methods	Observations	References
	Rubus fruticosus	Direct-solar drying, indirect-solar drying, and freeze-drying.	The freeze-dried product presents the highest anthocyanidin and total phenolic content, versus indirect-solar drying presents the highest total flavonoid content.	[52]
	Momordica cochinchinensis	Microwave and air drying.	Microwave improves the yield of oil and carotenoid.	[71]
		Ohmic heating and oven drying.	Ohmic heating yielded a higher amount of carotenoids and oil.	[72]
Pulp, pomace, or derivatives	*Aristotelia chilensis*	Spray-drying with silicon oxide as adjuvant.	The spray-dried aqueous extract contains a high content of anthocyanidins and high polyphenol content.	[56]
	Prunus cerasus	Oven-drying and freeze-drying.	The freeze-dried product presents the highest concentrations of ascorbic acid, total phenolics, total flavonoids, and total anthocyanidins.	[73]
	Prunus laurocerasus	Fan-oven drying, vacuum-oven, and freeze-drying.	Freeze-drying aqueous and methanolic extracts exhibit the highest antioxidant potential with high total phenolic content.	[60]
	Vaccinium ashei	Vacuum-belt drying and freeze-drying with maltodextrin as adjuvant.	Freeze-dried slur and 80°C vacuum-belt dried slurs without added maltodextrin present the highest total monomeric anthocyanidin content.	[74]
	Mangifera indica	Freeze-drying, hot air-drying, infrared drying, vacuum drying.	Freeze-drying presented the highest polyphenol content.	[75]
	Pistacia vera	Ultrasound-microwave drying, microwave convecting-drying, and infrared convecting-drying, solar drying, and fluidized bed-drying.	The ultrasound accelerates the drying rate.	[76]

time operation, total equipment, energy use, and optimization of extraction of compounds of interest with high content and low inalterability [77, 78] (Figure 3.5).

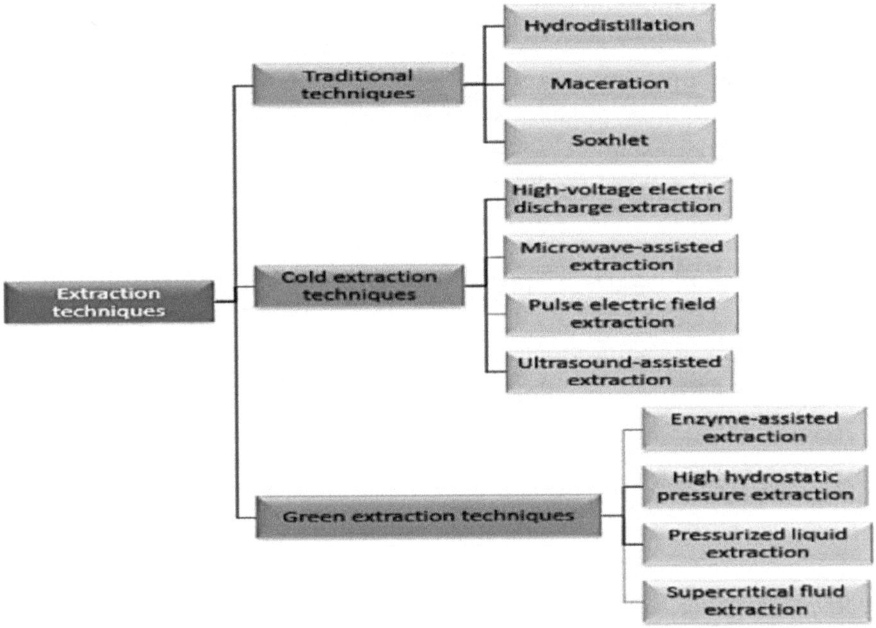

FIGURE 3.5 Classification of extraction techniques applied to fruits and their derivatives.
Source: Image of own making, information obtained from: Refs. [77, 78].

Recently, novel extraction techniques have been applied for rich-bioactive compound content fruit tissues such as mango and red prickle (*Opuntia ficus-indica* (L.) Mill.) peel, and olive kernel (*Olea europaea* L.), like the application of high-voltage electric discharges into an aqueous medium into a vessel with a needle electrode, where induction of electrohydraulic discharges with plasma channel, and a secondary formation of bubbles, turbulence, and shock waves into cells facilitate cell damage and liberation of bioactive compounds; and application of moderate-to-high pulsed electric field to the cell membrane induces electropolarization and/or electroporation that improves mass transfer and diffusion through cell membrane and walls, with an enhanced extraction of bioactive compounds and reducing the use of hazardous solvents [78–82]. High hydrostatic pressure extraction is a novel and useful extraction of fruit-related bioactive compounds such as flavonoids and lycopene from tomato pulp, and red anthocyanins from strawberries pulp and pomegranate juice, through the application of ultrahigh-pressure (100–1,000 MPa) of the mixture of fruit or their derivatives with extraction solvent into a thermoregulated vessel. This procedure allows disruption and deformation of cell walls, membranes, and organelles by alteration of intermolecular forces like hydrogen-bonding, hydrophobic, and ionic interactions that relate with the tridimensional cell structure; and additionally allow to protein denaturation for alteration of net formal charges by deprotonation [78, 83–85].

As a resume, Table 3.3 will be reviewed the type of compounds that can be extracted from these modern techniques.

3.5 ANALYSIS OF BIOACTIVE COMPOUNDS FROM FRUITS

Many fruits, including mangosteen, citrus, berries, pomegranate have attracted the attention of their health benefits due to the wide range of bioactivities [113]. These include targets for the treatment of diseases like diabetes, inflammation, cancer, neurological, and cardiovascular trouble, liver damage, bacterial, and fungus infection, etc. [114]. As mentioned above, the principal secondary metabolites produced by fruits are phenolic compounds and terpenoids [115]. Phenolic or polyphenol compounds include simple phenols and phenolic acids, hydroxycinnamic acid derivatives, and flavonoids. Many phenolic compounds in plants are good sources of natural antioxidants and antimicrobial agents [116–118]. On the other hand, terpenoids are recognized as a group of natural compounds possessing a broad range of biological activities such as antimicrobial, anticancer, anti-inflammatory, and hepatoprotective [119, 120]. The phytochemical profiles of botanical preparations can differ significantly, and they can be affected by geography, genetics, ontogenetic stage, plant materials used, and post-harvest processing methods among others [121]. Thermal and chemical treatment of the extracts from plants may cause the degradation of phytochemicals. During the preparation of botanical supplements, compositions, and levels of phytochemicals may change and affect the bioactivity of the preparations [121, 122].

In the past, the analysis of phytochemical preparations was largely based on thin-layer chromatography (TLC) or liquid chromatography (LC) in conjunction with ultraviolet (UV) spectrophotometric or fluorometric detection. Currently, LC in combination with MS is the principal and most used method to analyze phytochemical preparations [121]. Some examples of fruits

TABLE 3.3 Different Extraction Methods Performed on Spices

Extraction Method	Fruit and Extraction Conditions	Chemical Nature of Compounds	References
Solvent extraction	*Arbutus unedo* fruits with aqueous methanol or ethanol (80% v/v).	Phenolics (flavonoids, tannins, and their glycosides).	[86]
	Solanum lycopersicum Sun Black® variety peels and/or freeze-drying powder with methanol:water:trifluoracetic acid (70:29.5:0.5 v/v/v) for anthocyanins, methanol:ethanol:water:formic acid (35:35:28:2 v/v/v/v) for polyphenols, and hexane/ethanol/acetone (2/1/1 v/v/v) for carotenoids.	Anthocyanidins (petenin and negretein), polyphenols, and carotenoids.	[87]
	Euterpe oleracea ripe fruits with acetic acid-acidified ethanol 92%.	Anthocyanidins	[88]
	Hylocereus polyrhizus flesh (mesocarp) and peel (exocarp) with chilled methanol 70%.	Betacyanins, flavonoids, and phenolic acids.	[89]
	Momordica cochinchinensis with a mixture of hexane:acetone:methanol (2:1:1).	Carotenoids	[90]
	Momordica cochinchinensis with hexane.	Oil	[72]
Hydro distillation or steam distillation.	*Citrus* sp. Flavedo (exocarp) associated with extra-virgin olive oil.	Volatile mono- and sesquiterpenoids.	[91]
Soxhlet apparatus	*Juglans regia* shells with 60% methanol.	Saturated fatty acids and their éster derivatives.	[92]
	Vaccinium myrtillis milled fruits with 70% ethanol.	Phenolics (tannins, flavonoids, anthocyanins, and procyanidins).	[93]
	Mangifera indica with ethanol at boiling point for 3 hours.	Fatty acids and phenolic compounds.	[94]
Ultrasound-assisted extraction	*Plinia cauliflora (Myrciaria cauliflora)* peel and flesh puree at 200 Watts and 24 kHz.	Anthocyanidins glycosides.	[95]
	Sechium edule fruits and seeds with ethanol/water solutions in the percentage of 90%, 80%, 70%, 60%, and 50%.	Flavonoids	[96]
	Punica granatum with methanol at 35°C, 20 KHz for 15 minutes.	Phenolic compounds	[97]

TABLE 3.3 *(Continued)*

Extraction Method	Fruit and Extraction Conditions	Chemical Nature of Compounds	References
Microwave-assisted extraction.	*Berberis jaeschkeana* dried pulp with methanol.	Anthocyanidins and phenolic compounds.	[98]
	Opuntia ficus-indica peels complemented with hydrodistillation and hydro diffusion.	Betalains	[99]
	Mangifera indica at 75°C, ethanol 90% v/v, 22 minutes.	Hydrollisable and condensed tannins.	[100]
Pulse electric field extraction	*Vaccinium myrtillus* fresh fruits into different field strengths.	Flavonoids	[101]
	Citrus maxima dried peel mixture with ethanol:water (40:60 v/v).	Naringin	[102]
High-voltage electric discharge extraction.	*Punica granatum* peel suspension with water with constant peak pulse voltage of 9 kV and frequency of 100 Hz.	Phenolic compounds	[103]
	Vitis vinifera pomace mixed with water and 40 kV discharges pulse generator.	Phenolic compounds	[104]
CO_2 supercritical fluids.	*Byrsonima crassifolia* lyophilized pulp with 100% supercritical CO_2 at 50°C and 15–45 MPa, followed with 90% supercritical CO_2:10% ethanol at 60°C and 15–45 MPa.	Lutein and phenolic content.	[105]
	Hippophae rhamnoides pomace with 100% supercritical CO_2 at 35°C/45 MPa and 37.5°C/36.5 MPa.	Carotenoids, fatty acids, phytosterols, and polyphenols.	[106]
High hydrostatic extraction	*Morinda citrifolia* dried powder mixed with ethanol.	Alizarin, rutin, and scopoletin.	[107]
	Morus alba frozen-dried fruits complemented with enzyme-assisted extraction.	Anthocyanins	[108]
	Physalis peruviana frozen dried fruits mixed with 5% metaphosphoric acid aqueous solution and hexane:acetone:ethanol (50:25:25 v/v).	Ascorbic acid and β-carotene	[109]

TABLE 3.3 *(Continued)*

Extraction Method	Fruit and Extraction Conditions	Chemical Nature of Compounds	References
Enzyme-assisted extraction	*Citrullus lanatus* fresh fruits with Kemzyme® enzyme mixture with pH between 6 and 9 and temperature between 30°C and 70°C.	Phenolic acids (vanillic, sinapic, *p*-coumaric, chlorogenic, and *p*-hydroxybenzoic acid).	[110]
	Momordica balsamina dried pulp with Kemzyme® dry-plus, Natuzyme®, and Zympex-014® enzyme mixtures with pH between 6 and 9 and temperature between 30°C and 70°C.	Phenolic acids (ferulic, sinapic, *p*-coumaric, gallic, and *m*-coumaric acid) and flavonoids (quercetin).	[111]
	Rubus idaeus pomace with alcalase, bacterial proteolytic enzymes, papain, bacterial cellulase, bacterial pectinase, and bacterial xylanase with acetic acid buffer.	Tannins (lambertianins, sanguiins)	[112]

rich in phenolic compounds are berries, grapes, avocados, papaya, guava among others. Flavonoids, tannins, anthocyanins, carotenoids, ascorbic acid are some of the most identified and characterized phytochemicals in fruits. Here, we will describe some of the most recent techniques to identify these kinds of phytochemicals.

3.5.1 MANGO (MANGIFERA INDICA L.)

Mango (*Mangifera indica* L.) fruit is considered a good source of dietary antioxidants, such as ascorbic acid, carotenoids, and phenolic compounds [123], which have been demonstrated different health-promoting properties, mainly due to their remarkable antioxidant capacity [123, 124]. The antioxidant capacity of fruits and vegetables has been correlated to their total phenolic content (TPC) and composition [125] Phenolic compounds have the different antioxidant capacity, depending on their structure, number of hydroxyl groups and their distribution in the structure [126]. In a study carried out by Palafox-Carlos et al. [123] with "Ataulfo" mango pulp, they identified the major phenolic compound and evaluated its contribution to antioxidant activity. The identification of phenolic compounds was determined by high-performance liquid chromatography-mass spectrometry (HPLC-ESI-MS) analysis and was identified by comparing their retention time (RT) and UV-Vis data with references standards. Quantitative data for phenolic compounds were obtained by calibration curves constructed with known standards.

The results of these analyzes showed the major phenolic compounds in "Ataulfo" mango were phenolic acids gallic, chlorogenic, protocatechuic, and vanillic [123]. Chlorogenic acid was the most abundant polyphenol in "Ataulfo" mango. It had a range of concentrations from 301 mg/100 dry weight (dw) to 28 mg/100 g dw. Chlorogenic acid was identified as a [M-H]-deprotonated molecule (*m/z* 356) with an UV spectra (l_{max} = 286, 240 nm) in RT of 20.05 min. Gallic acid was the second most abundant polyphenol (98.7 to 94.6 mg/100 g dw) and it was also identified as a [M-H]-deprotonated molecule (*m/z* 169) with UV spectra (l_{max} = 220, 275 nm) in a RT of 15.5 min and yielded ion fragments at *m/z* 125, 126 and 170. Vanillic acid exhibited a lower concentration (24.4 to 16.9 mg/100 g dw). It was identified as a [M-H]-deprotonated molecule (*m/z* 168), with a RT of 20.75 min, and yielded fragments at 124 and 133. Finally, protocatechuic acid was identified according to its UV spectra (l_{max} = 230, 266 nm) as a [M-H]-deprotonated molecule (*m/z* 153) with a RT of 18.85 min and yielded ion fragments at *m/z* 108 and 155. It had a concentration from 1.1 to 0.48 mg/100 g dw. [123].

3.5.2 BLACK MULBERRY (MORUS NIGRA)

Black mulberry (*Morus nigra*) has been recognized due to its beneficial effects. The fruit is used to treat several diseases, such as diabetes, hypertension, arthritis, Alzheimer´s disease, and other disorders associated with chronic inflammation [127–129]. Moreover, has a tonic effect on kidney energy, and it is used as an antiphlogistic, diuretic, and expectorant [130]. Many studies have reported its importance for human diet and health as the result of its compounds such as alkaloids, carotenoids, anthocyanins, organic acids, phenolics, and sugar [130–132]. Also, fruits of black mulberry provide an inexhaustible source of vitamins (C, E, B2) and minerals (K, Ca, Mg, Na, Fe, Zn) which make fruit highly physiologically valuable [131]. Paunović et al. [131] evaluated and compared the chemical composition and biological activity of fruit of black mulberry and evaluated the potential use of the fruit as a natural nutrient highly beneficial for human health.

The determination of total and invert sugars was carried out by the Luff-Schorl method. Individual invert sugars (glucose, fructose, sucrose) were determined by HPLC analysis. Fructose was the dominant invert sugar in berry extracts, followed by glucose, while the amount of sucrose was very low [131]. These results are consistent with Özgen et al. [133] who found that fructose (about 52%) and glucose (about 48%) are dominant sugars in black mulberry fruits. The determination of vitamin C was assessed by UV/Vis spectrophotometry. The absorbance was measured at 665 nm, meanwhile, the determination of vitamins A, B_1, B_2, and B_3 was assessed by HPLC analysis. The main vitamin in berries was vitamin C (19.355 mg/100 g), followed by Vitamin B_3 (3.814 mg/100 g) and vitamin A (1.635 mg/100 g). To determine minerals (K, P, Ca, Mg, Na, Fe, Zn, Cu, Mo) a flame atomic absorption spectrophotometry was carried out. In this experiment, major elements were K, P, Ca, Mg, and Na, whereas Fe, Cu, Zn, and Mn were found in lower amounts. The highest amount was determined for phosphorus (222.32–231.05 mg/100 g) and potassium (206.27–207.67 mg/100 g). Hassimotto et al. [132] determined and characterized the anthocyanins of wild mulberry grown in Brazil. Flavonoids were separated by HPLC and two cyanidin derivatives and one quercetin derivative were identified. To elucidate the structure of two cyanidin derivatives UV-Vis spectra and mass spectrometry (MS-MS) were performed. The anthocyanidins and their glycosylated forms, anthocyanins, exhibited broad absorption maxima in the visible region (465–550 nm). In acidified methanolic solution, the maximum wavelength (l_{max}) for cyanidin derivatives was 531 nm y 529 nm. The acidic hydrolysis of both derivatives resulted in one major anthocyanidin with l_{max} at 537 nm.

From MS-MS analysis, a molecular ion [M^+] was found at *m/z* 595 and the fragmentation yielded two daughter ions an *m/z* 449, resulting from the loss of rhamnose moiety, and the other at *m/z* 287 that results from the loss of glucose residue. According to UV-Vis and MS/MS data, one of the cyanidin derivatives corresponded to cyanidin-3-glucosylrhamnoside which has directly attached a glucose residue to cyanidin. The cyanidin-3-glucoside was the other cyanidin derivative identified [132]. D'Urso et al. [134] investigated the metabolite compositions of ripe fruits of mulberries, using reversed-phase HPLC coupled to high-resolution mass spectrometry (LC-MS). Around 35 metabolites, mainly phenolic compounds, and 4 piperidine alkaloids were identified. Morusimic acids, B, C, and E (Figure 3.6) [135] were identified based on their exact molecular mass and fragmentation.

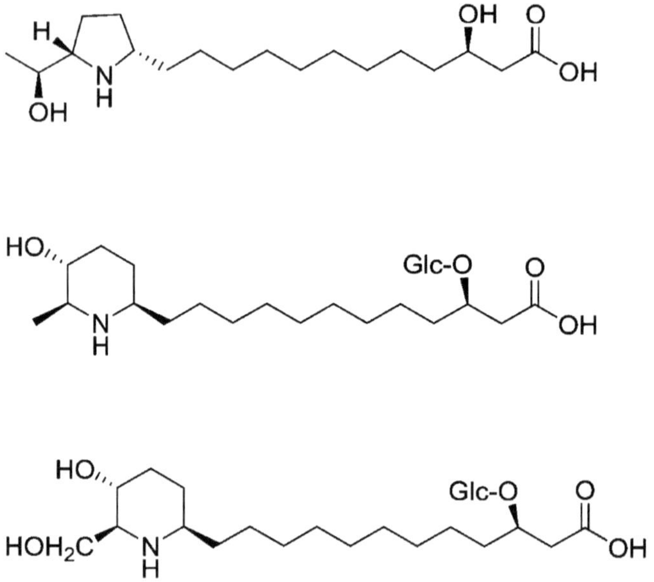

Morusimic acid B

Morusimic acid C

Morusimic acid E

FIGURE 3.6 Morusimic acids, B, C, and E (piperidine alkaloids) extracted from mulberries.

A series of flavonols with different substituents either were present in mulberry fruits. All these compounds showed the characteristic flavonol absorbance peaks at around 260 nm (resulting from the A-ring) and at around 350 nm (due to B-ring) and producing daughter ions of *m/z* 303.0496 (quercetin) or 287.0547 (kaempferol) [134].

3.5.3 APPLES (MALUS DOMESTICA)

Apples (*Malus domestica*) are one of the most consumed fruits worldwide, and they are recognized as an incredible source of bioactive compounds related to both functional and nutraceutical values (Figure 3.7) [136].

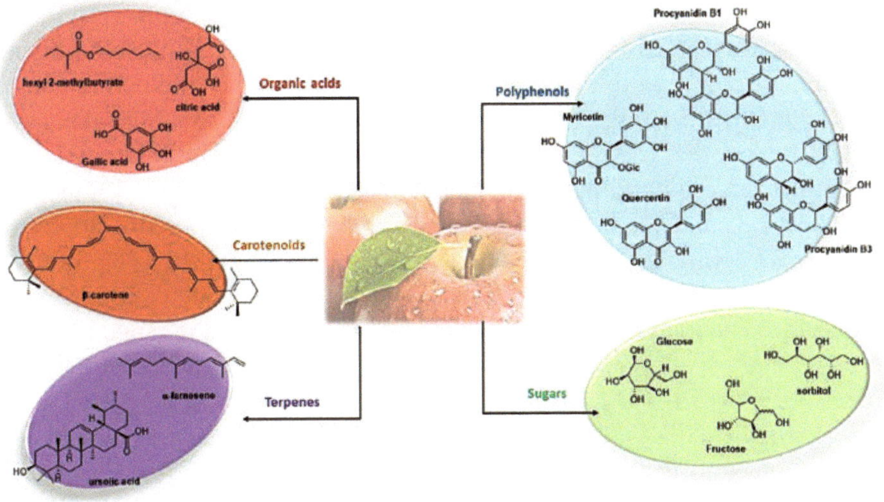

FIGURE 3.7 Main primary and secondary metabolites involved in functional and nutraceutical values of *Malus domestica*.

Source: Image of own making, information obtained from: Ref. [136].

Apples have a varied and well-balanced composition; more than 84% of their weight is represented by water. The remaining 16% is represented by minerals (K, Ca, Mg, and Na, mainly), complex polysaccharides (pectin, celluloses, hemicelluloses, and lignin), complex B vitamins, and polyphenols which are responsible for their well-known antioxidant properties [136–138]. For the quantification of polyphenols in apple fruits extract, colorimetric methods are widely used, due to their simplicity and sensitivity. Folin-Ciocalteu (analysis of total polyphenols content (TPC)) method and Aluminum ($AlCl_3$) chloride assay are the most used to determine flavonoids [136, 137]. In presence of polyphenols the Folin-Ciocalteu reagent, under basic conditions, reduces itself conferring to the sample a blue color (l_{max}= 495 nm) [136]. The principle involved in $AlCl_3$ colorimetric method is that $AlCl_3$ forms acid stables complexes with the C-4 keto groups and either the C-3 and C-5 hydroxyl group of flavones and flavonol, which show maximum absorbance at 415 nm [136].

More selective and sensitive methods, based on chromatographic techniques, to identify secondary metabolites from fruits are currently employed. These chromatographic methods are coupled with UV-Vis, fluorescence, or MS detection. Petrosvska et al. [139] elucidated the polyphenolic composition of different apple varieties using an efficient UHPLC-DAD-HESI-MS^n method. Six compounds were identified as phenolic acid derivatives. According to their UV (l_{max} = 209–220 nm) and MS (ion *m/z* 163), four of them were identified as hydroxycinnamic acids (gallic, chlorogenic, *p*-coumaric, and caffeic acid). The flavan-3-ols, procyanidin B1, and B3 were identified. A fragmentation pattern (*m/z* 425 and *m/z* 289) along with RT of 0.52 and 2.31 min, respectively, were reported. Eight flavonols were identified according to their UV and MS data. The ions at *m/z* 317, m/z 301, and *m/z* 285 indicated three aglycones with the corresponding molecular ions: myricetin, quercetin, and kaempferol. Quercetin had deprotonated molecular ion at *m/z* 463 and a fragment ion at *m/z* 301 indicating quercetin hexoside. The total polyphenolic content values were in the range from 5.59 to 14.7 mg/g lyophilized peel, from 4.52 to 16.3 mg/g lyophilized flesh, and from 8.07 to 15.1 mg/g dried leaves [139].

Routinely HPLC with UV-Vis, are used for the structural elucidation, separation, and quantification of carotenoids occurring in apples. Carotenoids pigments in the skin of apples contribute to fruit coloration, and therefore to their attractiveness. These compounds show absorption in the visible region (400 and 500 nm), due to long-conjugated double bond system [136, 140]. MS/MS is used to find out the characteristic/typical fragmentation pattern of carotenoids (molecular ion at *m/z* 601 and *m/z* 583). Most of the MS analysis of carotenoids is performed, by using positive ion mode. APCI (atmosphere pressure chemical ionization) has become the most widely used ionization technique for various carotenoids in apples, because of highly sensitive [141]. Recently, ultrahigh-performance liquid chromatography (UHPLC) technology has been used for the analyzes of carotenoids in *Malus* fruits [136]. Aromatic volatile compounds alongside sugar and organic acids are responsible for the taste and flavor of apples. Volatile organic compounds (VOCs) contribute to the overall sensory quality. These include carboxylic esters, alcohol, aldehydes, ketones, and terpenes in minor concentrations.

Terpenes such as α-farnesene, geranyl acetone, and farnesol have been identified in apples. α-farnesene, an acyclic branched sesquiterpene, is highly occurring in ripe fruits [136]. Ferreira et al. [142] investigated the volatile composition of different apple varieties of *Malus domestica*. The analysis was performed by headspace solid-phase microextraction followed by GC-MS. It was possible to identify about 100 volatile compounds among pulp, peel, and entire fruit. Ethyl esters, higher alcohols, and terpenes were found to be the most representative volatiles. α-farnesene, hexane-1-ol, and hexyl 2-methyl butyrate were the compounds found in the volatile profiles of studied apples with the largest area. The relative percent amount (%RPA) of VOCs of pulp were 45.57% for higher alcohols, 38.28% for ethyl esters, 7.64% for carbonyl compounds, and 1.18% for acids. For peel, the significant contribution for the total volatile profile arises from terpenes (50.94%) followed by ethyl esters (23.71%), higher alcohols (12.94%), carbonyl compounds (7.76%), and acid (1.18%). Finally, for the entire fruit samples terpenes, higher alcohols, ethyl esters, carbonyl compounds, and acids account for 34.61, 23.27, 19.16, 10.85, and 3.10%, respectively of the total GC peak area of the chromatograms.

Ferreira et al. [142], reported that the most prominent constituents in apple pulp were hexyl 2-methyl butyrate (49.05–29.77%), hexane-1-ol (40.2–29.77%). Relatively to apple peel VOCs and semi-volatile organic compounds (SVOCs) with the major contribution to their volatile profile were α-farnesene (30.57–67.33%), hexylacetate (0.76–12.66%), and hexyl-2-methylbutyrate (1.37–8.51%). Concerning apple entire fruit, α-farnesene (30.57%–16.87%) was the major volatile compound. In conclusion, the total apple's volatile fraction occurred in the peel and the pulp. As triterpenoids such as betulinic acid (BA) and UA are polar metabolites, they are not efficiently volatilized for analysis using GC. Thus, LC-MS is more suitable for analysis [136]. Thus, Butkevičiūtė et al. [143] developed a new technique for the rapid detection and analysis of triterpene compounds in apple extracts using HPLC. In total, four triterpene compounds were isolated and identified, corosolic acid, betulinic acid, oleanolic acid, and UA. The last one was the predominant compound in both, entire fruit, and peel. It comprised the greatest part (81.2–72.1%) of the total amount of all triterpene compounds detected in the extracts of the apple samples [143]. In previous studies developed and published by Grigoras et al. [144], UA is reported as the major triterpene component in apples (flesh, peel, and entire fruit). It comprises about 70% of the total amount of all triterpene compounds detected in extracts of apple samples. The amounts of acids, oleanolic, corosolic, may range from

0.16 g/100 g to 1.0 g/100 g, up to 0.51 g/100 g, respectively. The percentage of betulinic acid, compared to the other identified triterpenoids (ursolic, oleanolic, and corosolic acids) was significantly lower and comprised only 0.7–1.1% of the total amount of all the triterpene compounds identified in extracts of apple samples [143].

3.5.4 DATE PALM FRUITS (PHOENIX DACTYLIFERA)

Date palm fruits of the species *Phoenix dactylifera* have been related to several biological effects including antioxidant, antimutagenic, anti-inflammatory, anticancer, antimicrobial, and immunostimulant [145]. Polyphenols such as phenolic acids, hydroxycinnamates, flavonoids glycosides, and proanthocyanidin oligomers are the major metabolites present in date fruits, all likely to mediate for its beneficial effects [145, 146]. Several studies have shown that HPLC coupled to diode array detection (DAD) an MS presents a powerful tool for metabolite structure elucidation [145, 147].

Hong et al. [148] analyzed flavonoid extracts derived from Delget Noor dates (*P. dactylifera*) fruits using HPLC/DAD/MS technique. In total, 19 different flavonoids among procyanidin oligomers were identified. The procyanidin composition was determined by comparing RTs of date drupe procyanidins with the procyanidin standard. HPLC chromatograms and mass spectra indicated that Delget Noor dates are homogeneous in B-type oligomers and contain oligomers through decamers. Mass spectra of peaks showed trimers, hexamers, and nonamers at *m/z* 865, 864, and 863.6, respectively. Flavonoid glycosides were identified by reverse-phase LC-ESI/MS/MS in positive ion mode. The HPLC chromatogram demonstrated the presence of 19 peaks absorbing at 370 nm. These were consistent with conjugates arising from either quercetin, luteolin, or apigenin (Figure 3.8) [148] based upon calculated masses. Eight luteolin glycosides were identified. Ion of *m/z* 287 for the $[M+H]^+$ corresponding to aglycone of luteolin in positive ion mode. Conjugates of quercetin (three) were characterized using precursor ion scanning of *m/z* 303 for the aglycone of quercetin in conjunction with LC-ESI/MS/MS. Only one apigenin derivative was identified. Thus, UV-Vis spectra demonstrated l_{max} at 267 and 338 nm, which are characteristics of apigenin. Fragmentation of *O*-glycosides and *C*-glycosides resulted in ions at *m/z* 457, 427, 355, and 325 [148].

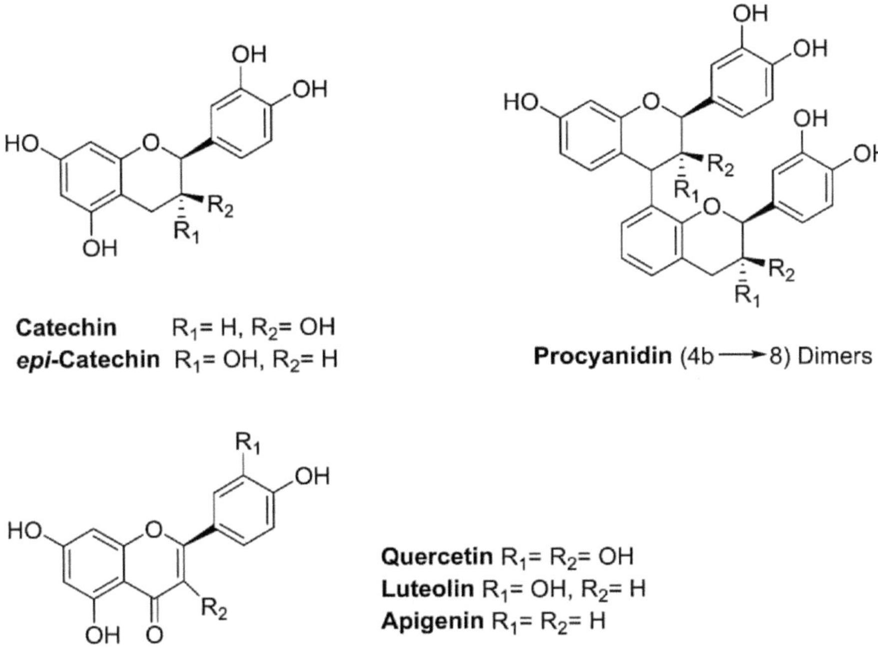

FIGURE 3.8 Representative structure of catechins, epicatechin, procyanidins B type dimers, quercetin, luteolin, and apigenin.

Source: Image of own making, information obtained form: Ref. [148].

Identification of fatty acids and sphingolipids present in date fruits also is possible using the high-resolution UPLC/PDA/(–)ESI-qTOF-MS technique. Thus, Marag et al. [145] elucidated fatty acids, i.e., oleic, linoleic, linolenic, and palmitic acids by their high-resolution masses 281.2486, 279.2335, 277.2114, and 255.2316, respectively. Lycopene (carotenoid) has been the object of extensive research which suggests its antioxidant role and antiproliferative activity in cancer cells [149, 150]. In addition, vitamins C (ascorbic acid) y E (tocopherol) have largely been known for their role as antioxidants in the body [151]. Yahia et al. [152], carried out a study to identify and quantify the presence of phenolic compounds, carotenoids, and vitamin E in fruits of black sapote (*Diospyros digyna* Jacq.). Black sapote belongs to the *Ebanaceae* family and is native to the south of Mexico and Central America. Identification and quantification of metabolites of black sapote were performed by HPLC/MS analysis. From this analysis, 10 phenolic compounds were identified. From these, five were hydroxycinnamic acids (sinapic, ferulic, caffeic, cinnamic, and *o*-coumaric); two were flavan-3-ols (catechin and epicatechin); two were hydroxybenzoic acids (*p*-benzoic and protocatechuic), and one flavonol (myricetin). The four phenolic compounds found in black sapote were sinapic acid with a concentration of 110.7 mg/100 g dw, followed by myricetin (85.0 mg/100 g dw), ferulic acid (82 mg/100 g dw), and catechin 79.9 mg/100 g dw). All these compounds were identified by the LC-ESI-MS technique. A peak with ions at *m/z* 578 and 289 suggested that catechin was present in black sapote as a dimer. Epicatechin was identified based on fragmentation profile (*m/z* 289 and 245) and RT (20.30 min). Ferulic acid showed a RT of 21.32 min and the presence of ions at *m/z* 194. Sinapic acid displayed fragment ions at *m/z* 385 and 224, due to it being present as sinapic acid hexoside, and RT of 25.50 min. Regarding identification and quantification of carotenoids and vitamin E, only lutein and β-carotene were detected in black sapote. Lutein monoepoxide was identified through its fragmentation profile and RT (7.3 min). The UV spectra of lutein epoxides were reported with three l_{max}: 416, 440, and 470 nm. The fragmentation pattern revealed an ion at *m/z* 571 (protonated molecule) and a base ion at *m/z* 553 because of the loss of water which is typical for lutein. Identification of β-carotene was carried out by comparison of its RT (16.0 min). The fragmentation pattern indicated it might correspond to a β-carotene isomer. α-tocoferol was no detected in black sapote, instead γ-tocoferol was extracted in low concentration (0.716 mg/100 g dw) [152].

3.5.5 OTHER ANALYTICAL TECHNIQUES OF INTEREST FOR FRUIT ANALYSIS

The analytical methods reviewed in the present chapter, even though their use is more common every day that passes, still have the disadvantage that some methods destroy the sample without having the chance to recover it such as MS and GC. In this regard, Cappellin et al. proposed a method based on proton transfer, derived from direct injection MS coupled to a time of flight detector because of a higher proton affinity to hydronium ions compared to water, which makes it more sensible to detect metabolites of very low concentration (below 2%), in which it has been used to perform metabolomics for identification and systematics of fresh fruits, as the authors applied it on different apple cultivars varieties, Gala, Braeburn, and Fuji complemented with multivariate analysis such as principal component analysis followed by classification methods such as random forest and discriminant analysis [153]. Another analytical technique that has the feature of being non-destructive is near-infrared spectroscopy (NIR), as reported by Alves Filho et al. based on reflectance and transmittance of samples on cashew apples for classification purposes by preprocessing the spectra obtained due to the overlapping of C-H, N-H, S-H, and O-H bonds related in different wavelengths and inferring the type of compound present followed by hierarchical clustering based on the genotype of fruits [154]. Another example of the use of NIR was observed on the quality of pomegranate in four different stages, where the authors proved the usefulness of NIR to predict quality features of pomegranate itself because of differences in the peak intensities, especially at 720 and 680 nm [155].

Ion mobility spectrometry (IMS) is another analytical technique that enhances previous methods such as GC, HPLC, MS, and CE, by improving the selectivity to differentiate between isomeric compounds as well as compounds of minor concentration that could overlap easily with noise and also with the collision cross-section can provide complementary information for more reliable compound identification. This technique consists similarly to electrophoresis, but instead of using a buffer to mobilize the molecules, it occurs in a gaseous phase at atmospheric pressure, where the collision cross-section provides information about the tridimensional state of ions on the gas phase. This technique has been used for terpene identification, as well as phenolic acids that coelute with chromatographic methods such as vanillic, salicylic, ferulic, and caffeic acids [156]. An application of this technique is reported by Liu et al. with the structural elucidation of anthocyanins present on blueberry, raspberry, strawberry, pomegranate, and blackberry, determining anthocyanins in the range from 400 to 800 m/z [157]. A more recent application can be seen on jujube fruits reported by Yang et al. where they used IMS coupled with GC for the characterization of volatile compounds in different times of storage [158].

Another technique that could be used to solve the coelution of compounds besides IMS is through multidimensional chromatography because sometimes HPLC or GC methods by using a single column might not provide enough resolution of compounds [159, 160], as happened in the case of carotenoids extracted form red mamey, where the authors compared the one dimensional HPLC (one column) against two dimensional HPLC (two columns) complemented with MS and PDA detectors, allowing this separation according to their polarity followed by their hydrophobicity. The authors observed with one column that five carotenoids coeluted in the same peak at 80 minutes while with two columns and both detectors they managed to identify 23 carotenoids [160]. Another interest in the analysis of natural products with potential nutraceutical use is the preprocessing technique of spectrometric as well as chromatographic data, because of the variability of the measurements that may rise between different runs because of instrumental factors, sample state as well as environmental factors, without mentioning the human factor [161]. An example is the one reported by Mishra et al. where the authors preprocessed near-infrared spectra of different mango fruit from four different harvests, by using the second derivative with Savitzky-Golay polynomial fitting, standard normal variate, and variable sorting of normalization, which were compared with partial least square regression followed by the use of a designed neural network, they saw this preprocessing affected the quality of the spectra and therefore the capability to predict the classification of the mango fruits, they also concluded that the use of neural networks was a better tool to model the spectral data than least partial square regression [162]. However, Khodabakhshian et al. demonstrated in pomegranate with different normalization techniques by the multiplicative scatter correction and standard normal variate followed by the implementation of principal component analysis and then the partial least squares regression and also considering the smoothing of data allowed the generation of better models [155]. These findings suggest that the preprocessing of spectral and chromatographic data needs to be worked to obtain a better prediction model.

3.6 CONCLUSIONS AND FUTURE PROSPECTS

With the intake of fresh fruits because of the phytochemical content reviewed along in this chapter, the use of residues such as bagasse, peels, and seeds to mention a few, are becoming novel sources of bioactive compounds because of their higher content, thus becoming an attractive research field with an ecological and economic impact in the society; where the development of alternative extraction techniques such as SFs are becoming also important for the extraction of different phytochemicals depending on the extraction conditions [163]. Fruits by-products have also direct applications as it is in the case of citrus fruits where the seeds for their vitamin C, sterols, and fatty acids have been used to obtain oils, flours, cosmetics, and ointments among others and the pith residue has been used for the production of ethanol and vinegar because of cellulose content, among other residues [164].

The exploration of exotic fruits such as noni, pitanga, camu-camu as well as their residues, is also an emerging trend because these fruits are being known as superfruits as a marketing strategy to increase their commercial status; they also possess important phytochemicals such as flavonoids, iridoids, and anthocyanins among others and biological activities such as cardioprotective, antitumoral, immunomodulatory, and gastroprotective among others [165]. Another concern that has been lately explored is the preservation of fruits, in this regard, lactic acid fermentation, even though is an ancient preservation method it has been recently explored again because of changes in the phytochemical content because of their conversion to compounds of high added value (e.g., polyphenols), thus increasing their biological activities, along with correlation of the interactions of the microbiota through metabolomic analysis [163, 166].

KEYWORDS

- **bioactive compounds**
- **biological activities**
- **chemical composition**
- **diet**
- **extraction techniques**
- **fruit analysis**
- **Mangifera indica**
- **metabolites**

REFERENCES

1. Das, L., Bhaumik, E., Raychaudhuri, U., & Chakraborty, R., (2012). Role of nutraceuticals in human health. *Journal of Food Science and Technology, 49*(2), 173–183.
2. Dutta, S., Ali, K. M., Dash, S. K., & Giri, B., (2018). Role of nutraceuticals on health promotion and disease prevention: A review. *Journal of Drug Delivery and Therapeutics, 8*(4), 42–47.
3. Tagde, P., Tagde, S., Tagde, P., Bhattacharya, T., Monzur, S. M., Rahman, M., et al., (2021). Nutraceuticals and herbs in reducing the risk and improving the treatment of COVID-19 by targeting SARS-CoV-2. *Biomedicines, 9*(9).
4. Prakash, V., (2018). Terpenoids as cytotoxic compounds: A perspective. *Pharmacognosy Reviews, 12*(24), 166–177.
5. Prakash, D., Gupta, C., & Sharma, G., (2012). Importance of phytochemicals in nutraceuticals. *Journal of Chinese Medicine Research and Development, 1*(3), 70–78.

6. Ramappa, V. K., Srivastava, D., Singh, P., Kumar, U., Kumar, D., Gosipatala, S. B., et al., (2020). Mulberries: A promising fruit for phytochemicals, nutraceuticals, and biological activities. *International Journal of Fruit Science, 20*(3), S1254–S1279.
7. Khan, M. P. Z., Ahmad, M., Zafar, M., Sultana, S., Ali, M. I., & Sun, H., (2015). Ethnomedicinal uses of edible wild fruits (EWFs) in Swat Valley, Northern Pakistan. *Journal of Ethnopharmacology,* 173.
8. Natić, M., Pavlović, A., Bosco, F. L., Stanisavljević, N., Zagorac, D. D., Akšić, M. F., et al., (2019). Nutraceutical properties and phytochemical characterization of wild Serbian fruits. *European Food Research and Technology, 245*(2), 469–478.
9. Bhatt, I. D., Rawat, S., Badhani, A., & Rawal, R., (2017). Nutraceutical potential of selected wild edible fruits of the Indian Himalayan region. *Food Chemistry,* 215.
10. Ghaffari, S., & Roshanravan, N., (2020). The role of nutraceuticals in prevention and treatment of hypertension: An updated review of the literature. *Food Research International,* 128.
11. Keservani, R. K., Sharma, A. K., & Kesharwani, R., (2016). Medicinal effect of nutraceutical fruits for the cognition and brain health. *Scientifica, 2016.*
12. Altemimi, A., Lakhssassi, N., Baharlouei, A., Watson, D., & Lightfoot, D., (2017). Phytochemicals: Extraction, isolation, and identification of bioactive compounds from plant extracts. *Plants, 6*(4).
13. Thavamoney, N., Sivanadian, L., Tee, L. H., Khoo, H. E., Prasad, K. N., & Kong, K., (2018). Extraction and recovery of phytochemical components and antioxidative properties in fruit parts of *Dacryodes rostrata* influenced by different solvents. *Journal of Food Science and Technology, 55*(7), 2523–2532.
14. Bernal, J., Mendiola, J. A., Ibáñez, E., & Cifuentes, A., (2011). Advanced analysis of nutraceuticals. *Journal of Pharmaceutical and Biomedical Analysis, 55*(4).
15. Ahmad, N., Zuo, Y., Anwar, F., Abbas, A., Shahid, M., Hassan, A. A., et al., (2021). Ultrasonic-assisted extraction as a green route for hydrolysis of bound phenolics in selected wild fruits: Detection and systematic characterization using GC–MS–TIC method. *Process Biochemistry,* 111.
16. Delgado, A. M., Issaoui, M., & Chammem, N., (2019). Analysis of main and healthy phenolic compounds in foods. *Journal of AOAC International, 102*(5), 1356–1364.
17. Yahia, E. M., (2010). The contribution of fruit and vegetable consumption to human health. In: De La Rosa, L. A., Alvarez-Parrilla, E., & González-Aguilar, G. A., (eds.), *Fruit and Vegetable Phytochemicals: Chemistry, Nutritional Value, and Stability* (pp. 3–51). Iowa, USA: Blackwell Publishing.
18. Kazi, N. A., Yadav, J. P., & Agale, M., (2015). Nutritional value of fruits. *Scholarly Research Journal for Interdisciplinary Studies, 16,* 2937–2943.
19. Vicente, A., Manganaris, G., Sozzi, G., & Crisosto, C., (2009). Nutritional quality of fruits and vegetables. In: Florkowski, W., Shewfelt, R., Brueckner, B., & Prussia, S., (eds.), *Postharvest Handling: A Systems Approach* (2nd edn., pp. 69–122). Elsevier-Academic Press.
20. Angelino, D., Godos, J., Ghelfi, F., Tieri, M., Titta, L., Lafranconi, A., et al., (2019). Fruit and vegetable consumption and health outcomes: An umbrella review of observational studies. *International Journal of Food Sciences and Nutrition, 70*(6), 652–667.
21. Yahia, E. M., Maldonado, C. M. E., & Svendsen, M., (2018). The contribution of fruit and vegetable consumption to human health. In: Yahia, E. M., (ed.), *Fruit and*

Vegetables Phytochemicals, Chemistry, and Human Health (2nd edn., pp. 3–52). Blackwell Publishing.
22. Rejman, K., Górska-Warsewicz, H., Kaczorowska, J., & Laskowski, W., (2021). Nutritional significance of fruit and fruit products in the average polish diet. *Nutrients, 13*(6), 2079.
23. Sadler, M. J., Gibson, S., Whelan, K., Ha, M. A., Lovegrove, J., & Higgs, J., (2019). Dried fruit and public health – what does the evidence tell us? *International Journal of Food Sciences and Nutrition, 70*(6), 675–687.
24. Suchánková, M., Kapounová, Z., Dofková, M., Ruprich, J., Blahová, J., & Kouřilová, I., (2016). Selected fruits and vegetables: Comparison of nutritional value and affordability. *Czech Journal of Food Sciences, 33*(3), 242–246.
25. PBH Foundation, (2017). *Health and Wellness Resource Guide for Fruit & Vegetables* (pp. 14–26). PBH Foundation.
26. Li, Z., Yang, H., Li, P., Liu, J., Wang, J., & Xu, Y., (2013). Fruit biomechanics based on anatomy: A review. *International Agrophysics, 27*(1), 97.
27. Cerri, M., & Reale, L., (2020). Anatomical traits of the principal fruits: An overview. *Scientia Horticulturae, 270.*
28. Dardick, C., & Callahan, A., (2014). Evolution of the fruit endocarp: Molecular mechanisms underlying adaptations in seed protection and dispersal strategies. *Frontiers in Plant Science,* 5.
29. Oranusi, S., Onibokun, A., Afolabi, O., Okpalajiaku, C., Seweje, A., Olopade, B., et al., (2020). Chemical, microbial and antioxidant activity of *Cola lepidota* K. Schum fruits. *Czech Journal of Food Sciences, 38*(1), 11–19.
30. Cervantes-Hernández, F., Alcalá-González, P., Martínez, O., & Ordaz-Ortiz, J., (2019). Placenta, pericarp, and seeds of tabasco chili pepper fruits show a contrasting diversity of bioactive metabolites. *Metabolites, 9*(10), 206.
31. Lachowicz, S., Seliga, Ł., & Pluta, S., (2020). Distribution of phytochemicals and antioxidative potency in fruit peel, flesh, and seeds of Saskatoon berry. *Food Chemistry,* 305.
32. Spjut, R. W., & Thieret, J., (1989). Confusion between multiple and aggregate fruits. *The Botanical Review, 55*(1).
33. Jiang, Y., & Song, J., (2017). Fruits and fruit flavor: Classification and biological characterization. In: Hui, Y. H., (ed.), *Handbook of Fruit and Vegetable Flavors* (pp. 3–24). Hoboken, New Jersey: John Wiley & Sons, Inc.
34. Mohapatra, D., Mishra, S., Giri, S., & Kar, A., (2013). Application of hurdles for extending the shelf life of fresh fruits. *Trends in Post Harvest Technology, 1*(1), 37–54.
35. Pott, D. M., Osorio, S., & Vallarino, J., (2019). From central to specialized metabolism: An overview of some secondary compounds derived from the primary metabolism for their role in conferring nutritional and organoleptic characteristics to fruit. *Frontiers in Plant Science,* 10.
36. Lima, M. C., Paiva De, S. C., Fernandez-Prada, C., Harel, J., Dubreuil, J. D., & De Souza E. L., (2019). A review of the current evidence of fruit phenolic compounds as potential antimicrobials against pathogenic bacteria. *Microbial Pathogenesis, 130,* 259–270.
37. Suleria, H. A. R., Barrow, C. J., & Dunshea, F., (2020). Screening and characterization of phenolic compounds and their antioxidant capacity in different fruit peels. *Foods, 9*(9), 1206.

38. Khan, M. S. A., Khundmiri, S. U. K., Khundmiri, S. R., Al-Sanea, M. M., & Mok, P., (2018). Fruit-derived polysaccharides and terpenoids: Recent update on the gastroprotective effects and mechanisms. *Frontiers in Pharmacology, 9.*
39. Wahdaningsih, S., Wahyuono, S., Riyanto, S., & Murwanti, R., (2020). Terpenoid-lupeol of red dragon fruit (*Hylocereus polyrhizus*) and its immunomodulatory activity. *Pakistan Journal of Pharmaceutical Sciences,* 2.
40. Mele, M. A., Kang, H. M., Lee, Y. T., & Islam, M., (2021). Grape terpenoids: Flavor importance, genetic regulation, and future potential. *Critical Reviews in Food Science and Nutrition, 61*(9), 1429–1447.
41. Marsol-Vall, A., Sgorbini, B., Cagliero, C., Bicchi, C., Eras, J., & Balcells, M., (2017). Volatile composition and enantioselective analysis of chiral terpenoids of nine fruit and vegetable fibers resulting from juice industry by-products. *Journal of Chemistry, 2017,* 8675014.
42. Yazaki, K., Arimura, G., & Ohnishi, T., (2017). 'Hidden' terpenoids in plants: Their biosynthesis, localization, and ecological roles. *Plant and Cell Physiology, 58*(10).
43. Shofian, N. M., Hamid, A. A., Osman, A., Saari, N., Anwar, F., Dek, M. S. P., et al., (2011). Effect of freeze-drying on the antioxidant compounds and antioxidant activity of selected tropical fruits. *International Journal of Molecular Sciences, 12*, 4678–4692 [Internet]. Available from: https://www.mdpi.com/1422–0067/12/7/4678/htm (accessed on 26 September 2023).
44. Sun, Y., Zhang, M., & Mujumdar, A., (2019). Berry drying: Mechanism, pretreatment, drying technology, nutrient preservation, and mathematical models. *Food Engineering Reviews, 11*(2), 61–77 [Internet]. Available from: https://link.springer.com/article/10.1007/s12393-019-9188-3 (accessed on 26 September 2023).
45. Castagna, A., Chiavaro, E., Dall'Asta, C., Rinaldi, M., Galaverna, G., & Ranieri, A., (2013). Effect of postharvest UV-B irradiation on nutraceutical quality and physical properties of tomato fruits. *Food Chemistry, 137*(1–4), 151–158.
46. Assumpção, C. F., Hermes, V. S., Pagno, C., Castagna, A., Mannucci, A., Sgherri, C., et al., (2018). Phenolic enrichment in apple skin following post-harvest fruit UV-B treatment. *Postharvest Biology and Technology, 138*, 37–45.
47. Omolola, A. O., Jideani, A. I. O., & Kapila, P., (2016). *Quality Properties of Fruits as Affected by Drying Operation, 57*(1), 95–108. https://doi.org/101080/104083982013859563 [Internet]. Available from: https://www.tandfonline.com/doi/abs/10.1080/10408398.2013.859563 (accessed on 26 September 2023).
48. Moses, J. A., Norton, T., Alagusundaram, K., & Tiwari, B., (2014). Novel drying techniques for the food industry. *Food Engineering Reviews, 6*(3), 43–55. [Internet]. Available from: https://link.springer.com/article/10.1007/s12393-014-9078-7 (accessed on 26 September 2023).
49. Bhople, S., Singh, M., & Kumar, V. D., (2019). Application of drying techniques in food processing. *Int. J. Curr. Microbiol. App. Sci., 8*(8), 556–558. [Internet]. Available from: https://doi.org/10.20546/ijcmas.2019.808.066 (accessed on 26 September 2023).
50. Bhople, S., Singh, M., & Verma, D., (2019). Application of drying techniques in food processing. *International Journal of Current Microbiology and Applied Sciences, 8*(8), 556–558.
51. Battocchio, C., Bruni, F., Di Nicola, G., Gasperi, T., Iucci, G., Tofani, D., et al., (2021). *Solar Cookers and Dryers: Environmental Sustainability and Nutraceutical Content in*

Food Processing. Foods, *10*(10), 2326. [Internet]. Available from: https://www.mdpi.com/2304–8158/10/10/2326/htm (accessed on 26 September 2023).
52. López-Vidaña, E. C., Pilatowsky, F. I., Antonio, M. E. G., Navarro-Ocaña, A., Hernández-Vázquez, L., & Santiago-Urbina, J., (2021). Solar drying kinetics and bioactive compounds of blackberry (*Rubus fruticosus*). *Journal of Food Process Engineering, 42*(4), e13018. [Internet]. Available from: https://onlinelibrary.wiley.com/doi/full/10.1111/jfpe.13018 (accessed on 26 September 2023).
53. Hamid, Thakur, N. S., Thakur, A., & Kumar, P., (2020). Effect of different drying modes on phenolics and antioxidant potential of different parts of wild pomegranate fruits. *Scientia Horticulture, 274*, 109656.
54. McSweeney, M., & Seetharaman, K., (2014). *State of Polyphenols in the Drying Process of Fruits and Vegetables, 55*(5), 660–669. http://dx.doi.org/101080/104083982012670673. [Internet]. Available from: https://www.tandfonline.com/doi/abs/10.1080/10408398.2012.670673 (accessed on 26 September 2023).
55. Lim, K., Ma, M., & Dolan, K., (2011). Effects of spray drying on antioxidant capacity and anthocyanidin content of blueberry by-products. *Journal of Food Science, 76*(7), H156–H164.
56. Garrido, M. F., Gallo, L., Sette, P., Salvatori, D., & Bucalá, V., (2020). Nutraceutical tablets from maqui berry (*Aristotelia chilensis*) spray-dried powders with high antioxidant levels. *Drying Technology, 38*(9), 1231–1242.
57. Pereira, S. A. C., Deyse, G. P., & Damasceno, F. M. L., (2017). Maltodextrin, pectin, and soy protein isolate as carrier agents in the encapsulation of anthocyanins-rich extract from jaboticaba pomace. *Food and Bioproducts Processing*, 102.
58. Murugesan, R., & Orsat, V., (2012). Spray drying for the production of nutraceutical ingredients: A review. *Food and Bioprocess Technology, 5*, 3–14.
59. Benseddik, A., Benahmed-Djilali, A., Azzi, A., Zidoune, M. N., Bensaha, H., Lalmi, D., et al., (2020). Effect of drying processes on the final quality of potimarron pumpkin (*Cucurbita maxima*) powders. *Journal of Dispersion Science and Technology, 43*(1), 136–146.
60. Celik, O. F., Demirkol, M., Durmus, Y., & Tarakci, Z., (2020). Effects of drying method on the phenolics content and antioxidant activities of cherry laurel (*Prunus laurocerasus* L.). *Journal of Food Measurement and Characterization, 14*(1).
61. Horuz, E., & Maskan, M., (2015). Hot air and microwave drying of pomegranate (*Punica granatum* L.) arils. *Journal of Food Science and Technology, 52*(1).
62. Alkaltham, M. S., Salamatullah, A. M., Özcan, M. M., Uslu, N., Hayat, K., & Mohamed, A. I. A., (2021). Influence of different drying methods on antioxidant activity, total phenol, and phenolic compounds of myrtle (*Myrtus communis* L.) fruits. *Journal of Food Processing and Preservation, 45*(4), E15308.
63. Bernaert, N., Van, D. B., Van, P. E., & De Ruyck, H., (2019). Innovative refractance window drying technology to keep nutrient value during processing. *Trends in Food Science and Technology*, *84*, 22–24.
64. Waghmare, R., (2021). Refractance window drying: A cohort review on quality characteristics. *Trends in Food Science and Technology, 110,* 652–662.
65. Mphahlele, R. R., Fawole, O. A., Makunga, N. P., & Opara, U., (2016). Effect of drying on the bioactive compounds, antioxidant, antibacterial and anti-tyrosinase activities of pomegranate peel. *BMC Complementary and Alternative Medicine, 16*(1), 1–12.

66. Rahman, N. F. A., Shamsudin, R., Ismail, A., Shah, N. N. A. K., & Varith, J., (2018). Effects of drying methods on total phenolic contents and antioxidant capacity of the pomelo (*Citrus grandis* (L.) Osbeck) peels. *Innovative Food Science and Emerging Technologies*, 50.
67. Xu, M., Tian, G., Zhao, C., Ahmad, A., Zhang, H., Bi, J., et al., (2017). Infrared drying as a quick preparation method for dried tangerine peel. *International Journal of Analytical Chemistry*, *2017*, 6254793.
68. Vu, H. T., Scarlett, C. J., & Vuong, Q. V., (2017). Effects of drying conditions on physicochemical and antioxidant properties of banana (*Musa cavendish*) peels. *Drying Technology, 35*(9), 1141–1151.
69. Donno, D., Mellano, M. G., Raimondo, E., Cerutti, A. K., Prgomet, Z., & Beccaro, G., (2016). Influence of applied drying methods on phytochemical composition in fresh and dried goji fruits by HPLC fingerprint. *European Food Research and Technology, 242*(11), 1961–1974.
70. López-Vidaña, E. C., Rojano, B. A., Figueroa, I. P., Zapata, K., & Cortés, F., (2016). Evaluation of the sorption equilibrium and effect of drying temperature on the antioxidant capacity of the jaboticaba (*Myrciaria cauliflora*). *Chemical Engineering Communications, 203*(6), 809–821.
71. Kha, T. C., Nguyen, M. H., Roach, P. D., & Stathopoulos, C., (2014). Effect of drying pre-treatments on the yield and bioactive content of oil extracted from Gac aril. *International Journal of Food Engineering, 10*(1), 103–112.
72. Aamir, M., & Jittanit, W., (2017). Ohmic heating treatment for Gac aril oil extraction: Effects on extraction efficiency, physical properties, and some bioactive compounds. *Innovative Food Science & Emerging Technologies, 41*, 224–234.
73. Ciccoritti, R., Paliotta, M., Centioni, L., Mencarelli, F., & Carbone, K., (2018). The effect of genotype and drying condition on the bioactive compounds of sour cherry pomace. *European Food Research and Technology, 244*(4), 635–645.
74. Kim, M., & Kerr, W., (2013). Vacuum-belt drying of rabbit eye blueberry (*Vaccinium ashei*) slurries: Influence of drying conditions on physical and quality properties of blueberry powder. *Food and Bioprocess Technology, 6*(11), 3227–3237.
75. Sogi, D. S., Siddiq, M., Greiby, I., & Dolan, K., (2013). Total phenolics, antioxidant activity, and functional properties of 'Tommy Atkins' mango peel and kernel as affected by drying methods. *Food Chemistry, 141*(3), 2649–2655.
76. Jahanbakhshi, A., Kaveh, M., Taghinezhad, E., & Rasooli, S. V., (2020). Assessment of kinetics, effective moisture diffusivity, specific energy consumption, shrinkage, and color in the pistachio kernel drying process in microwave drying with ultrasonic pretreatment. *Journal of Food Processing and Preservation, 44*(6), E14449.
77. Sut, S., Baldan, V., Faggian, M., Peron, G., & Dall'Acqua, S., (2016). Nutraceuticals: A new challenge for medicinal chemistry. *Current Medicinal Chemistry, 23*(28), 3198–3223.
78. Soquetta, M. B., Terra, L. De. M., & Bastos, C., (2018). Green technologies for the extraction of bioactive compounds in fruits and vegetables. *CYTA – Journal of Food, 16*, 400–412.
79. Boussetta, N., & Vorobiev, E., (2014). Extraction of valuable biocompounds assisted by high voltage electrical discharges: A review. *Comptes Rendus Chimie, 17*(3).
80. Roselló-Soto, E., Barba, F. J., Parniakov, O., Galanakis, C. M., Lebovka, N., Grimi, N., et al., (2015). High voltage electrical discharges, pulsed electric field, and ultrasound

assisted extraction of protein and phenolic compounds from olive kernel. *Food and Bioprocess Technology, 8*(4), 885–894.
81. Koubaa, M., Barba, F. J., Grimi, N., Mhemdi, H., Koubaa, W., Boussetta, N., et al., (2016). Recovery of colorants from red prickly pear peels and pulps enhanced by pulsed electric field and ultrasound. *Innovative Food Science and Emerging Technologies*, 37.
82. Parniakov, O., Barba, F. J., Grimi, N., Lebovka, N., & Vorobiev, E., (2016). Extraction assisted by pulsed electric energy as a potential tool for green and sustainable recovery of nutritionally valuable compounds from mango peels. *Food Chemistry*, 192.
83. Ferrari, G., Maresca, P., & Ciccarone, R., (2011). The effects of high hydrostatic pressure on the polyphenols and anthocyanins in red fruit products. *Procedia Food Science*, 1.
84. Huang, H. W., Hsu, C. P., Yang, B. B., & Wang, C., (2013). Advances in the extraction of natural ingredients by high pressure extraction technology. *Trends in Food Science and Technology, 33*, 54–62.
85. Briones-Labarca, V., Giovagnoli-Vicuña, C., & Cañas-Sarazúa, R., (2019). Optimization of extraction yield, flavonoids, and lycopene from tomato pulp by high hydrostatic pressure-assisted extraction. *Food Chemistry*, 278.
86. Salem, I. B., Ouesleti, S., Mabrouk, Y., Landolsi, A., Saidi, M., & Boulilla, A., (2018). Exploring the nutraceutical potential and biological activities of *Arbutus unedo* L. (*Ericaceae*) fruits. *Industrial Crops and Products*, 122.
87. Blando, F., Berland, H., Maiorano, G., Durante, M., Mazzucato, A., Picarella, M. E., et al., (2019). Nutraceutical characterization of anthocyanin-rich fruits produced by "sun black" tomato line. *Frontiers in Nutrition*, 6.
88. Silva, H. R. Da., Assis, D. Da. C. De., Prada, A. L., Silva, J. O. C., Sousa, M. B. De., Ferreira, A. M., et al., (2019). Obtaining and characterization of anthocyanins from *Euterpe oleracea* (açaí) dry extract for nutraceutical and food preparations. *Revista Brasileira de Farmacognosia, 29*(5).
89. Tenore, G. C., Novellino, E., & Basile, A., (2012). Nutraceutical potential and antioxidant benefits of red pitaya (*Hylocereus polyrhizus*) extracts. *Journal of Functional Foods, 4*(1), 129–136.
90. Petchsak, P., & Sripanidkulchai, B., (2015). *Momordica cochinchinensis* aril extract induced apoptosis in human MCF-7 breast cancer cells. *Asian Pacific Journal of Cancer Prevention, 16*(13), 5507–5513.
91. Ascrizzi, R., Taglieri, I., Sgherri, C., Flamini, G., Macaluso, M., Sanmartin, C., et al., (2019). Nutraceutical oils produced by olives and citrus peel of Tuscany varieties as sources of functional ingredients. *Molecules, 24*(1), 65.
92. Jadhav, R. S., Palamthodi, S., & Saha, B., (2019). Nutraceutical potential and antipsychotic activity of local dry fruit waste – *Juglans regia. European Journal of Medicinal Plants*, pp. 1–7.
93. Brasanac-Vukanovic, S., Mutic, J., Stankovic, D. M., Arsic, I., Blagojevic, N., Vukasinovic-Pesic, V., et al., (2018). Wild bilberry (*Vaccinium myrtillus* l., Ericaceae) from Montenegro as a source of antioxidants for use in the production of nutraceuticals. *Molecules, 23*(8), 1864.
94. Awolu, O. O., & Manohar, B., (2019). Quantitative and qualitative characterization of mango kernel seed oil extracted using supercritical CO_2 and solvent extraction techniques. *Heliyon, 5*(12), e03068.
95. Fernández-Barbero, G., Pinedo, C., Espada-Bellido, E., Ferreiro-González, M., Carrera, C., Palma, M., et al., (2019). Optimization of ultrasound-assisted extraction of bioactive

compounds from jabuticaba (*Myrciaria cauliflora*) fruit through a box-Behnken experimental design. *Food Science and Technology, 39*(4), 1018–1029.
96. Díaz-de-Cerio, E., Verardo, V., Fernández-Gutiérrez, A., & Gómez-Caravaca, A., (2019). New insight into phenolic composition of chayote (*Sechium edule* (Jacq.) Sw.). *Food Chemistry*, 295.
97. Lantzouraki, D. Z., Sinanoglou, V. J., Zoumpoulakis, P., & Proestos, C., (2016). Comparison of the antioxidant and antiradical activity of pomegranate (*Punica granatum* L.) by ultrasound-assisted and classical extraction. *Analytical Letters, 49*(7), 969–978.
98. Belwal, T., Giri, L., Bhatt, I. D., Rawal, R. S., & Pande, V., (2017). An improved method for extraction of nutraceutically important polyphenolics from *Berberis jaeschkeana* C.K. Schneid. fruits. *Food Chemistry*, 230.
99. Ciriminna, R., Fidalgo, A., Avellone, G., Danzì, C., Timpanaro, G., Locatelli, M., et al., (2019). Integral extraction of *Opuntia ficus*-indica peel bioproducts via microwave-assisted hydrodiffusion and hydrodistillation. *ACS Sustainable Chemistry and Engineering, 7*(8), 7884–7891.
100. Torres-León, C., Rojas, R., Serna-Cock, L., Belmares-Cerda, R., & Aguilar, C., (2017). Extraction of antioxidants from mango seed kernel: Optimization assisted by microwave. *Food and Bioproducts Processing, 105*, 188–196.
101. Bobinaitė, R., Pataro, G., Lamanauskas, N., Šatkauskas, S., Viškelis, P., & Ferrari, G., (2015). Application of pulsed electric field in the production of juice and extraction of bioactive compounds from blueberry fruits and their by-products. *Journal of Food Science and Technology, 52*(9), 5898–5905.
102. Niu, D., Ren, E. F., Li, J., Zeng, X. A., & Li, S., (2021). Effects of pulsed electric field-assisted treatment on the extraction, antioxidant activity and structure of naringin. *Separation and Purification Technology*, 265.
103. Xi, J., He, L., & Yan, L. G., (2017). Continuous extraction of phenolic compounds from pomegranate peel using high voltage electrical discharge. *Food Chemistry*, 230.
104. Boussetta, N., Vorobiev, E., Deloison, V., Pochez, F., Falcimaigne-Cordin, A., & Lanoisellé, J., (2011). Valorization of grape pomace by the extraction of phenolic antioxidants: Application of high voltage electrical discharges. *Food Chemistry, 128*(2), 364–370.
105. Seabra, P. F. C., de Souza E. S., A. P., Rodriguez, S. M. A., da Costa, W. A., Cardoso da Costa, H. S., Santos, L. A., Rogez, H., & de Carvalho, J. R. N., (2019). Determination of process parameters and bioactive properties of the murici pulp (*Byrsonima crassifolia*) extracts obtained by supercritical extraction. *Journal of Supercritical Fluids*, 146.
106. Mihalcea, L., Turturică, M., Cucolea, E. I., Dănilă, G. M., Dumitrașcu, L., Coman, G., et al., (2021). CO_2 supercritical fluid extraction of oleoresins from sea buckthorn pomace: Evidence of advanced bioactive profile and selected functionality. *Antioxidants, 10*(11), 1681.
107. Jamaludin, R., Kim, D. S., Md. Salleh, L., & Lim, S. B., (2020). Optimization of high hydrostatic pressure extraction of bioactive compounds from noni fruits. *Journal of Food Measurement and Characterization, 14*(5), 2810–2818.
108. Jung, S., Lee, M. S., Choi, A. J., Kim, C. T., & Kim, Y., (2019). Anti-inflammatory effects of high hydrostatic pressure extract of mulberry (*Morus alba*) fruit on LPS-stimulated RAW264.7 cells. *Molecules, 24*(7), 1425.
109. Briones-Labarca, V., Giovagnoli-Vicuña, C., Figueroa-Alvarez, P., Quispe-Fuentes, I., & Pérez-Won, M., (2013). Extraction of β-carotene, vitamin C and antioxidant compounds

from *Physalis peruviana* (cape gooseberry) assisted by high hydrostatic pressure (cape gooseberry) assisted by high hydrostatic pressure. *Food and Nutrition Sciences, 4*(8), 35284.

110. Mushtaq, M., Sultana, B., Bhatti, H. N., & Asghar, M., (2015). RSM based optimized enzyme-assisted extraction of antioxidant phenolics from underutilized watermelon (*Citrullus lanatus* Thunb.) rind. *Journal of Food Science and Technology, 52*(8), 5048–5056.
111. Qadir, R., Anwar, F., Batool, F., Mushtaq, M., & Jabbar, A., (2019). Enzyme-assisted extraction of *Momordica balsamina* L. fruit phenolics: Process optimized by response surface methodology. *Journal of Food Measurement and Characterization, 13*(1), 697–706.
112. Saad, N., Louvet, F., Tarrade, S., Meudec, E., Grenier, K., Landolt, C., et al., (2019). Enzyme-assisted extraction of bioactive compounds from raspberry (*Rubus idaeus* L.) pomace. *Journal of Food Science, 84*(6).
113. Dembitsky, V. M., Poovarodom, S., Leontowicz, H., Leontowicz, M., Vearasilp, S., Trakhtenberg, S., et al., (2011). The multiple nutrition properties of some exotic fruits: Biological activity and active metabolites. *Food Research International, 44*(7), 1671–1701.
114. Che, C. T., & Zhang, H., (2019). Plant natural products for human health. *International Journal of Molecular Sciences, 20*(4), 830.
115. Pott, D. M., Osorio, S., & Vallarino, J., (2019). From central to specialized metabolism: An overview of some secondary compounds derived from the primary metabolism for their role in conferring nutritional and organoleptic characteristics to fruit. *Frontiers in Plant Science,* 10.
116. Daglia, M., (2012). Polyphenols as antimicrobial agents. *Current Opinion in Biotechnology, 2*, 174–181.
117. Ho, C. T., (1992). Phenolic compounds in food. In: Huang, M. T., Ho, C. T., & Lee, C. Y., (eds.), *Phenolic Compounds in Food and Their Effects on Health II* (pp. 2–7).
118. Rice-Evans, C., Miller, N., & Paganga, G., (1997). Antioxidant properties of phenolic compounds. *Trends in Plant Science, 4*, 152–159.
119. Wang, G., Tang, W., & Bidigare, R., (2005). Terpenoids as therapeutic drugs and pharmaceutical agents. In: *Natural Products* (pp. 197–227). Totowa, NJ: Humana Press.
120. Ludwiczuk, A., Skalicka-Woźniak, K., & Georgiev, M., (2017). Terpenoids. In: *Pharmacognosy* (pp. 233–266). Elsevier.
121. Alcazar, M. A., Wright, K., Vaswani, A., Caruso, M., Reed, R. L., Bailey, C. F., et al., (2020). Integration of mass spectral fingerprinting analysis with precursor ion (MS1) quantification for the characterization of botanical extracts: Application to extracts of *Centella asiatica* (L.) urban. *Phytochemical Analysis, 31*(6), 722–738.
122. Magana, A. A., Wrobel, K., Elguera, J. C. T., Escobosa, A. R. C., & Wrobel, K., (2015). Determination of small phenolic compounds in tequila by liquid chromatography with ion trap mass spectrometry detection. *Food Analytical Methods, 8*(4), 864–872.
123. Palafox-Carlos, H., Yahia, E. M., & González-Aguilar, G., (2012). Identification and quantification of major phenolic compounds from mango (*Mangifera indica*, cv. Ataulfo) fruit by HPLC–DAD–MS/MS-ESI and their individual contribution to the antioxidant activity during ripening. *Food Chemistry, 135*(1), 105–111.

124. Kim, Y., Brecht, J. K., & Talcott, S., (2007). Antioxidant phytochemical and fruit quality changes in mango (*Mangifera indica* L.) following hot water immersion and controlled atmosphere storage. *Food Chemistry, 105*(4), 1327–1334.
125. Corral-Aguayo, R. D., Yahia, E. M., Carrillo-Lopez, A., & González-Aguilar, G., (2008). Correlation between some nutritional components and the total antioxidant capacity measured with six different assays in eight horticultural crops. *Journal of Agricultural and Food Chemistry, 56*(22), 10498–10504.
126. Heo, H. J., Kim, Y. J., Chung, D., & Kim, D. O., (2007). Antioxidant capacities of individual and combined phenolics in a model system. *Food Chemistry, 104*(1), 87–92.
127. Kim, H. G., Ju, M. S., Shim, J. S., Kim, M. C., Lee, S. H., Huh, Y., et al., (2010). Mulberry fruit protects dopaminergic neurons in toxin-induced Parkinson's disease models. *British Journal of Nutrition, 104*(1), 8–16.
128. Chen, P. N., Chu, S. C., Chiou, H. L., Kuo, W. H., Chiang, C. L., & Hsieh, Y. S., (2006). Mulberry anthocyanins, cyanidin 3-rutinoside and cyanidin 3-glucoside, exhibited an inhibitory effect on the migration and invasion of a human lung cancer cell line. *Cancer Letters, 235*(2), 248–259.
129. Sass-Kiss, A., Kiss, J., Milotay, P., Kerek, M. M., & Toth-Markus, M., (2005). Differences in anthocyanin and carotenoid content of fruits and vegetables. *Food Research International, 38*(8, 9), 1023–1029.
130. Koyuncu, F., (2004). Organic acid composition of native black mulberry fruit. *Chemistry of Natural Compounds, 40*(4), 367–369.
131. Paunović, S. M., Mašković, P., & Milinković, M., (2020). Determination of primary metabolites, vitamins and minerals in black mulberry (*Morus nigra*) berries depending on altitude. *Erwerbs-Obstbau, 62*(3), 355–360.
132. Hassimotto, N. M. A., Genovese, M. I., & Lajolo, F., (2007). Identification and Characterization of anthocyanins from wild mulberry (*Morus nigra* L.) growing in Brazil. *Food Science and Technology International, 13*(1), 17–25.
133. Özgen, M., Serçe, S., & Kaya, C., (2009). Phytochemical and antioxidant properties of anthocyanin-rich Morus nigra and *Morus rubra* fruits. Scientia *Horticulture, 119*(3), 275–279.
134. D'Urso, G., Mes, J. J., Montoro, P., Hall, R. D., & De Vos, R. C. H., (2019). Identification of bioactive phytochemicals in mulberries. *Metabolites, 10*(1), 7.
135. Kusano, G., Orihara, S., Tsukamoto, D., Shibano, M., Coskun, M., Guvenc, A., et al., (2002). Five new nortropane alkaloids and six new amino acids from the fruit of *Morus alba* Linne growing in turkey. *Chemical and Pharmaceutical Bulletin, 50*(2), 185–192.
136. Acquavia, M., Pascale, R., Foti, L., Carlucci, G., Scrano, L., Martelli, G., et al., (2021). Analytical methods for extraction and identification of primary and secondary metabolites of apple (*Malus domestica*) fruits: A review. *Separations, 8*(7), 91.
137. Figuerola, F., Hurtado, M. L., Estévez, A. M., Chiffelle, I., & Asenjo, F., (2005). Fiber concentrates from apple pomace and citrus peel as potential fiber sources for food enrichment. *Food Chemistry, 91*(3), 395–401.
138. Huber, G. M., & Rupasinghe, H. P. V., (2009). Phenolic profiles and antioxidant properties of apple skin extracts. *Journal of Food Science, 74*(9), C693–700.
139. Petkovska, A., Gjamovski, V., Stanoeva, J. P., & Stefova, M., (2017). Characterization of the polyphenolic profiles of peel, flesh and leaves of malus domestica cultivars using UHPLC-DAD-HESI-MS. *Natural Product Communications, 12*(1), 1934578X1701200.

140. Saini, R. K., Nile, S. H., & Park, S., (2015). Carotenoids from fruits and vegetables: Chemistry, analysis, occurrence, bioavailability and biological activities. *Food Research International, 76*, 735–750.
141. Arathi, B. P., Sowmya, P. R. R., Vijay, K., Baskaran, V., & Lakshminarayana, R., (2015). Metabolomics of carotenoids: The challenges and prospects: A review. *Trends in Food Science & Technology, 45*(1), 105–117.
142. Ferreira, L., Perestrelo, R., Caldeira, M., & Câmara, J., (2009). Characterization of volatile substances in apples from *Rosaceae* family by headspace solid-phase microextraction followed by GC-qMS. *Journal of Separation Science, 32*(11), 1875–1888.
143. Butkevičiūtė, A., Liaudanskas, M., Kviklys, D., Zymonė, K., Raudonis, R., Viškelis, J., et al., (2018). Detection and analysis of triterpenic compounds in apple extracts. *International Journal of Food Properties, 21*(1), 1716–1727.
144. Grigoras, C. G., Destandau, E., Fougère, L., & Elfakir, C., (2013). Evaluation of apple pomace extracts as a source of bioactive compounds. *Industrial Crops and Products, 49*, 794–804.
145. Farag, M. A., Mohsen, M., Heinke, R., & Wessjohann, L., (2014). Metabolomic fingerprints of 21 date palm fruit varieties from Egypt using UPLC/PDA/ESI–qTOF-MS and GC–MS analyzed by chemometrics. *Food Research International, 64*, 218–226.
146. Al-Turki, S., Shahba, A. M., & Stushnoff, C., (2010). Diversity of antioxidant properties and phenolic content of date palm (*Phoenix dactylifera* L.) fruits as affected by cultivar and location. *Journal of Food, Agriculture & Environment, 8*(1), 253–260.
147. Bedair, M., & Sumner, L., (2008). Current and emerging mass-spectrometry technologies for metabolomics. *TrAC Trends in Analytical Chemistry, 27*(3), 238–250.
148. Hong, Y. J., Tomas-Barberan, F. A., Kader, A. A., & Mitchell, A., (2006). The flavonoid glycosides and procyanidin composition of Deglet Noor dates (*Phoenix dactylifera*). *Journal of Agricultural and Food Chemistry, 54*(6), 2405–2411.
149. Heber, D., (2004). Vegetables, fruits, and phytoestrogens in the prevention of diseases. *Journal of Postgraduate Medicine, 50*(2), 145–149.
150. Obermüller-Jevic, U. C., Olano-Martin, E., Corbacho, A. M., Eiserich, J. P., Van, D. V. A., Valacchi, G., et al., (2003). Lycopene inhibits the growth of normal human prostate epithelial cells *in vitro*. *The Journal of Nutrition, 133*(11), 3356–3360.
151. Huang, D., Ou, B., & Prior, R., (2005). The chemistry behind antioxidant capacity assays. *Journal of Agricultural and Food Chemistry, 53*(6), 1841–1856.
152. Yahia, E. M., Gutierrez-Orozco, F., & Leon, C., (2011). Phytochemical and antioxidant characterization of the fruit of black sapote (*Diospyros digyna* Jacq.). *Food Research International, 44*(7), 2210–2216.
153. Cappellin, L., Soukoulis, C., Aprea, E., Granitto, P., Dallabetta, N., Costa, F., et al., (2012). PTR-ToF-MS and data mining methods: A new tool for fruit metabolomics. *Metabolomics, 8*(5), 761–770.
154. Alves, Filho, Silva, Lima, Ribeiro, Silva, Zocolo, et al., (2019). Metabolomic variability of different genotypes of cashew by LC-MS and correlation with near-infrared spectroscopy as a tool for fast phenotyping. *Metabolites, 9*(6), 121.
155. Khodabakhshian, R., Emadi, B., Khojastehpour, M., Golzarian, M. R., & Sazgarnia, A., (2017). Non-destructive evaluation of maturity and quality parameters of pomegranate fruit by visible/near infrared spectroscopy. *International Journal of Food Properties, 20*(1), 41–52.

156. Hernández-Mesa, M., Ropartz, D., García-Campaña, A. M., Rogniaux, H., Dervilly-Pinel, G., & Le Bizec, B., (2019). Ion mobility spectrometry in food analysis: Principles, current applications, and future trends. *Molecules, 24*(15), 2706.
157. Liu, W., Zhang, X., Siems, W. F., Hill, H. H., & Yin, D., (2015). Rapid profiling and identification of anthocyanins in fruits with Hadamard transform ion mobility mass spectrometry. *Food Chemistry, 177*, 225–232.
158. Yang, L., Liu, J., Wang, X., Wang, R., Ren, F., Zhang, Q., et al., (2019). Characterization of volatile component changes in jujube fruits during cold storage by using headspace-gas chromatography-ion mobility spectrometry. *Molecules, 24*(21), 3904.
159. Rasheed, D. M., Serag, A., Abdel, S. Z. T., & Farag, M., (2021). Novel trends and applications of multidimensional chromatography in the analysis of food, cosmetics and medicine bearing essential oils. *Talanta, 223*, 121710.
160. Cacciola, F., Giuffrida, D., Utczas, M., Mangraviti, D., Dugo, P., Menchaca, D., et al., (2016). Application of comprehensive two-dimensional liquid chromatography for carotenoid analysis in Red Mamey (*Pouteria sapote*) fruit. *Food Analytical Methods, 9*(8), 2335–2341.
161. Mishra, P., Biancolillo, A., Roger, J. M., Marini, F., & Rutledge, D., (2020). New data preprocessing trends based on ensemble of multiple preprocessing techniques. *TrAC Trends in Analytical Chemistry, 132*, 116045.
162. Mishra, P., Rutledge, D. N., Roger, J. M., Wali, K., & Khan, H., (2021). Chemometric pre-processing can negatively affect the performance of near-infrared spectroscopy models for fruit quality prediction. *Talanta, 229*, 122303.
163. Ruiz, R. L. G., Zamora, G. V. M., Pescuma, M., Van, N. C., Mozzi, F., & Sánchez, B. J. A., (2021). Fruits and fruit by-products as sources of bioactive compounds. Benefits and trends of lactic acid fermentation in the development of novel fruit-based functional beverages. *Food Research International, 140.*
164. Mahato, N., Sharma, K., Sinha, M., & Cho, M., (2018). Citrus waste derived Nutra-/pharmaceuticals for health benefits: Current trends and future perspectives. *Journal of Functional Foods, 40.*
165. Chang, S. K., Alasalvar, C., & Shahidi, F., (2019). Superfruits: Phytochemicals, antioxidant efficacies, and health effects – A comprehensive review. *Critical Reviews in Food Science and Nutrition, 59*(10), 1580–1604.
166. Septembre-Malaterre, A., Remize, F., & Poucheret, P., (2018). Fruits and vegetables, as a source of nutritional compounds and phytochemicals: Changes in bioactive compounds during lactic fermentation. *Food Research International, 104.*

CHAPTER 4

Technological and Analytical Aspects of Bioactive Compounds and Nutraceuticals from Roots and Tubers Sources

ANTIMA GUPTA, HANUMAN BOBADE, RAJAN SHARMA, and SAVITA SHARMA

Department of Food Science and Technology, Punjab Agricultural University, Ludhiana, Punjab, India

ABSTRACT

Roots and tubers are the second most important source of carbohydrates after cereals. They include cassava, potato, sweet potato, yam, carrot, beetroot, radish, ginger, turnip, and many more. Apart from being a good source of carbohydrate, they do contain certain health beneficial compounds like phenolic substances, carotenoids, saponins, and sapogenins, bioactive proteins like dioscorin, glycoalkoloids, and others. To isolate these health-beneficial compound, conventional technologies like Soxhlet, maceration, percolation, etc., have been used since so many years, but these technologies are time-consuming and require a lot of solvents. However, they are being used as a reference method to check the correctness of the newly developed technologies for extraction. Various technologies are used these days to increase the extractive effectiveness of bioactive compounds. They include enzyme-assisted extraction (EAE), microwave-assisted extraction, pulsed electric field extraction, ultrasound-assisted extraction, pressurized liquid extraction (PLE), and supercritical fluid extraction (SCFE). These are being preferred because, they are less time consuming, results in better yield and quality extract with less use of synthetic and organic chemicals. Generally, chromatographic techniques like high-pressure performance

Bioactive Compounds and Nutraceuticals from Plant Sources: Extraction Technology, Analytical Techniques, and Potential Health Prospects. Charanjit Singh Riar, Parmjit S. Panesar (Eds.)

liquid chromatography (HPPLC), high-performance liquid chromatography (HPLC), gas-liquid chromatography (GLC), etc., are being used to quantify these compounds. Bioactives present in root and tuber crops exhibit many pharmacological activities and therefore offer several health benefits. The biological activities possessed by nutraceuticals of root and tuber crops include antioxidant activity, anticancer activity, anti-microbial activity, anti-diabetic activity, anti-obesity property and many other activities.

4.1 INTRODUCTION

Roots and starchy tubers are the second most important source of carbohydrates after cereals, which belong to different botanical families (Figure 4.1). Root crops particularly include carrot, swede, beetroot, ginger, radish, turnip, and parsnip. On the other hand, tuber crops are potato, taro, yam, sweet potato, tapioca, Colocasia, kumara, etc. As all of these produce underground edible parts, they are grouped together. Owing to their favorable adaption to diversified environmental conditions, this vegetable group is consumed as a staple food in many countries. Besides a considerable part of the supplied world, food roots and tubers are important sources of processed products for various food items for industrial purposes, human consumption, and animal feed. Tubers and starchy roots are edible plants that are originated from modified or diversified botanical sources, store starch in corms, rhizomes, roots, subterranean stems and tubers. Cocoyam and taro are derived from corms, swollen hypocotyls, and underground stems, whereas potatoes and yams are tubers. In comparison to cereals, roots, and tubers produce more quantity of energy value per day. In terms of annual volume of production, the most important food crops globally are yam, potato, cassava, and sweet potato. In developing countries, in terms of production top 10 crops are sweet potato, cassava, and potato. At present scenario most widely cultivated are developed by natural pollination and crossbreeding. According to an estimation, world potato demand has been raised to double by the year 2020. As potato starch can be used for controlled release of functional ingredients and for encapsulation in biopharmaceuticals. Owing to its high moisture content about 60–90%; root and tuber crops are bulky in nature and have short shelf life, high transportation cost thus associated to limited market margin in developing countries.

The general characteristics of R&T crops are the following:

- Root and tuber crops are characterized by high moisture content of about 70–80%.
- These crops have large unit sizes, typically ranging from 100 g to 15 kg.
- These crops are perishable owing to its less shelf-life ranges from a few days to a month.
- Degradation or spoilage is caused by senescence, rotting caused by fungi or bacteria, bruising, and sprouting.
- These crops are characterized by high respiration rate as compared to that of cereals.
- In case of Nutritional composition, richest source of dietary fibers (DFs) and deficient in protein and lipids.
- High nutritive value in terms of dry weight basis.

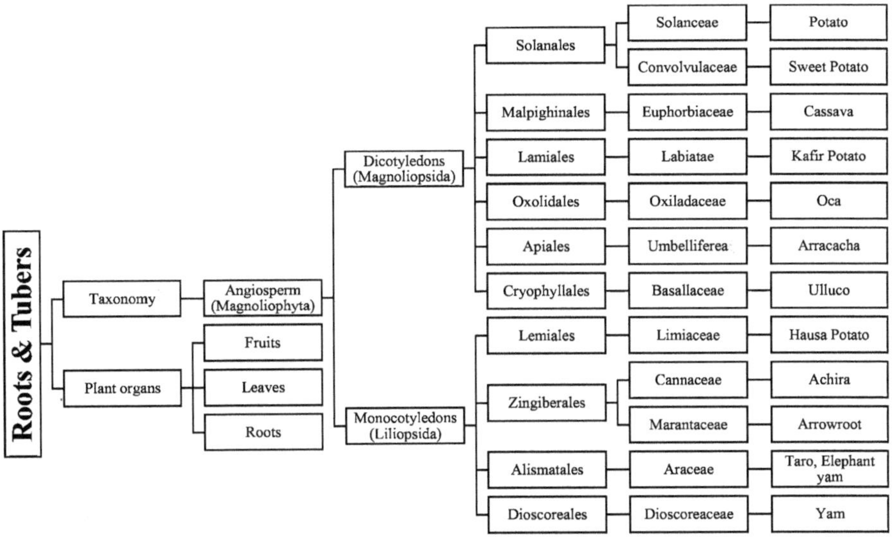

FIGURE 4.1 Taxonomic classification of roots and tubers.

4.2 ORIGIN AND PRODUCTION

Tuber and root crops were originated in majorly in three regions independently:

- Africa – Madagascar;
- Southeast Asia and its geographic continuation – the Sunda Islands, Papua New Guinea, Oceania; and
- Tropical America (Table 4.1).

Other than these were Japan and southern China, where a few species were domesticated. There was no system of exchange of techniques and materials between the old and the new worlds in the early 1500s. In Oceania and Tropical America, the sweet potato was the only known crop. The land between the South China Sea, the Deccan peninsula, and the Indo-Malayan region could be considered the primary domestication region of these plant resources and materials in relation to the utilization and invention of these plant materials. Owing to the natural barriers of the Himalayas, these regions are limited in the northeast region, basically towards China. On the other hand, towards the west, the Indian desert region is open. The rest of the part is surrounded by water bodies. Natural continuation could be considered as Papua New Guinea and the larger Sunda Islands region. The coastal areas of Deccan Peninsula are mainly comprised by the Indian Section. Five racial groups in these regions who were belonging to the Negrito stock, lived only by consumption of these root and tuber crops. Even in present scenario, small groups of these races who have survived in these regions for many centuries owing to unavailability of food resources and reduction in their population size; collect aroids and yams which is present in abundance in these regions. The Negrito stock who was settled in these regions from 2,500 years ago was pushed by other immigrant towards eastwards. The new mixed population to whom Malays or Indo-Malays names have applied were the population group who starts domestication as well as cultivation of roots and tubers. Using combinations of simple practices and simple tools, agricultural practices are possible in this region. The most considerable expansion of tuber and root crops in eastwards is from the Southeast Asian region which was carried out by the Proto-Malays, the mixed population. The expansion of crop is limited towards west owing to desert and dry regions in northeast India. Some yams and taro are pushed towards the northeast India into the subtropical area of China towards to Japan. The origin of some crops is as follows.

Potato, and tuber crop has been domesticated and originated in the high Andes of South America, thus becoming a popular food crop in South America, eastern and Central Africa, and Asia. Central America is the origin of sweet potato and is the seventh largest food crop grown in subtropical, tropical, and warm temperate regions. Taro is most popular crop in Africa, Asia, Caribbean Islands and Pacific furthermore Colocasia is originated in southeast Asia and India.

TABLE 4.1 Common and Botanical Name of Tuber and Root Crops with Their Origin and Present Distribution

Common Name	Botanical Name	Origin	Distribution Presently
Yams	*Dioscora* spp.	America, Asia, Africa	Africa, Asia, Caribbean Pacific Islands
Sweet potato	*Ipomea batatas*	North-western, central, and south America	Warmer parts of temperate region and all tropical and sub-tropical region.
Giant taro	*Alocasia macrorrhiza*	Asia	Southeast Asia, Sri Lanka, Pacific Islands
Peruvian carrot	*Arracada xanthorrhiza* Bancr.	Bolivia	Particularly South America, Latin America
Taro, Dasheen	*Colocasia esculenta*	Southeast Asia India	All parts of tropical region
Edible canna	*Canna edulis* Ker.	Tropical America	Asia, Pacific Islands, Latin America
Tannia	*Xanthosoma* spp.	Tropical America	Pacific Islands, Southeast Asia, Africa, Tropical America
Arrowroot	*Marantha arundinacea*	Tropical America	Tropical Asia and Tropical America
Ulluco	*Ullucus tuberosus*	Andean region	Andean region
New Zealand Yam, Oca	*Oxalis tuberosa* Mol *O. crenata*	Andean region	New Zealand, Mexico, Andean region
Irish potato	*Solanum tuberosum* L.	Andean region	All temperate, subtropical, and tropical regions
Ahipa	*Pachyrrh ahipa*	Andean region	Bolivia, Peru
Yacon	*Polymnia sonçhifolia*	Andean region	Southeast Asia and South America

Production In 2019, the production of roots and tubers crops globally was 861 million tons. On the other hand, in India, it recorded 56.3 million tons, which is an increase from 10.7 in the last 4 decades. From 1970 to 2019, world production of root and tuber crops increased from 559 to 861 million tons, growing at the average annual rate of 0.95%. On the other hand, at the national level at the rate of 3.81%. With the production of 368 million tons, potato is the fourth important food crop, followed by maize, wheat, and rice globally. Carrots are fifth leading food crop with 40% processing share and 60% market share. With global production of 80%, covering 52% of world's area; Asia is considered as largest region under production of sweet potato. After that is Africa which produce 17% of total world production and cover 42% of world's area reported by FAO statistics (2013) [1]. Other than these, eastern Asia includes major production; China alone covers 43% of the world's area, and production is about 72%.

4.3 CHEMICAL COMPOSITION

In general, roots and tuber crops are considered as poor source of Nutrition as compared to other crops based on fresh weight. R&T crops provide approximately 25% of available energy to 17 countries of the world, according to FAO statistics. Nutritionally, tubers and root crops, in the form of carbohydrates, provide an economical source of dietary fiber and thus have great importance (Table 4.2). Owing to higher moisture content of tuber crops, energy obtained from these crops is equivalent to one-third weight of wheat and rice. Furthermore, high yield of tubers and roots provides more energy value per day per land unit as compared to cereal grains. On dry weight basis, protein content in root and tuber crops is low ranges from 1–2%. Around 50% to 75% of daily requirement of energy and 80% of protein requirement is fulfilled by Root and tubers in infants and younger children. As compared to other tuber crops, potato, and yams are rich source of protein. Tuber crops lack Cysteine and methionine, which are sulfur-containing amino acids. Yam, cassava, and yellow variety of sweet potato contains β-carotene and sweet potatoes, cassava, yam, and potatoes contain a little amount of vitamin C. Among these crops, taro is a good source of potassium. As compared to cereals, root and tuber crops comprise good and advantageous Ca/P ratio and higher calcium content. R&T is a good source of DFs; on the other hand, it is deficient in minerals and vitamins.

Owing to its nutraceutical components, sweet potato is considered a popular health food. Sweet potato is considered as functional food and antidiabetic effects due to law-calorific value. It includes vitamin C, vitamin B complex, xanthophylls, carotenes, phenolic acids, amino acids, protein, anthocyanins, and minerals. Sweet potato produces more than 30% of starch/unit area as compared to that of maize [2, 3]. Cassava is considered as richest source of vitamin C and calcium and provides significant amount of riboflavin, thiamine, and nicotinic acid. Cassava leaves contains a considerably high amount of protein owing to the presence of essential amino acids like methionine, cystine, and cysteine, whereas the storage part, which is the root, lack protein content. And every part of cassava is edible. Carrot is important source of β-carotene which is precursor of Vitamin A (retinol). Different colored carrots contain different colored carotenoids like α- and β-carotene, and other colored carrots have different carotene compositions [4]. Carrots is constituted about 300–700 mg/kg of vitamin C in Dry matter [5].

However, cooking affects the nutritive value of roots and tubers, causing a reduction in nutrients, including minerals, protein, vitamins, and carbohydrates. This nutritional reduction occurs in two different manners: by some chemical changes like oxidation or degradation by some sort of destruction, and other is a leaching process. Free amino acids can react with free sugars and form complex and may be leach down. Minerals are only degraded by leaching process. Vitamins like vitamin C get easily degraded by leaching; during cooking, on the other hand, vitamin B complex shows considerable loss during boiling up to 25%. Peeling after boiling is advisable to prevent leaching loss in root and tuber crops. Like other crops, root and tuber vary in nutritional composition depending upon agronomical practices, soil type, and environmental factors.

TABLE 4.2 Nutritional Composition of Major and Minor Root and Tuber Crops

R&T Crops (Major)	Fibers (g/100 g)	Carbohydrate (g/100 g)	Protein (g/100 g)	Fat (g/100 g)	Energy k/Cal 100 g (DM)
Sweet potato	10.2	78.2	5.3	1.95	351
Cassava	7.9	86.9	2.7	0.62	364
Irish potato	9.3	66.7	9.2	0.50	316
Yam	17.9	72.8	6.4	0.42	318
R&T Crops (Minor)					
Arrowroot	4.4	91.1	5.6	0.23	367
Cocoyam	3.1	88.2	6.3	0.79	362
Achira	1.5	94.3	2.7	0.30	392
Giant Taro	6.2	75.6	7.2	0.34	361
Arracacha	2.2	92.2	3.0	0.74	385
Ulluco	4.3	88.6	7.1	Traces	362
Oca	4.9	85.2	6.2	3.70	389

4.4 EXTRACTION TECHNOLOGY

Extraction techniques play a significant role in correctly estimating the bioactive compounds qualitatively and quantitatively. They are also known as "sample preparation techniques." Generally, this part of the study is neglected and usually done by untrained research personnel, which is the major cause of wrong results. Majors [6] reported that most of the researchers believe in the importance of sample preparation during any analytical study. In earlier days, Soxhlet is being used as a conventional method to extract bioactive compounds from roots and tubers. This method of extraction is time consuming, and a lot of solvent is required, but because of its accuracy and repeatability, still it is considered as one of the reference method to check the correctness of newly developed extraction technique. During the last 50 years, many green technologies have been developed to extract the bioactive compounds, they include, enzyme assisted extraction (EAE), microwave assisted extraction, pulsed electric field extraction, ultrasound assisted extraction, pressurized liquid extraction (PLE) and supercritical fluid (SF) assisted fluid extraction [7]. These are being preferred for extraction because, they are less time consuming, results in better yield and quality extract with less use of synthetic and organic chemicals. Due to the difference in the chemical nature of bioactive compounds and the presence of massive numbers of roots and tubers species, there is a need to develop a standard and integrated approach to extracting these compounds. The factor affecting the extraction process includes properties of roots and tubers, process time, temperature, pressure, and solvent used for extraction [8].

4.4.1 CONVENTIONAL TECHNIQUE FOR THE EXTRACTION OF BIOACTIVE COMPOUNDS FROM ROOTS AND TUBERS

Soxhlet extraction, maceration, and hydro-distillation are some of the commonly used conventional techniques. These are the classical techniques known for their accuracy and repeatability of results. Efficiency of these conventional method chiefly depends on the selection of solvents which further connects with the targeted bioactive compound.

The Soxhlet extraction technique, when proposed in early 1879 by Franz Ritter Von Soxhlet, was mainly designed to extract lipids; later, its application was expanded to the extraction of bioactive compounds as well. It is used as a reference method to check the correctness of newly developed technologies for extraction. In Soxhlet, dry sample is placed in thimble which is then placed for refluxing with solvent of particular interest. During refluxing, bioactives dissolves in solvent and carried with them into the bulk solvent. This process runs repeatedly till all the compounds are extracted. Halwani et al. [9] extracted the betalains from beetroot using different solvent, they reported concentration of betalains was higher in water extract in comparison to citric acid (2%) solution. Liu et al. [10] optimized the extraction process of anthocyanin from purple sweet potato and the result showed that ethanol concentration of 25%, ammonium sulfate concentration 22% with solid-liquid ratio 1:45 and pH 3.3, the optimum extraction of anthocyanin was 90.02% which was very close to the predicted extraction rate, i.e., 90.12%. Vanja et al. [11] extracted the carotenoid and polyphenol using different solvent, i.e., acetone, ethanol, hexane, and ethyl acetate. They reported hexane and ethyl acetate resulted in a highest yield of carotenoid, i.e., 18.2 and 15.73 mg/100 g, respectively whereas acetone and ethanol were found to have slightly lower levels of carotenoid content, i.e., 14.52 and 11.45 mg/100 g, respectively. Maceration is an inexpensive method to isolate bioactives and essential oils from roots and tubers. It is practiced at small scale level, and it consists of several steps. Firstly, roots and tubers are grinded into small particles for increasing the surface area. Secondly, solvent of interest is added to the ground sample in a closed vessel. Thirdly, the solution is pressed and filtered to collect the extract. To increase the extraction, yield occasional shaking and changing of concentrated solvent can be done. Hydro-distillation is the traditional way of extracting bioactives and essential oils from roots and tubers without any organic solvent. Hydro-distillation involves three processes that are hydrodiffusion, hydrolysis, and decomposition by heat. There are three types of hydro-distillation process such as steam distillation, water, and steam distillation, and water distillation [12]. In this process, sample is packed in a still section and steam/water-steam/water is injected into the sample to free the bioactives from roots and tubers tissue. Later, it is cooled to condense the vapor mixture of water and bioactive compound, which is then passed to a separator, where bioactives are automatically separate from the water [13]. Hydro-distillation involves heat that limits it usage to extract heat labile compounds.

4.4.2 NON-CONVENTIONAL TECHNIQUE FOR THE EXTRACTION OF BIOACTIVE COMPOUNDS FROM ROOTS AND TUBERS

Conventional techniques pertain certain limitations such as high purity and costly solvents, longer extraction time, low extraction selectivity, evaporation of large amount of solvent, decomposition of heat sensitive compounds. Therefore, to overcome these limitations, new techniques are developed for extraction. Some of the most promising non-conventional techniques are EAE, microwave-assisted extraction, pulsed electric field extraction, ultrasound-assisted extraction, PLE, and supercritical fluid extraction (SCFE). These are also referred to as "Green-technologies" for extracting bioactives from roots and tubers.

4.4.2.1 ENZYME ASSISTED EXTRACTION (EAE)

Several bioactives are present in the cell cytoplasm and some are retained in the polysaccharide-lignin network that are not available to be extracted by conventional extraction process. Thereby, certain enzymes like cellulases, pectinases, and α-amylases, that breaks the cell wall and hydrolyze the structural polysaccharide to release the bound bioactives followed by Soxhlet extraction [14]. Various factors are responsible for the effectiveness of this method, which include enzyme concentration and composition, moisture content and particle size of the sample, water-to-solid ratio, and time of hydrolysis [15]. Han [16] used α-amylase and pectinase for extraction of anthocyanin from purple sweet potato, they reported optimum extraction was 50°C in temperature, 1:15 in sample: solvent ratio, pH 5.5, 0.25% α-amylase and 0.10% pectinase dosage resulting in the yield of 2.71 mg/g.

4.4.2.2 MICROWAVE ASSISTED EXTRACTION

In microwave assisted extraction, microwaves are generated by electromagnetic field, are used with the frequency range of 300 MHz to 300 GHz. When microwaves pass through the sample, the polar molecules tend to align themselves with the electric field. But the electric field reverses 300 million times per second. The molecules attempting to oscillate at such frequencies generate intermolecular friction, which cause heat generation. This increase in temperature and pressure causes separation of bioactives from active sites of the sample matrix, thereafter solvent diffusion occurs that releases the bioactive compounds from the sample matrix [17]. Microwave assisted extraction results in rapid and better recovery of these compounds in comparison with conventional method. Xu et al. [18] optimized the extraction of anthocyanin from purple sweet potato and reported the optimized condition: power 700 W (temperature 70°), time 6.37 min, solvent: sample ratio 15:1, hydrochloric acid concentration 0.3% resulting the yield of anthocyanin of 4.66 mg/100 g. Cardoso-Ugarte et al. [19] employed microwaves with power 400 W with 100 duty cycle for 90–140 s for the extraction of betalains were reported to produce higher yield. Addition of ascorbic acid (0.040 mol/L) into the ethanol (50%) stabilized the pigment. They also conducted two stage extraction process at 400 W with 100% duty cycle for different time interval 130 s and 100 s, leads to higher extraction of the pigment. Elik and his co-worker [20] extracted the carotenoid from carrot juice processing waste using microwave technology and flaxseed oil as solvent. They reported optimum conditions for extraction were microwave power of 165 W, extraction time of 9.39 min, and oil: waste ratio of 8.06:1 g/g, resulting in a carotenoid recovery of 77.48%.

4.4.2.3 PULSED ELECTRIC FIELD EXTRACTION

Pulsed electric field extraction is another newly developed method of extraction in which cell membrane was destructed to increase the extraction rate. The pulsed electric field technique involves the application of an electric field pulse of 0.1–50 kV/cm for a short duration μ second to the material placed between the electrodes [21, 22]. The principal mechanism of pulsed electric field involves passing of electric current through the membrane of cells of roots and tubers, and based on the dipole nature of the membrane, electric potential separates the molecules based on their charges in the cell membrane. When the critical value exceeds the critical limit, i.e., 1 V of transmembrane potential, repulsion occurs between the charge carrying molecules in the membrane that forms pores in the membrane and causes leaching of bioactives [23]. Factors that improve the extraction efficiency include energy input, field strength, treatment temperature, pulse number, and properties of the roots and tubers [24]. Electric field of 500 and 1,000 V/cm for 10^{-4}–10^{-2} s is reported to damage cell membrane of tissue with slight increase in temperature [25]. This technique is suitable for heat-sensitive bioactive compounds also. Pulse electric field (PEF) can be used solely or be utilized as a pre-treatment with conventional method to increase the efficiency. Lopez et al. [26] systematically studied the effect of pulsed electric field on the recovery of betanin from red beetroot. The beetroot was sliced and subjected to the different electric field strength of 1–9 kV/cm and pulses 5–100. They reported 5 pulses of 2 μs at 7 kV/cm results in release of betanine fivefolds faster than conventional method. Kannan [27] given a prior treatment of pulsed electric field (1.5 kV/cm, 0.66 μF capacitance and 20 pulses), there after juice was extracted and analyzed for total phenolics, betalains, and antioxidant activity. They reported higher concentration of phenolics, betalains with higher antioxidant activity in the juice which was given a pre-treatment of pulsed electric field.

4.4.2.4 ULTRASOUND-ASSISTED EXTRACTION

In ultrasound-assisted extraction, sound waves are used with a frequency range of 20 kHz to 100 kHz, which is beyond human hearing. During this process, when ultrasound waves are passes through the liquid medium, it creates microbubbles that grow and collapse rapidly, this phenomenon is referred to as cavitation [28]. Due to this cavitation process, a large amount of energy is produced, due to which the bubble has a temperature of about 5,000 K and pressure of 1,000 atm. This technology is only suited for liquid and liquid-containing solid materials. The reoccurrence of the cavitation and decavitation process produces microjets and shock waves, which in turn intensify the heat and mass transfer that accelerates the migration of solvent from outside to inside the membrane [29, 30]. The mechanism involved in ultrasound-assisted extraction is first the diffusion across the cell wall and then the leaching of the content of the cell [31]. There are multiple factor that affects the effective extraction of bioactives such as moisture content, degree of milling, particle size, type of solvent, temperature, time, pressure, and frequency of sonication. This technique is used in assistance with conventional technology to increase extraction efficiency. Shivkumar et al. [28] extracted the betalains from beetroot with sample: ethanol (50%) ratio of 1:50 with 80 W ultrasonic power for 3 h. Maran & Priya [30] optimized the extraction condition with higher extraction efficiency using distilled water as a solvent. They reported maximum extraction at 1:19 sample: deionized water ratio with ultrasonic power of 89 W at 53°C for 35 min. This variation in the Shivkumar et al. [28] and Maran & Priya [30] study was due to the difference in the raw material in terms of its structure, composition, cell integrity, compactness, and other intrinsic factor. Yang et al. [32] optimized the extraction of pectin from potato pulp at 50 W, pH 2.0 and at a temperature of 93° for 50 min. Samarin et al. [33] extracted the potato peel using five different solvent, i.e., water, ethanol, hexane, methanol, and acetone and using two extraction techniques, i.e., solvent and ultrasound assisted. They reported that the yield of extract was higher in water, but the amount of phenolics was found to be higher in methanol. Moreover, when conventional and ultrasound-assisted technologies were used, ultrasound-assisted extraction improved the phenolic compounds percentage in the potato peel extract.

4.4.2.5 PRESSURIZED FLUID EXTRACTION (PFE)

Pressurized fluid extraction (PFE) is introduced in early 1996 by Richter and his co-worker, this is also known as accelerated fluid extraction (AFE), high pressure solvent extraction and enhanced solvent extraction (ESE) [34]. High pressure and temperature facilitate the enhanced extraction process with reduced time by promoting the solubility of analyte into the solvent [35]. This technology is an alternate to conventional as well SCFE as it increases the extractability efficiency with reduced requirement of time and solvent. Gizir et al. [36] successfully extracted the anthocyanin-rich extract from the black carrot using PFE, they reported no destruction during the extraction as this technique overcome the disadvantage of high temperature employed in the extraction. Saponjac et al. [37] optimized the conditions for extraction of carotenoid and polyphenols using response surface methodology and compared different solvents for extraction such as acetone, ethanol, and with and without addition of 20% water. They observed in the case of 20% water inclusion, 49% of acetone and 31% of ethanol is appropriate while in absence of water pure ethanol was the best option. They suggested addition of water clearly affects the extraction efficiency and optimal solvent to extract carotenoids from carrot by accelerated solvent extraction (ASE) is by pure ethanol.

4.4.2.6 SUPERCRITICAL FLUID EXTRACTION (SCFE)

Supercritical fluid extraction (SCFE) is an alternative extraction technology to conventional technology because it extracts the bioactive in a very less time with better quality and left no residue in the extract. SF possesses gas and liquid-like properties. It is the process of separating one component from another using SFs as the extracting solvent. Generally, carbon dioxide is used as the fluid for extraction of bioactive compounds. The liquid carbon dioxide is pumped to a heating zone, where it is heated to the supercritical stage, followed by passing it to the extraction vessel, where it rapidly diffuses into the solid sample and dissolves the bioactive compound. The dissolved compound is swept from the extraction cell into a separator at lower pressure and extracted compounds settles out. The carbon dioxide collected can then be cooled, re-compressed, and recycled. Various factor influences the extraction process such as temperature, time, pressure, particle size, moisture content of the sample, flow rate of carbon dioxide and solvent: sample ratio [35, 38]. Stefan et al. [39] extracted the bioactives from carrot pomace using SCFE and later checked the application as antioxidant to retard the rancidity in linseed oil. They reported extract was obtained at 30 MPa and 60°C with a carbon dioxide flow rate of 2 L/h for 120 min resulting in the yield of 1–2%. Later on, the extract was microencapsulated at a pressure of 10, 20, or 30 MPa at 65°C. They reported powder prepared at 10 MPa showed twofold more oxidative stability than those prepared at 30 MPa. Moreover, they have also observed addition of carrot extract improved the oxidative stability to the powder by 8 times longer than those of without carrot extract.

4.5 ANALYTICAL TECHNIQUES FOR QUALITATIVE AND QUANTITATIVE ANALYSIS

The bioactive compounds of the plants represent the secondary metabolites with certain therapeutic or toxicological effects in biological system. These bioactive compounds in root and tuber crops primarily constitute the phenolic substances, carotenoids, saponins, and sapogenins, bioactive proteins like dioscorin, glycoalkoloids, and others.

4.5.1 TECHNIQUES FOR QUALITATIVE ANALYSIS OF BIOACTIVE COMPOUNDS FROM ROOT AND TUBER CROPS

The bioactives such as polyphenols and B group vitamins in the juice of beetroot has been selectively determined by the hydrophilic interaction liquid chromatography (LC) coupled to mass spectrometry (MS) [40]. HPLC coupled with UV absorption spectra was used for characterization of the bioactive compounds, hopoxoside, and rooperol, from African potato. The isolated molecules were identified by comparing the UV spectra with previously published data. Further, the identification of these compounds was verified with the help of bidimensional heteronuclear single quantum coherence using ^{1}H and NMR using ^{13}C isotope. The UV spectra of these bioactives so obtained in methyl alcohol was coherent with the observations of Kruger et al. [41] found with the help of diode array detection. The UV absorption spectrum for hypoxoside exhibited three maxima at 212 nm, 258 nm and 298 nm whereas for rooperol the three maxima displayed were at 212 nm, 260 nm and 302 nm. The UV absorption spectrum obtained for both the bioactives of African potato didn't differ significantly from each other, however, the spectra of rooperol showed an ipsochromic transferral due to the occurrence of glucoside moity [42].

4.5.2 *TECHNIQUES FOR QUANTITATIVE ANALYSIS OF BIOACTIVE COMPOUNDS FROM ROOT AND TUBER CROPS*

The bioactive compounds present in root and tuber crops can be quantified by many techniques. However, notably, the high-performance LC and spectrophotometric methods are of particular choice and preferred quantification of bioactive compounds such as phenolic acids, flavonoids, anthocyanins, and others. The profiling of polyphenolic compounds from the extract of potato peel was carried out with the help of high-performance LC equipped with poroshell C18 column. The mobile phase was comprised of aqueous formic acid (0.1%) and formic acid (0.1%) in acetonitrile. The flow rate was maintained at 0.3 ml/min, and the detection was carried out at 280 nm and 325 nm wavelength. The compounds were recognized in relation to the retention time (RT) of the standards and quantified with the help of calibration curves obtained by using standards [43]. The phenolic acids and flavonoids from the colored potato cultivars were identified and quantified using high performance liquid chromatography (HPLC) coupled with C-18 column. The wavelengths used for detection of phenolic acids and flavonoids were 250 and 370 nm, respectively [44].

The bioactive compounds like β-carotene, phenolic acids, and anthocyanins from sweet potato flour subjected to high hydrostatic pressure were quantified by using high pressure LC (Agilent 1100), which was equipped with diode array detector (DAD). For quantification of β-carotene, the extract obtained from sample was reconstituted with methanol after drying with the help of nitrogen gas. The acetonitrile and methanol in the ratio of 1:9 (v/v) was used as mobile phase and the flow rate was maintained at 1.5 ml/min. The C18 reverse phase column with the dimensions of 25 cm × 4.6 mm was used for separation of β-carotene and detection was performed at 450 nm wavelength. For quantitative analysis of phenolic acids and anthocyanins, the mobile phases used were 1% acetic acid and pure acetonitrile and the flow rate was maintained at 0.4 ml/min. The same column, which was used for β-carotene, was also adopted for separation of phenolic acids and anthocyanins. The detection of phenolic acids and anthocyanins was performed at 320 nm and 520 nm wavelengths, respectively [45].

The hydrophilic interaction LC, which is comparatively recent chromatographic technique innovated in response to the traditional reverse phase chromatography [46], has been used for determination of bioactive compounds like polyphenols and B group vitamins from beet juice [40]. The hydrophilic interaction LC is reported to enhance the effectiveness of ionization in the ion source of mass spectrophotometer due to large amount of organic solvent in mobile phase. This results in obtaining considerably increased sensitivity [47]. The mechanism of hydrophilic interaction LC involves segregation between water layer formed near to solid surface and the mobile phase [48, 49]. Similarly other factors like adsorption through hydrogen bond formation and electrostatic as well as dipole-dipole interactions also play major role in mechanism of this chromatography [50].

The chromatographic methods have been adopted in isolation of various polyphenolic compounds like 4-ortho caffeoylquinic acid and 3,5-di-caffeolylquinic acid from methanolic and water-methanolic extract of sweet potato [51]. The norlignans compounds from the African potato were quantified using the high-pressure LC technique. The separation was performed in high performance liquid chromatogram having ultraviolet-visible DAD. The mobile phase consisted of acetonitrile and 0.05 M potassium dihydrogen phosphate. The flow rate was maintained at 0.8 ml/min and detection of hypoxoside and rooperol was carried out at 25 °C and at 260 nm wavelength [42]. The spectrophotometric based methods also find wide applications in the quantification of various bioactive compounds. The Folin-Ciocalteu method is popularly used in determination of phenolic content from various root and tuber crops like potato [52], beetroot [53], carrot [54]. The Folin-Ciocalteu method involves suspending the sample (preferably solvent extracted) in aqueous methanol followed by addition of tenfold diluted Folin-Ciocalteu reagent. The reaction mixture is added with sodium carbonate after 5 minutes. It is then incubated after making appropriate volume with distilled water in dark at room temperature for 2 h. The absorbance of the solution is then read at 765 nm wavelength using spectrophotometer against the reaction blank. The optical density is compared with the standard curve obtained with gallic acid at various known concentrations. The results are usually expressed as mg gallic acid equivalent (GAE) per g of dry material or per 100 g of dry material. Moreover, the spectrophotometric absorbance recording is used in quantification of flavonoids from carrot [54], beetroot [53] as well as anthocyanins in carrot [54].

4.6 POTENTIAL HEALTH PERSPECTIVES

The various bioactive compounds present in root and tuber crops exhibit many pharmacological activities and therefore offer several health benefits. The biological activities possessed by nutraceuticals of root and tuber crops include antioxidant activity, anticancer activity, anti-microbial activity, anti-diabetic activity, anti-obesity property and many other activities (Figure 4.2). Some specific health benefits of root and tuber crops with their bioactive compounds are summarized in Table 4.3 and are discussed in detail in subsections.

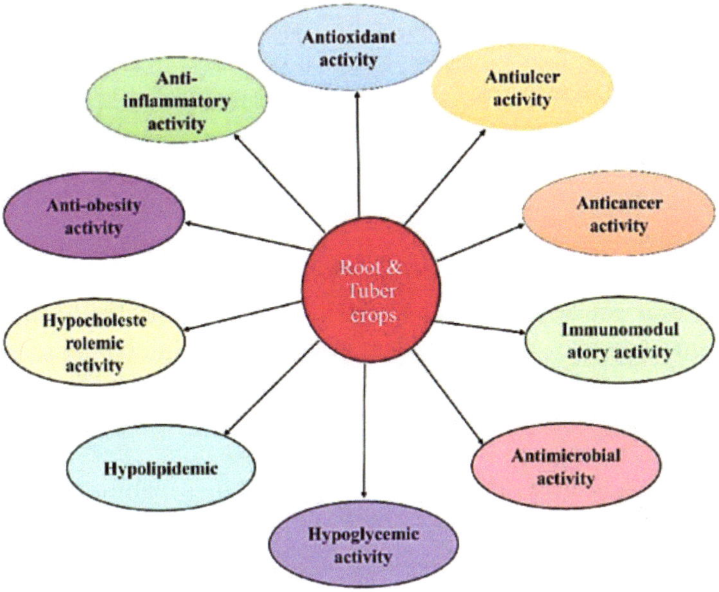

FIGURE 4.2 Health benefits of roots and tubers.

4.6.1 *ANTIOXIDANT ACTIVITIES*

The oxidative stress in biological system leads to the development of many chronic diseases like arthritis, certain types of cancers, cardiovascular diseases, diabetes, and disorders relating to neurodegeneration. The onset of aging is the chief ailment associated with oxidative stress. The internal antioxidant system through either certain enzymes (glutathione peroxidase, superoxide dismutase) or other molecules (tocopherol, glutathione, ascorbic acid, uric acid) help prevent the development of oxidative stress. However, the internal antioxidant system may get overpowered by the excessive exposure of materials inducing oxidative stress. Therefore, the body may require external dose of antioxidant through diet to combat the attack of oxidative substances.

Several research studies have established the potential of bioactive compounds from the root and tuber crops as an antioxidant [55–57]. The total phenolic content (TPC) of methanolic extract obtained from potato ranges from 16.6 to 32 mg GAE per 100 g dry sample. This methanolic extract exhibited strong antioxidant activity and has EC_{50} value of 94 mg/ml when evaluated using 2,2-diphenyl-1-picrylhydrazyl (DPPH) for radical scavenging activity [57]. The antioxidant activity of potato, thus, could be attributed to its phenolic and flavonoid content as Chu & Chang [58] reported that the flavonoid and flavones extracted from potato showed considerable scavenging activity against oxygen radical. Huang et al. [59] also reported that the antioxidant activity is associated with the phenolics and flavonoid content. The chlorogenic acid is the principal phenolic compound of potato and constitutes 90% of the total potato phenolics. Furthermore, the hydrophilic oxygen radical absorbance capacity (ORAC) and lipophilic ORAC of potato ranges from 100 to 1,400 μmol Trolox equivalent and 5 to 30 nanomol α-tocopherol equivalents per 100 fresh weights, respectively. Ji et al. [60] determined the TPC and antioxidant activity of 20 different potato clones. It was found that the peel of purple and red potato had higher TPC and antioxidant activity compared to unpigmented potato peels and potato peels show more DPPH radical scavenging activity than the potato flesh.

Many studies have reported the wound healing capability of sweet potato peel which is thought to be due to radical scavenging activity of the bioactive constituents and their inhibition of lipid oxidation [61, 62]. This healing effect of fibers of sweet potato against burns and wounds in rats has been observed through wound size reduction and its quality [63]. The study reported that the rats exposed with the fibers of sweet potato had reduced wound area against control. The sweet potato extract in petroleum ether also has a healing effect [64]. The methanolic extract of the sweet potato peel assessed for wound healing effect through incision and excision wound models on Wistar rat showed increase in hydroxyproline content which assist in collagen synthesis and thus hastening the wound healing [62]. Besides, this study also revealed a decrease in malondialdehyde content, which depicts the inhibitory effect of sweet potato peel for lipid oxidation.

The TPC of purple and lesser yam ranged between 69.9 and 421.8 mg GAE per 100 g dry matter. The *Daking* variety of purple yam had highest TPC and antioxidant activity as determined by DPPH radical scavenging activity, reducing power, and ferrous ion chelation ability [64]. Among different tuber crops (sweet potato, potato, coco yam, and Dioscorea yam, the water yam has highest (about 96%) DPPH radical scavenging activity

[65]. The study on antioxidant activity of ethanolic and aqueous extract of yam peel on mouse liver cells treated with tert-butyl hydroperoxide showed that these extracts possess significant antioxidant activity, with ethanolic extract showing protective effect against oxidative stress compared to aqueous yam extract. Moreover, the ethanolic extract resulted in increased catalase activity, while the aqueous extract declined it [66]. The organically grown cassava were found to contain more TPC and flavonoids compared to cassava grown with inorganic fertilizers. The antioxidant activity of organically grown cassava is also significantly higher than the inorganically grown cassava roots [67].

It was observed that the raw yam had more amount of TPC than the cooked yam. The heating of yam adversely affects TPC, antioxidant activity, and stability of bioactive compounds. The decline in DPPH radical scavenging activity is proportional to increase in temperature [55]. This suggests that the processing methods involving heat, like cooking, reduce the TPC of yam.

4.6.2 ANTICANCER ACTIVITIES

The cancer has remained one of the principal causes of global mortality and is linked to the unfit lifestyle coupled with unhealthy food habits. Cancer is a kind of multiphase illness condition and therefore hindering it at any early phase can facilitate mitigation of the disease condition. Hence, it necessitates finding interventions for preventing and reducing the risk of cancer through dietary guidelines. The bioactive compounds of root and tuber crops origin have established the anticancer activity in many kinds of carcinoma cell lines. The water and ethanolic extract of sweet potatoes showed considerable anticancer activities. The water extract was found to prevent multiplication of carcinoma cells more effectively than the ethanolic sweet potato extract. The antiproliferative activity of sweet potato extracts is linked to its various bioactive compounds which also contribute to antioxidant activity [59]. Specifically, 3-(6,6′-caffeylferulylsophoroside)-5-glucoside of cyanidin and peonidin which are anthocyanin pigments contribute hugely to the anticancer activities [68]. The phytochemicals, namely quercetin and chlorogenic acid, present in sweet potatoes, also contribute to anticancer activity. The sweet potatoes are also good sources of carotenoids, particularly, β-carotene, which intercept the free radicals, lower or hinder the mutagenesis in cells, and reduce the level of low-density lipoprotein cholesterol. The red pigmented potato cultivar exhibited strong inhibition of carcinogenesis compared to white potato cultivars in rats induced with breast cancer [69]. This may be due high content of anthocyanins and chlorogenic acid derivatives. However, another study Madiwale et al. [70] reported that compared with white fleshed potato, the purple potato cultivar has higher potential for controlling the multiplication and elated apoptosis [70]. Further, it was observed that the storage of potato has effect on anticancer as well as antioxidant and TPC. Though, the extracts from both fresh as well as stored potatoes have anticancer activity, but storage of potato results in decreasing the antiproliferative capacity. The duration of storage was found to have strong negative correlation with the apoptosis stimulation. The storage duration, however, had a positive correlation with antioxidant activity, suggesting that the antioxidant activity and TPC of potato increases with an increase in the duration of storage.

The anticancer activity of root and tuber crops is majorly attributed to the presence of phenolics. Besides phenolics, the saponins present in root and tuber crops are also responsible for their anticancer activities. The many saponin groups, including ammaranes, lupanes, cycloartanes, steroids, and oleananes, exhibit powerful antitumor activity against several types of cancers. For example, the cycloartane group of saponins are effective against the human colon cancer. Similarly, for oleananes group of saponins, several pathways like immunostimulation, chemoprevention, and antimetastasis are responsible for its antitumor effect [71]. Furthermore, some glycoalkaloids like alpha-solanine and alpha-chaconine found in many root and tuber crops also possess significant anticancer activity. The glycoalkaloids are reported to demonstrate antiproliferation against human colon and liver cancer cells [72].

4.6.3 ANTIMICROBIAL ACTIVITIES

The phenolic compounds present in yam tuber possess potential antimicrobial efficacy. The methanolic extract of *Dioscorea* yam are reported to have significant antimicrobial activity as determined by agar diffusion method and pour plate method. The methanolic extract of *D. hirtiflora* cultivar exhibit considerable antimicrobial activity against many microorganisms including *E. coli, Salmonella typhi, Staphylococcus aureus, Aspergillus Niger*, and *Bacillus subtilis* [73]. The antimicrobial potential of sweet potatoes varies depending on several factors including the genotypes, solvent used for extraction, bacterial strain under consideration, and the methods adopted for evaluating the antimicrobial activity [74, 75]. The antimicrobial activity of aqueous extract of sweet potato was evaluated against selected gram-positive as well as gram-negative bacteria. The results indicated that the aqueous extract of sweet potato demonstrated largest inhibition zone and is highly effective over *S. aureus* and *S. epidermidis* [76]. Furthermore, the ethanol and methanol extracts of orange and white cultivars of sweet potatoes were evaluated for its antimicrobial activity. The study revealed that the ethanol and methanol extracts of the orange cultivar of sweet potato showed high efficacy against gram-positive bacteria, *P. multocida*; however, neither ethanolic nor methanolic extracts could inhibit the growth of *E. coli* [77]. The biologically active constituents like caffeoylquinic acid and 3,5-di-caffeolylquinic acid present in sweet potato possess anti-bacterial activity [78].

4.6.4 ANTI-OBESITY ACTIVITIES

Recently, the ever-growing incidences of overweight and obese-ness among people of all age group is becoming a chief concern of community health. In 2016, about 1,900 million adults were identified as overweight, reflecting more than 34% population as obese [79]. Obesity is a kind of medical disorder that stems from the build-up of abnormal or surplus body fat due to disproportionate calorie intake, which may lead to many adverse health consequences. These days, obesity has translated into one of the leading preventable causes of death at the global level, associated with a higher levels of endangering of producing many metabolic disorders. The plant sources phytochemicals like phenolic acids, flavonoids, terpenoids, and phytosterols have potential to reduce the risk of obesity [80].

The purple sweet potato cultivars can be used to manage and prevent the obesity [81]. The bioactive component responsible for prevention of obesity was reported to be anthocyanin found in purple sweet potato. The mice fed with purple sweet potato for 4 weeks at dose concentration of 200 mg/kg body weight per day showed reduction in weight gain and hepatic accumulation of triacylglycerol as well as improvement in serum lipid parameters. This study also found that the anthocyanin rich fraction of sweet potato may recuperate the fatty liver disease caused by high fat diet and control the hepatic lipid metabolism. Jeong et al. [82] also mentioned the anti-obesity activities of yam extract in obese mice. The mice fed with high-fat diet containing yam extract in butanol at concentration of 100 mg/kg body weight for 8 weeks experienced considerable reduction in body weight as well as adipose tissue weight. Moreover, these mice also lost triglycerides and low-density lipoprotein cholesterol from the blood. The tubers, *D. steriscus* [83] and *D. nipponica* [84] also had anti-obesity potential and are effective against gain in body weight and adipose tissues. The anti-obesity potential of *D. steriscus* is much higher than the commercially available anti-obesity medicine, herbex. The anti-obesity activity of *D. steriscus* is attributed to presence of bioactive compounds in the tuber that are responsible for inhibition of α-amylase and lipase enzymes and thereby contribute in production of compounds for anti-obesity effect.

4.6.5 HYPOGLYCEMIC ACTIVITIES

The increased level of glucose sugar in the blood beyond normal causes the development of diabetes mellitus, which is one of the chronic disorder and can produce life-threatening complications that eventually may result in the death of a human being. The hypoglycemic and anti-diabetic activity of plant extracts originate from their capacity to reinstate the working of pancreatic tissues which possibly results in increased insulin output, malabsorption of glucose in intestine, and reconditioned enabling of insulin induced metabolites [85]. The sweet potato is considered as potent antidiabetic and its antidiabetic activity is rated more than the diabense, a standard drug used for treating diabetes [86]. The sweet potato peel extract is known to reduce the plasma glucose level in patients suffering from diabetes. The oral administration of white peeled sweet potato extract in rats reduced the hyper-insulinemia by about 50% after 8 weeks. Besides, the extract also found to reduce the triacylglycerol, free fatty acid, and lactate levels in blood. The pancreas of rats fed with sweet potato extract showed significant regranulation of islet B-cells after 8 weeks, as determined by histological examination. This suggests that the white peeled sweet potato extract facilitates the recovery of abnormal glucose and lipid metabolism in diabetic patients, which is also confirmed by a reduction in the fasting blood glucose level and low-density lipoprotein cholesterol level after ingestion of sweet potato extract for 6 weeks. However, these changes were dose-dependent and were not observed at low dose levels [87]. Ludvik et al. [88] also observed these beneficial effects of sweet potato extract concerning glucose and cholesterol levels in diabetic patients.

The ethanolic extract obtained from *Dioscorea alata* demonstrated antidiabetic activity in diabetic rats treated with alloxan. This antidiabetic activity is related to the presence of various bioactive compounds like dioscorin, alatanins, cyanidin, tocopherols, catechin in ethanolic extract of *D. alata*. The yam extract administration in diabetic rats considerably lowered the creatinine level. This may be due to enhancement in renal function as a result of lowering of plasm glucose level and consequent glycosylation of the renal membranes [89]. Moreover, the *D. alata* tuber exhibited decrease in plasma glucose in streptozotocin induced rats and decreased body weight in alloxan induced high body weight rats [89, 90]. The yam cultivar *D. dumetorum*, which is popularly recognized as bitter yam, has long been known in folk medicine to play a key role in management of diabetes owing to its hypoglycemic potential [91]. The water extract of *D. dumetorum*, which is rich in alkaloids like dioscoretin, is also effective in management of other medical conditions such as hyperketonemia, hypercholesterolemia, and hyperlipidemia.

4.6.6 HYPOCHOLESTEROLEMIC ACTIVITIES

The level of cholesterol is critical in the development of cardiovascular diseases, which is also one of the principal causes of global deaths. The cholesterol homeostasis, in effect, is governed by the dietary components. The continuous intake of dietary components having cholesterol lowering ability result in useful effect impact on reducing the risk and effective management of the cardiovascular diseases. Yams are known to possess hypolipidemic activity due to the presence of diosgenin, which is a kind of steroidal saponin. The diosgenin content in wild and cultivated yam varies from 0.12–0.48% and 40 ppm, respectively. The study involving feeding of rats for 6 weeks with high cholesterol diet having diosgenin demonstrated reduction in cholesterol level, inhibition of lipase activity, protection to liver and protective effect for oxidative damage resulting from polyunsaturated fatty acids [92]. The inclusion of yam in mice diet reduced plasma as well as hepatic cholesterol level and increased fecal excretion [93]. The yam-originated steroidal saponins are also used for drug processing in pharmaceutical industries. The yams are recognized to contain various saponins like dioscin (2.7%), gracillin, and prosapogenins. The cholesterol-lowering effect of yam saponins is attributed to its inhibitory activity for absorption of cholesterol [94]. The hypocholesterolemic activity of yam saponin, diosgenin, is also mentioned by Cayen & Dvornik [95]. The rats with high levels of cholesterol fed with yam extract exhibited a decrease in cholesterol absorption and an increase in bile cholesterol secretion, which is thought to be due to the presence of diosgenin in the yam extract. This antihypercholesterolemic effect of diosgenin is also confirmed by many studies who observed that the diosgenin result in decreased intestinal absorption of cholesterol and increased fecal bile acid secretion [96, 97]. The diosgenin-induced biliary secretion of cholesterol is related to the metabolism of hepatic cholesterol [98]. However, Chen et al. [93] indicated that the cholesterol lowering effect may not be attributed to diosgenin. The hypocholesterolemic effect could be due to dietary fiber and viscous mucilage of yam. Further, it was suggested that the hypolipidemic effect could be related either to dietary fiber, mucilage, plant sterols or synergistic effect of these bioactive components.

The *Dioscorea* species are reported to be effective for the control of abnormal cholesterol levels. The lipid lowering effect, particularly reduction in cholesterol level, of *Dioscorea* species such as *D. polygonoids* is linked to presence of sapogenin and diosgenin as has been found through studies on animals like rat and mice [99]. *D. alata* fed to rats at a concentration of 40% in high cholesterol diet causes substantial decrease in triglyceride level of plasma as well as cholesterol [100]. The hypolipidemic activity of sweet potato is thought to be due to the presence of carotenoids, which reduces the low-density lipoprotein cholesterol. Root and tuber crops, besides above discussed health benefits, also possess many other biological activities. Sweet potato can be used for treatment and prevention of cardiovascular diseases. The mechanism responsible for vasorelaxation activity of sweet potato is quite similar to the mechanism of action of acetylcholine, a popular pharmaceutical drug [101]. The sweet potato tuber is also used for treating and healing the burns and wounds. The anti-ulcer and gastro protective property of sweet potato tuber (methanolic extract) is well reported in cold stress induced mice. These activities are marked by lowering of lipid peroxidation and elevated levels of many enzymes like catalase, glutathione reductase, superoxide dismutase, etc., at various concentration of dose [102].

4.7 INFLUENCE OF PROCESSING ON NUTRACEUTICALS OF ROOT AND TUBER CROPS

The root and tuber crops are rarely consumed in their raw form, in which they are mostly unpalatable. These are subjected to one or other processing technique like boiling, roasting, baking, frying, drying or fermentation. The cultural and regional practices, as well as the type of root and tuber crop, govern the kind of processing these root and tuber crops undergo. Generally, most of the root and

TABLE 4.3 Potential Health Benefits of Various Root and Tuber Crops and Their Bioactive Compounds/Mechanism Responsible

Root/Tuber Crop	Health Benefit	Bioactive Compound/Mechanism Responsible	References
Potato	Anti-inflammatory	Phenolic acids, carotenoids, anthocyanins. Reduction of cytokine and C-reactive proteins.	Kaspar et al. [103]
	Antioxidant	TPC, flavonoids, flavones	Hesam et al. [57]; Chu et al. [58]
	Neuroprotective activity	TPC, chlorogenic acid, pelargonidin, petunidin glycoside.	Ji et al. [60]
Sweet potato	Anticancer	Phenolics, saponins, glycoalkaloids, cyanidin, peonidin. Inhibition of reverse mutation induced by mutagenic pyrolysates.	Yoshimoto et al. [68]; Auyeung et al. [71]; Lee et al. [72]
	Anti-obesity	Sporamin, anthocyanins. Prevention of hepatic lipid build-up by initiation of AMP stimulated protein kinase signaling pathways.	Xiong et al. [104]; Hwang et al. [81]
	Antioxidant	TPC, flavonoids	Dilworth et al. [56]
	Antiulcer	Increased activity of catalase, glutathione reductase, superoxide dismutase, and glutathione peroxidase.	Panda & Sonkamble [102]
	Hypoglycemic activity	Reduction in plasma glucose level, improvement in abnormal glucose and lipid metabolism.	Ludvik et al. [87]
	Wound healing	Antioxidant compounds increase hydroxyproline content, which enhances collagen synthesis.	Panda & Sonkamble [62]
Yam	Anticancer	Phenolics, saponins, mucilage polysaccharides. Inhibition and damage of DNA and protection of human lymphoblastoid cells.	Wang et al. [105]
	Anti-diabetic activity	Phenolic compounds, dioscorin, catechin, tocopherols, hydro Q9 chromene, cyanidin-3-glucoside, feruloyl glycerol, alatanins. Decrease in plasma glucose level coupled with glycosylation of renal basement membranes.	Maithili et al. [89]

TABLE 4.3 *(Continued)*

Root/Tuber Crop	Health Benefit	Bioactive Compound/Mechanism Responsible	References
	Anti-inflammatory	Saponins, β-sitosterol, stigmasterol, glycosides, diosgenin.	Olayemi & Ajaiyeoba [106]
	Antimicrobial	Inhibition of *Bacillus subtilis, Aspergillus niger*, *Staphylococcus aureus*, *Salmonella typhi*, *E coli*, and *Penicillium chrysogenum*.	Sonibare & Abegunde [73]
	Antioxidant	Dioscorin TPC.	Chen & Lin [55]
		Enhanced activity of endogenous antioxidant enzymes.	
	Hypolipidemic activity	Diosgenin.	Son et al. [92]
		Inhibition of pancreatic lipase, protection against damaging effects of polyunsaturated fatty acids.	
	Hypocholesterolemic activity	Saponins (dioscin, gracillin, prosapogenins).	Ma et al. [94]
		Inhibition of cholesterol absorption.	
	Immunomodulatory	Dioscorin.	Liu et al. [107]
		Stimulation of cytokine production and enhancement of phagocytosis.	
Cassava	Antioxidant	TPC, flavonoid	Omar et al. [67]

tuber crops are boiled to soften their flesh and increase the digestibility. Several processing techniques like peeling, drying, and others, depending on their intensity, influence the physicochemical characteristics and nutritional value of sweet potato [90]. Processing of sweet potato in flour preparation reduces its β-carotene content as a function of temperature. The peeled and unpeeled sweet potatoes not subjected to sulfitation also showed reduction in the TPC in their respective flour. Amongst the unpeeled and peeled sweet potato flour, the flour produced from unpeeled sweet potatoes had higher amount of total phenolics than the peeled sweet potato flour. This could be due to the contribution of phenolics from the peel of sweet potatoes [108]. Furthermore, the sweet potatoes subjected to sulfitation retained higher amount of phenolics as well as ascorbic acid in their flour. This could be attributed to the inhibition of polyphenol oxidase enzyme by the sulfitation treatment [109].

The effect of extrusion and drying (freeze and hot air) on physiological characteristics and bioactive composition of two sweet potato varieties, namely Tainong 57, which is yellow in color, and Tainong 66, which is orange in color, is studied by Shih et al. [110]. This study reported that the antioxidant contents as well as antioxidant activity of these sweet potato varied significantly due to these processing techniques. The freeze-drying process retained the highest amount of phenolics, anthocyanins, and β-carotene and showed higher free radical scavenging activity. Amongst the two varieties studied, the freeze-dried orange sweet potatoes had more amount of nutraceuticals and bioactivity as compared to the yellow sweet potatoes. The extrusion of sweet potato resulted in a significant increase in TPC and antioxidant activity; however, it caused a reduction in β-carotene and anthocyanin content. The positive effect of extrusion on phenolics and thereby antioxidant activity could be related to the liberation of bound phenolic compounds and their derivatives from the complex cell walls of plant matrices.

The processing methods like boiling, microwave cooking, and steaming reduce the β-carotene content of sweet potatoes with the progression in time [111]. Boiling of sweet potatoes for 50 minutes period reduces the β-carotene content by about 33%. Further, it is observed that the steaming had more pronounced effect on decreasing the β-carotene content of sweet potato compared to boiling. However, the highest losses in the content of β-carotene were found to be incurred by the microwave cooking of sweet potatoes. The effect of various processing methods on the bioactive compounds of yam is reported in Figure 4.3. All the processing methods studied (blanching, freezing, and vacuum frying) resulted in reducing the bioactive compounds like phenolics, anthocyanins, sinapic acid, and ferulic acid. It has been noticed that the blanching and vacuum frying destroy about 60 and 40% of anthocyanin in yam. The highest destruction of bioactive compounds was observed by the vacuum frying of yams whereas the blanching and freezing were found at par in decreasing the contents of various bioactive compounds in the yam [112].

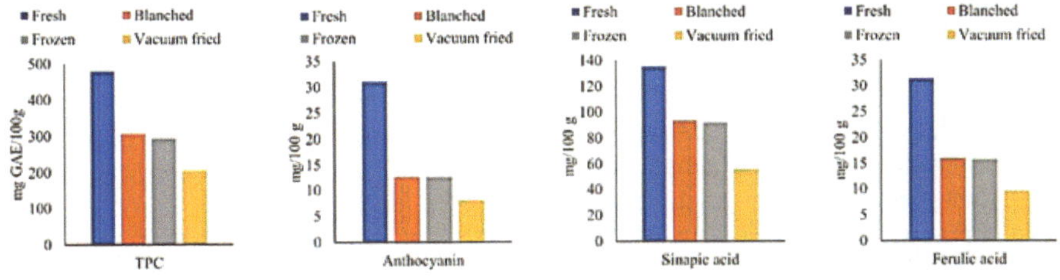

FIGURE 4.3 Effect of different processing on the bioactives of yam.

Source: Reprinted/Adapted with permission from Ref. [112]. Copyright © 2019 Elsevier Ltd.

4.8 CONCLUSION

Roots and tubers are the important species after cereals and known to be rich source of carbohydrate and many phytochemical compounds that are beneficial to the health. Different techniques for isolating these compounds are already discussed in this chapter. Many other technologies need to be identified to increase the extractability of these bioactive without affecting their chemical structure and activity. The bioactives from roots and tubers are having potential application. So, there is a need for more exploration of different varieties, different extraction techniques, and toxicology studies.

KEYWORDS

- **bioactive compounds**
- **extraction technology**
- **hypocholesterolemic activities**
- **microwave-assisted extraction**
- **non-conventional technique**
- **qualitative analysis**
- **quantitative analysis**
- **tuber crops**
- **ultrasound-assisted extraction**

REFERENCES

1. FAOSTAT, (2013). http://faostat3.fao.org (accessed on 26 September 2023).
2. Ecocrop, (2010). *Ecocrop Database*. FAO.
3. Chittaranjan, K., (2007). *Pulses, Sugar, and Tuber Crops: Genome Mapping and Molecular Breeding in Plants*. Springer-Verlag, Berlin Heidelberg, New York.
4. Hammershøj, M., Kidmose, U., & Steenfeldt, S., (2010). Deposition of carotenoids in egg yolk by short-term supplement of colored carrot (*Daucus carota*) varieties as forage material for egg-laying hens. *J. Sci. Food Agric.*, *90*, 1163–1171.
5. Frias, J., Penas, E., Ullate, M., & Vidal-Valverde, C., (2010). Influence of drying by convective air dryer or power ultrasound on the vitamin C and beta-carotene content of carrots. *J. Agric. Food Chem.*, *58*, 10539–10544.

6. Majors, R. E., (1999). An overview of sample preparation methods for solids. *LC-GC Europe, 17*, 8–13.
7. Azmir, J., Zaidul, I. S. M., Rahman, M. M., Sharif, K. M., Mohamed, A., Sahena, F., Jahurul, M. H. A., et al., (2013). Techniques for extraction of bioactive compounds from plant materials: A review. *J. Food Eng., 117*, 426–436.
8. Hernandez, Y., Lobo, M. G., & Gonzalez, M., (2009). Factors affecting sample extraction in the liquid chromatographic determination of organic acids in papaya and pineapple. *Food Chem., 114*, 734–741.
9. Halwani, A. F., Sindi, H. A., & Jambi, H., (2018). Characterization of physical properties of red beet pigments. *J. Biochem. Technol., 9*, 10–14.
10. Liu, X., Mu, T., Sun, H., Zhang, M., & Chen, J., (2013). Optimization of aqueous two-phase extraction of anthocyanins from purple sweet potatoes by response surface methodology. *Food Chem., 141*, 3034–3041.
11. Vanja, N. S., Gordana, S. C., Čanadanović-Brunet, J. S., Tumbas, Š. V. T., Vulić, J. J., & Stajčić, S., (2017). Extraction and encapsulation of bioactive compounds from carrots. *APTEFF, 48*, 1–323.
12. Vankar, P. S., (2004). Essential oils and fragrances from natural sources. *Resonance, 9*, 30–41.
13. Silva, L. V., Nelson, D. L., Drummond, M. F. B., Dufosse, L., & Gloria, M. B., (2005). Comparison of hydrodistillation methods for the deodorization of turmeric. *Food Res. Int. 38*, 1087–1096.
14. Singh, R. K., Sarker, B. C., Kumbhar, B. K., Agrawal, Y. C., & Kulshreshtha, M., (1999). Response surface analysis of enzyme-assisted oil extraction factors for sesame, groundnut, and sunflower seeds. *J. Food Sci. Technol., 36*, 511–514.
15. Niranjan, K., & Hanmoungjai, P., (2004). Enzyme-aided aqueous extraction. In: Dunford, N. T., & Dunford, H., (eds.), *Nutritionally Enhanced Edible Oil Processing*. AOCS Publishing.
16. Han, Y., (2007). *Extraction, Composition Analysis, Stability and Antioxidant Activity of Anthocyanins from Purple Sweet Potato*. Ph.D. Thesis, Nanjing Agricultural University, Nanjing, China.
17. Alupului, A., (2012). Microwave extraction of active principles from medicinal plants. *U.P.B. Science Bulletin, 74*, 129–142.
18. Xu, Z., Gao, Y., Shi, S., & Liu, X., (2005). Study on the extraction of pigment from purple sweet potato powder by microwave-assisted technique. *Food Sci., 26*, 234–239.
19. Cardoso-Ugarte, G. A., Sosa-Morales, M. E., Ballard, T., Liceaga, A., & San Martín-Gonz′alez, M. F., (2014). Microwave-assisted extraction of betalains from red beet (*Beta vulgaris*). *LWT – Food Sci. Technol., 59*, 276–282.
20. Elik, A., Yanık, D. K., & Göğüş, F., (2020). Microwave-assisted extraction of carotenoids from carrot juice processing waste using flaxseed oil as a solvent. *LWT – Food Sci. Tech., 123*, 109100–109107.
21. Nirmal, N. P., Mereddy, R., & Maqsood, S., (2021). Recent developments in emerging technologies for beetroot pigment extraction and its food applications. *Food Chem., 356*, 129611–129622.
22. Nowacka, M., Tappi, S., Wiktor, A., Rybak, K., Miszczykowska, A., Czyzewski, J., & Tylewicz, U., (2019). The impact of pulsed electric field on the extraction of bioactive compounds from beetroot. *Foods, 8*, 244.

23. Bryant, G., & Wolfe, J., (1987). Electromechanical stress produced in the plasma membranes of suspended cells by applied electrical fields. *J. Membr. Biol., 96*, 129–139.
24. Heinz, V., Toepfl, S., & Knorr, D., (2003). Impact of temperature on lethality and energy efficiency of apple juice pasteurization by pulsed electric fields treatment. *Innov. Food Sci. Emerg. Technol., 4*, 167–175.
25. Fincan, M., & Dejmek, P., (2002). In situ visualization of the effect of a pulsed electric field on plant tissue." *J. Food Eng., 55*, 223–230.
26. Lopez, N., Puertolas, E., Condon, S., Raso, J., & Alvarez, I., (2009). Enhancement of the extraction of betanine from red beetroot by pulsed electric fields. *J. Food Eng., 90*, 60–66.
27. Kannan, V., (2009). *Extraction of Bioactive Compounds from Whole Red Cabbage and Beetroot using Pulsed Electric Fields and Evaluation of Their Functionality* (p. 11). Dissertations, Theses, & Student Research in Food Science and Technology.
28. Sivakumar, V., Anna, J. L., Vijayeeswarri, J., & Swaminathan, G., (2009). Ultrasound assisted enhancement in natural dye extraction from beetroot for industrial applications and natural dyeing of leather. *Ultrason. Sonochem., 16*, 782–789.
29. Sivakumar, V., Vijaeeswarri, J., & Anna, J., (2011). Effective natural dye extraction from different plant materials using ultrasound. *Ind. Crops Prod., 33*, 116–122.
30. Maran, J. P., & Priya, B., (2016). Multivariate statistical analysis and optimization of ultrasound-assisted extraction of natural pigments from waste red beet stalks. *J. Food Sci. Technol., 53*, 792–799.
31. Mason, T. J., Paniwnyk, L., & Lorimer, J., (1996). The uses of ultrasound in food technology. *Ultrason. Sonochem., 3*, 253–260.
32. Yang, J. S., Mu, T. H., & Ma, M., (2019). Optimization of ultrasound- microwave-assisted acid extraction of pectin from potato pulp by response surface methodology and its characterization. *Food. Chem., 289*, 351–359.
33. Samarin, A. M., Poorazarang, H., Hematyar, N., & Elhamirad, A., (2012). Phenolics in potato peels, extraction, and utilization as natural antioxidants. *World Appl. Sci. J., 18*, 191–195.
34. Richter, B. E., Jones, B. A., Ezzell, J. L., Porter, N. L., Avdalovic, N., & Pohl, C., (1996). Accelerated solvent extraction: A technology for sample preparation. *Anal. Chem., 68*, 1033–1039.
35. Ibanez, E., Herrero, M., Mendiola, J. A., & Castro-Puyana, M., (2012). Extraction and characterization of bioactive compounds with health benefits from marine resources: Macro and microalgae, cyanobacteria, and invertebrates. In: Hayes, M., (ed.), *Marine Bioactive Compounds: Sources, Characterization and Applications* (pp. 55–98). Springer.
36. Gizir, A. M., Turker, N., & Artuvan, E., (2008). Pressurized; acidified water extraction of black carrot *Daucus carota ssp. sativus var. atrorubens Alef.* anthocyanins. *Eur. Food Res. Technol., 226*, 363–370.
37. Šaponjac, V. T., Kovaˇcevi´c, S., Šeregelj, V., Šovljanski, O., Mandi´c, A., Cetkovic, G., Vulic, J., et al., (2021). Improvement of carrot accelerated solvent extraction efficacy using experimental design and chemometric techniques. *Processes, 9*, 1652–1668.
38. Temelli, F., & Guclu-Ustundag, O., (2005). *Supercritical Technologies for Further Processing of Edible Oils: Bailey's Industrial Oil and Fat Products*. John Wiley & Sons, Inc.

39. Stefan, K., Ferrentino, G., Zendehbad, S. H., Morozova, K., & Scampicchio, M., (2021). Bioactive compounds from carrot pomace as natural antioxidants to enhance the oxidative stability of linseed oil encapsulated by particles from gas saturated solutions technique. *Chem. Eng. Trans.*, *87*, 145–150.
40. Sentkowska, A., & Pyrzyńska, K., (2018). Zwitterionic hydrophilic interaction liquid chromatography coupled to mass spectrometry for analysis of beetroot juice and antioxidant interactions between its bioactive compounds. *LWT – Food Sci. Technol. 93*, 641–648.
41. Kruger, P. B., Albrecht, C. D. V., Liebenberg, R. W., & Van, J. P., (1994). Studies on hypoxoside and rooperol analogues from *Hypoxis rooperi* and *Hypoxis latifolia* and their biotransformation in man by using high-performance liquid chromatography with in-line sorption enrichment and diode-array detection. *J. Chromatogr. B Biomed. Appl.*, *662,* 71–78.
42. Laporta, O., Pérez-Fons, L., Mallavia, R., Caturla, N., & Micol, V., (2007). Isolation, characterization, and antioxidant capacity assessment of the bioactive compounds derived from *Hypoxis rooperi* corm extract (African potato). *Food Chem., 101*, 1425–1437.
43. Martinez-Fernandez, J. S., Seker, A., Davaritouchaee, M., Gu, X., & Chen, S., (2021). Recovering valuable bioactive compounds from potato peels with sequential hydrothermal extraction. *Waste Biomass Valor.*, *12*, 1465–1481.
44. Vaitkevičienė, N., Kulaitienė, J., Jarienė, E., Levickienė, D., Danillčenko, H., Średnicka-Tober, D., et al., (2020). Characterization of bioactive compounds in colored potato (*Solanum tuberosum* L.) cultivars grown with conventional, organic, and biodynamic methods. *Sustainability*, *12*, 2701.
45. Cui, R., & Zhu, F., (2019). Physicochemical properties and bioactive compounds of different varieties of sweet potato flour treated with high hydrostatic pressure. *Food Chem.*, *299*, 125129.
46. McCalley, D. V., (2017). Understanding and manipulating the separation in hydrophilic interaction liquid chromatography. *J. Chromatogr. A.*, *1523*, 49–71.
47. Yang, Y., Boysen, R. I., & Hearn, M., (2013). Selectivity differences of water-soluble vitamins separated on hydrophilic interaction stationary phases. *J. Sep. Sci.*, *36*, 1897–1903.
48. Guo, Y., (2015). Recent progress in the fundamental understanding of hydrophilic interaction chromatography (HILIC). *Analyst.*, *140*, 6452–6466.
49. Salas, D., Borrull, F., Fontanals, N., & Marcé, R., (2017). Hydrophilic interaction liquid chromatography coupled to mass spectrometry-based detection to determine emerging organic contaminants in environmental samples. *Trend. Anal. Chem.*, *94*, 141–149.
50. Zuo, R., Zhou, S., Zuo, Y., & Deng, Y., (2015). Determination of creatinine, uric and ascorbic acid in bovine milk and orange juice by hydrophilic interaction HPLC. *Food Chem.*, *182*, 242–245.
51. Jung, J. K., Lee, S. U., Kozukue, N., Levin, C. E., & Friedman, M., (2011). Distribution of phenolic compounds and antioxidative activities in parts of sweet potato (*Ipomoea batata* L.) plants and in home processed roots. *J. Food Compost. Anal.*, *24*, 29–37.
52. Lemos, M. A., Aliyu, M. M., & Hungerford, G., (2015). Influence of cooking on the levels of bioactive compounds in purple majesty potato observed via chemical and spectroscopic means. *Food Chem.*, *173,* 462–467.

53. Ramos, J. A., Furlaneto, K. A., Lundgren, G. A., Mariano-Nasser, F. A. D. C., Mendonca, V. Z., Nasser, M. D., & Vieites, R., (2017). Stability of bioactive compounds in minimally processed beet according to the cooking methods. *Food Sci. Technol.*, *38*, 643–646.
54. Koley, T. K., Singh, S., Khemariya, P., Sarkar, A., Kaur, C., Chaurasia, S. N. S., & Naik, P., (2014). Evaluation of bioactive properties of Indian carrot (*Daucus carota* L.): A chemometric approach. *Food Res. Int.*, *60*, 76–85.
55. Chen, Y. T., & Lin, K., (2007). Effects of heating temperature on the total phenolic compound, antioxidative ability and the stability of dioscorin of various yam cultivars. *Food Chem.*, *101*, 955–963.
56. Dilworth, L., Brown, K., Wright, R., Oliver, M., Hall, S., & Asemota, H., (2012). Antioxidants, minerals and bioactive compounds in tropical staples. *Afr. J. Food Sci. Technol.*, *3*, 90–98.
57. Hesam, F., Balali, G. R., & Tehrani, R., (2012). Evaluation of antioxidant activity of three common potato (*Solanum tuberosum*) cultivars in Iran. *Avicenna J. Phytomedicine*, *2*, 79–85.
58. Chu, Y. H., Chang, C. L., & Hsu, H., (2000). Flavonoid content of several vegetables and their antioxidant activity. *J. Sci. Food Agric.*, *80*, 561–566.
59. Huang, D. J., Chun-Der, L. I. N., Hsien-Jung, C. H. E. N., & Yaw-Huei, L. I., (2004). Antioxidant and antiproliferative activities of sweet potato (*Ipomoea batatas* [L.] LamTainong 57') constituents. *Bot. Bull. Acad. Sinica*, *45*, 465–470.
60. Ji, X., Rivers, L., Zielinski, Z., Xu, M., MacDougall, E., Stephen, J., Zhang, S. H., Wang, Y., et al., (2012). Quantitative analysis of phenolic components and glycoalkaloids from 20 potato clones and in vitro evaluation of antioxidant, cholesterol uptake, and neuroprotective activities. *Food Chem.*, *133*, 1177–1187.
61. Chimkode, R., Patil, M. B., & Jalalpure, S., (2009). Wound healing activity of tuberous root extracts of *Ipomoea batatas*. *Adv. Pharmacol. Toxicol.*, *10*, 69–72.
62. Panda, V., & Sonkamble, M., (2012a). Anti-ulcer activity of *Ipomoea batatas* tubers (sweet potato). *Funct. Food Health Dis.*, *2*, 48–61.
63. Suzuki, T., Tada, H., Sato, E., & Sagae, Y., (1996). Application of sweet potato fiber to skin wound in rat. *Biol. Pharm. Bull.*, *19*, 977–983.
64. Cornago, D. F., Rumbaoa, R. G. O., & Geronimo, I., (2011). Philippine yam (*Dioscorea* spp.) tubers phenolic content and antioxidant capacity. *Philipp J. Sci.*, *140*, 145–152.
65. Dilworth, L., Brown, K., Wright, R., Oliver, M., Hall, S., & Asemota, H., (2012). Antioxidants, minerals and bioactive compounds in tropical staples. *Afr. J. Food Sci.*, *3*, 90–98.
66. Hsu, C. K., Yeh, J. Y., & Wei, J. H., (2011). Protective effects of the crude extracts from yam (*Dioscorea alata*) peel on tert-butylhydroperoxide-induced oxidative stress in mouse liver cells. *Food Chem.*, *126*, 429–434.
67. Omar, N. F., Hassan, S. A., Yusoff, U. K., Abdullah, N. A. P., Wahab, P. E. M., & Sinniah, U., (2012). Phenolics, flavonoids, antioxidant activity, and cyanogenic glycosides of organic and mineral-base fertilized cassava tubers. *Mol.*, *17*, 2378–2387.
68. Yoshimoto, M., Okuno, S., Yoshinaga, M., Yamakawa, O., Yamaguchi, M., & Yamada, J., (1999). Antimutagenicity of sweetpotato (*Ipomoea batatas*) roots. *Biosci. Biotechnol. Biochem.*, *63*, 537–541.
69. Thompson, M. D., Thompson, H. J., McGinley, J. N., Neil, E. S., Rush, D. K., Holm, D. G., & Stushnoff, C., (2009). Functional food characteristics of potato cultivars (*Solanum*

tuberosum L.): Phytochemical composition and inhibition of 1-methyl-1-nitrosourea induced breast cancer in rats. *J Food Compost Anal., 22*, 571–576.

70. Madiwale, G. P., Reddivari, L., Holm, D. G., & Vanamala, J., (2011). Storage elevates phenolic content and antioxidant activity but suppresses antiproliferative and pro-apoptotic properties of colored-flesh potatoes against human colon cancer cell lines. *J. Agric. Food Chem., 59*, 8155–8166.
71. Auyeung, K. K. W., Law, P. C., & Ko, J. K., (2009). Astragalus saponins induce apoptosis via an ERK-independent NF-κB signaling pathway in the human hepatocellular HepG2 cell line. *Int. J. Mol. Med., 23*, 189–196.
72. Lee, K. R., Kozukue, N., Han, J. S., Park, J. H., Chang, E. Y., Baek, E. J., Chang, J. S., & Friedman, M., (2004). Glycoalkaloids and metabolites inhibit the growth of human colon (HT29) and liver (HepG2) cancer cells. *J. Agric. Food Chem., 52*, 2832–2839.
73. Sonibare, M. A., & Abegunde, R., (2012). In vitro antimicrobial and antioxidant analysis of *Dioscorea dumetorum* (Kunth) Pax and *Dioscorea hirtiflora* (Linn.) and their bioactive metabolites from Nigeria. *J. Appl. Biosci., 51*, 3583–3590.
74. Mbaeyi-Nwaoha, I. E., & Emejulu, V., (2013). Evaluation of phytochemical composition and antimicrobial activity of sweet potato *(Ipomoea batatas)* leaf. *Pak. J. Nutr., 12*, 575–586.
75. Pochapski, M. T., Fosquiera, E. C., Esmerino, L. A., Dos, S. E. B., Farago, P. V., Santos, F. A., & Groppo, F., (2011). Phytochemical screening, antioxidant, and antimicrobial activities of the crude leaves' extract from *Ipomoea batatas* (L.) Lam. *Pharmacogn. Mag., 7*, 165–170.
76. Lee, K., Park, J. H., & Kim, Y., (2020). Antioxidant, cytotoxic and antimicrobial activities of *Dendropanax morbifera* and sweet potato extracts for production of health-oriented food materials. *African Journal of Biotechnology, 19*, 301–306.
77. Naz, S., Naqvi, S. A. R., Khan, Z. A., Mansha, A., Ahmad, M., Zahoor, A. F., & Hussain, Z., (2017). Antioxidant, antimicrobial and antiproliferative activities of peel and pulp extracts of red and white varieties of *Ipomoea batatas* (L) Lam. *Trop. J. Pharm. Res., 16*, 2221–2229.
78. Meira, M., Silva, E. P. D., David, J. M., & David, J., (2012). Review of the genus ipomoea: Traditional uses, chemistry, and biological activities. *Rev. bras. Farmacogn, 22*, 682–713.
79. Singh, M., Thrimawithana, T., Shukla, R., & Adhikari, B., (2020). Managing obesity through natural polyphenols: A review. *Future Foods*, 100002.
80. Mir, S. A., Shah, M. A., Ganai, S. A., Ahmad, T., & Gani, M., (2019). Understanding the role of active components from plant sources in obesity management. *Journal of the Saudi Society of Agricultural Sciences, 18*, 168–176.
81. Hwang, Y. P., Choi, J. H., Han, E. H., Kim, H. G., Wee, J. H., Jung, K. O., Jung, K. H., et al., (2011). Purple sweet potato anthocyanins attenuate hepatic lipid accumulation through activating adenosine monophosphate–activated protein kinase in human HepG2 cells and obese mice. *Nutr. Res., 31*(12), 896–906.
82. Jeong, E. J., Jegal, J., Ahn, J., Kim, J., & Yang, M., (2016). Anti-obesity effect of *Dioscorea oppositifolia* extract in high-fat diet-induced obese mice and its chemical characterization. *Biol. Pharm. Bull., 39*, 409–414.
83. Dzomba, P., & Musekiwa, C., (2014). Anti-obesity and antioxidant activity of dietary flavonoids from *Dioscorea serious* tubers. *J. Coast. Life Med., 2*, 465–470.

84. Kwon, C. S., Sohn, H. Y., Kim, S. H., Kim, J. H., Son, K. H., Lee, J. S., Lim, J. K., & Kim, J., (2003). Anti-obesity effect of *Dioscorea nipponica* Makino with lipase-inhibitory activity in rodents. *Biosci. Biotechnol. Biochem.*, *67*, 1451–1456.
85. Malviya, N., Jain, S., & Malviya, S. A. P. N., (2010). Antidiabetic potential of medicinal plants. *Acta Pol. Pharm.*, *67*, 113–118.
86. Oke, J. M., Oladosu, B., & Okunola, M., (1999). Sweet potato (*Ipomea batatas*) tuber-potential oral anti-diabetic agent. *Afr. J. Biomed. Res.*, *2*, 13–17.
87. Ludvik, B. H., Mahdjoobian, K., Waldhaeusl, W., Hofer, A., Prager, R., Kautzky-Willer, A., & Pacini, G., (2002). The effect of *Ipomoea batatas* (Caiapo) on glucose metabolism and serum cholesterol in patients with type 2 diabetes: A randomized study. *Diabetes Care*, *25*, 239–240.
88. Ludvik, B., Neuffer, B., & Pacini, G., (2009). Efficacy of *Ipomoea batatas* (Caiapo) on diabetes control in type 2 diabetic subjects treated with diet. *Diabetes Care*, *27*, 436–440.
89. Maithili, V., Dhanabal, S. P., Mahendran, S., & Vadivelan, R., (2011). Antidiabetic activity of ethanolic extract of tubers of *Dioscorea alata* in alloxan-induced diabetic rats. *Indian J. Pharmacol.*, *43*, 455–459.
90. Ahmed, Z., Chishti, M. Z., Johri, R. K., Bhagat, A., Gupta, K. K., & Ram, G., (2009). Antihyperglycemic and antidyslipidemic activity of aqueous extract of *Dioscorea bulbifera* tubers. *Diabetol. Croat.*, *38*, 63–72.
91. Iwu, M. M., Okunji, C. O., Ohiaeri, G. O., Akah, P., Corley, D., & Tempesta, M., (1990). Hypoglycaemic activity of dioscoretine from tubers of *Dioscorea dumetorum* in normal and alloxan diabetic rabbits. *Planta. Medica.*, *56*, 264–267.
92. Son, I. S., Kim, J. H., Sohn, H. Y., Son, K. H., Kim, J. S., & Kwon, C., (2007). Antioxidative and hypolipidemic effects of diosgenin, a steroidal saponin of yam (*Dioscorea* spp.), on high-cholesterol fed rats. *Biosci. Biotechnol. Biochem.*, *71*, 3063–3071.
93. Chen, H. L., Wang, C. H., Chang, C. T., & Wang, T., (2003). Effects of Taiwanese yam (*Dioscorea japonica Thunb* var. *Pseudo japonica* Yamamoto) on upper gut function and lipid metabolism in Balb/c mice. *Nutr. 19*, 646–651.
94. Ma, H. Y., Zhao, Z. T., Wang, L. J., Wang, Y., Zhou, Q. L., & Wang, B., (2002). Comparative study on anti-hypercholesterolemia activity of diosgenin and total saponin of *Dioscorea panthaica*. *China J. Chinese Materia. Medica.*, *27*, 528–531.
95. Cayen, M. N., & Dvornik, D., (1979). Effect of diosgenin on lipid metabolism in rats. *J. Lipid Res.*, *20*, 162–174.
96. Thewles, A., Parslow, R. A., & Coleman, R., (1993). Effect of diosgenin on biliary cholesterol transport in the rat. *Biochem. J., 291*, 793–798.
97. Uchida, K., Takse, H., Nomura, Y., Takeda, K., Takeuchi, N., & Ishikawa, Y., (1984). Effects of diosgenin and B-sisterol on bile acids. *J. Lipid Res.*, *25*, 236–245.
98. Temel, R. E., Brown, J. M., Ma, Y., Tang, W., Rudel, L. L., Ioannou, Y. A., Davies, J. P., & Yu, L., (2009). Diosgenin stimulation of fecal cholesterol excretion in mice is not NPC1L1 dependent. *J. Lipid Res.*, *50*, 915–923.
99. McKoy, M. L., Thomas, P. G., Asemota, H., Omoruyi, F., & Simon, O., (2014). Effects of Jamaican bitter yam (*Dioscorea polygonoides*) and diosgenin on blood and fecal cholesterol in rats. *J. Med. Food.*, *17*, 1183–1188.
100. Yeh, Y. H., Lee, Y. T., & Hwang, D., (2007). Yam (*Dioscorea alata*) inhibits hypertriglyceridemia and liver enlargement in rats with hypercholesterol diet. *J. Chin. Med. Assoc.*, *18,* 65–74.

101. Runnie, I., Salleh, M. N., Mohamed, S., Head, R. J., & Abeywardena, M., (2004). Vasorelaxation induced by common edible tropical plant extracts in isolated rat aorta and mesenteric vascular bed. *J. Ethnopharm.*, *92*, 311–316.
102. Panda, V., & Sonkamble, M., (2012b). Phytochemical constituents and pharmacological activities of *Ipomoea batatas* L. (Lam): A review. *Int. J. Res. Phytochem. Pharmcol.*, *2*, 25–34.
103. Kaspar, K. L., Park, J. S., Brown, C. R., Mathison, B. D., Navarre, D. A., & Chew, B., (2011). Pigmented potato consumption alters oxidative stress and inflammatory damage in men. *J. Nutr.*, *141*, 108–111.
104. Xiong, Z. D., Li, P. G., & Mu, T., (2009). The differentiation-and proliferation-inhibitory effects of sporamin from sweet potato in 3T3-L1 preadipocytes. *Agric. Sci. China.*, *8*, 671–677.
105. Wang, T. S., Lii, C. K., Huang, Y. C., Chang, J. Y., & Yang, F., (2011). Anticlastogenic effect of aqueous extract from water yam (*Dioscorea alata* L.). *J. Med. Plants Res.*, *5*, 6192–6202.
106. Olayemi, J. O., & Ajaiyeoba, E., (2007). Anti-inflammatory studies of yam (*Dioscorea esculenta*) extract on Wistar rats. *Afr. J. Biotechnol.*, *6*, 1913–1915.
107. Liu, Y. W., Shang, H. F., Wang, C. K., Hsu, F. L., & Hou, W., (2007). Immunomodulatory activity of dioscorin, the storage protein of yam (*Dioscorea alata* cv. Tainong No. 1) tuber. *Food Chem. Toxicol.*, *45*, 2312–2318.
108. Mondy, N. I., & Gosselin, B., (1988). Effect of peeling on total phenols, total glycoalkaloids, discoloration and flavor of cooked potatoes. *J. Food Sci.*, *53*, 756–759.
109. Sapers, G. M., Cooke, P. H., Heidel, A. E., Martin, S. T., & Miller, R., (1997). Structural changes related to texture of pre-peeled potatoes. *J. Food Sci.*, *62*, 797–803.
110. Shih, M. C., Kuo, C. C., & Chiang, W., (2009). Effects of drying and extrusion on color, chemical composition, antioxidant activities and mitogenic response of spleen lymphocytes of sweet potatoes. *Food Chem.*, *117*, 114–121.
111. Wu, X., Sun, C., Yang, L., Zeng, G., Liu, Z., & Li, Y., (2008). β-carotene content in sweet potato varieties from China and the effect of preparation on β-carotene retention in the Yanshu No. 5. *Innov. Food Sci. Emerg. Technol.*, *9*, 581–586.
112. Fang, Z., Wu, D., Yü, D., Ye, X., Liu, D., & Chen, J., (2011). Phenolic compounds in Chinese purple yam and changes during vacuum frying. *Food Chem.*, *128*, 943–948.

CHAPTER 5

Technological and Analytical Approaches for Extraction of Bioactive Compounds and Nutraceuticals from Fruit and Vegetable Processing Byproducts

BRAHMEET KAUR,[1] DIPAK DAS,[1] GAURAV PANESAR,[1] PARMJIT S. PANESAR,[1] CHARANJIT S. RIAR,[1] and ZAINUL A. ZAKARIA[2]

[1]Department of Food Engineering and Technology, Sant Longowal Institute of Engineering and Technology, Longowal, Punjab, India

[2]School of Chemical and Energy Engineering, Faculty of Engineering, Universiti Teknologi Malaysia, UTM Johor Bahru, Johor, Malaysia

ABSTRACT

Huge quantity of by-products is generated by the fruit and vegetable processing industries mostly in the form of peels, seeds, rind, and pomace. These by-products carry high nutritional and economical value owing to their rich composition of bioactive constituents. Utilization of various green extraction methods for the recovery of high-value compounds and their quantification using new and innovative technologies can help achieve complete valorization of fruit and vegetable by-products. The bioactive compounds present in fruit and vegetable by-products contribute to the prevention of several diseases and are therefore, beneficial to human health. Its incorporation in food products as functional food ingredient can be a novel strategy to meet the growing concerns of health and nutrition. This chapter highlights the bioactive potential of fruit and vegetable processing by-products, methods of their extraction and quantification along with its health benefits.

Bioactive Compounds and Nutraceuticals from Plant Sources: Extraction Technology, Analytical Techniques, and Potential Health Prospects. Charanjit Singh Riar, Parmjit S. Panesar (Eds.)

5.1 INTRODUCTION

The food processing industry has witnessed a phenomenal growth since the last decade and is emerging as one of the fastest growing sectors around the world. However, processing of food is accompanied by a significant quantity of waste in the form of food by-products. Apart from causing deleterious effects on environment, food waste has severe socioeconomic impacts especially in low-income and developing countries [1]. The reports indicate approximately 25–30% of the waste is obtained from fruits and vegetables (F&Vs) industry. The major by-products obtained after its processing include peels, seeds, rind, stones, core, pomace, pods, etc. [2]. In the developing countries, harvesting and processing of fruits and vegetables are the two most important stages which contributes to waste generation with a ca. 50% share of the initial food production [3]. The by-products of fruits and vegetables generated during its processing alone constitutes 25–30% of the horticulture product loss [4]. According to an estimate, around 55 MMT waste is released during every processing step, which includes 5.5 MMT during processing of F&Vs, 6 MMT in canning as well as freezing, 5–9 MMT after wine processing and others.

The large quantities of these waste residues have a great deal of biochemical oxygen demand (BOD) and chemical oxygen demand (COD) contents. Moreover, inadequate biological stability and presence of pathogens leads to increase in microbial activity. The most common adopted methods like incineration, composting, anaerobic fermentation, landfilling or using food residues as animal feed or fertilizer have been consistently utilized in the past as crucial strategies for waste minimization and valorization. Attempts have been made widely in the past to develop appropriate methods for utilization of fruit and vegetable wastes therapeutically. The increasing demand of consumers for natural and plant-based foods has fueled the interest in fruit and vegetable by-products as source of functional and bioactive components. These naturally occurring bioactive compounds have great significance in terms of health enhancing and disease reduction effects. Their efficient interaction with DNA, proteins, and other biological molecules can help in designing natural therapeutic agents [5]. A number of fruits such as apple, oranges, mango, etc., and vegetables like potatoes, onions, carrots, asparagus, etc., after processing, have been utilized to produce value added products. The by-products of such F&Vs, often discarded as waste are rich in phytochemicals and secondary metabolites entrapped within tissue [6, 7]. Many studies indicate presence of bioactive compounds concentration even more in these by-products than their edible counterparts [8, 9]. Some of the major health benefits associated with these compounds include anti-inflammatory, anti-oxidant, cardio protective, anti-allergic, and anti-microbial effects. Moreover, the natural extracts derived from plant sources have potential in development of useful drugs and prevention of diseases like atherosclerosis, cancer, obesity, etc. [10].

The molecular pool of bioactive compounds possesses substantial number of properties for applications as functional foods, nutraceuticals, and food additives [11]. F&Vs represent the category of simplest form of functional foods as they comprise of several bioactive constituents. Carotenoids and polyphenols present in fruit and vegetable by-products have antioxidant activity and prevent certain types of cancer [12]. The phytochemicals like tocopherols, carotenes, terpenes, sterols, etc., contain significant antioxidant activities. These valuable components derived from the F&V by-products can be isolated and used as natural antioxidants for the formulation of functional foods and shelf-life extension of perishable foods. Bioactive peptides from the residual F&Vs waste are the new generation of biologically active regulators which can prevent microbial and oxidative degradation of foods and can work as disease preventing agents [13]. These can be recovered and purified on large-scale using suitable techniques for increased applications in food and pharmaceutical sectors.

Selective extraction and modification of the targeted components in F&Vs by-products can be achieved through combined approaches of biochemical, chemical, and physical means. Ideal extraction methods can help determine type and quantity of bioactive compounds obtained from F&Vs by-products [14]. Characterization of bioactive components can be achieved after identification from stem, leaves, flowers, and fruits. Various compounds, such as phytosterols, carotenoids, curcumin, omega-3 fatty acids, oil-soluble vitamins have been extracted using various extraction techniques [10]. Solvent extraction had been applied to recover target substances and separate them from the vegetal matrix. Mechanical pressing had been employed for vegetable oil extraction and volatile compounds have been recovered using hydro or steam distillation. Solvent extraction of biologically useful compounds from F&Vs by-products can be carried out using conventional techniques such as maceration, decoction, infusion, or Soxhlet extraction. To overcome the drawbacks encountered using conventional methods, reduce the energy consumption and negative environmental impacts, several alternative green techniques have been developed. An increasing focus is laid on the use of nonthermal processes to avoid the risk of matrix overheating [15].

5.2 CHEMICAL COMPOSITION AND BIOACTIVE POTENTIAL OF FRUIT AND VEGETABLE PROCESSING BY-PRODUCTS

Fruit and vegetable wastes generated from the processing industries mainly comprise of the peels, seeds, and pomace. These are an excellent resource for biologically useful compounds like proteins, polysaccharides, dietary fibers (DFs), etc. Moreover, the rich composition of fruit and vegetable by-products in antioxidants, polyphenols, and antimicrobial compounds has been reported in previous studies [16]. Major bioactive compounds present in fruits and vegetables processing by-products described in Figure 5.1. The natural pigments present in these by-products possess coloring as well as pharmacological properties with applications in nutraceuticals and functional foods development. In the last few decades, increased emphasis has been laid on the identification, extraction, quantification, and evaluation of physiochemical action of bioactive compounds from fruits and vegetables. A significant growth in health and medical related research has been observed pertaining to the reduced risk of cardiometabolic diseases and cancer [17]. The F&Vs by-products mainly constituted of the peels, seed, and pomace fractions has a significant potential as a rich source of high value bioactive compounds like polysaccharides, DFs, proteins, flavor compounds, etc. [4, 18]. With remarkable health enhancing effects, bioactive compounds are known to lower the risk of diseases like cancer, Parkinson, cataracts, etc. The abundance of phytochemicals like phenolic compounds and DFs in F&Vs waste exhibiting antioxidant activity are responsible for providing several health benefits. The residue obtained after extraction of juices, oils, starch, and sugars are rich in compounds of nutraceutical importance. These include dietary fiber, fatty acids, carotenoids, isoprenoids, lipids, proteins, phenolic compounds, phytoestrogens, and saponins [19].

5.2.1 POLYPHENOLS

Polyphenols represent one of the largest classes of phytochemicals with significant biological functions in human body [20]. The beneficial factor of phenolic compounds is associated with their antioxidant activity. Phenolic compounds naturally are present as conjugates with mono and polysaccharides and can also be found as esters and methyl esters [21]. The structure of phenolic compounds is composed of an aromatic ring with one or more hydroxyl substituents, ranging from simple phenolic molecules to highly polymerized compounds. The peels, seeds, and rind of fruits and vegetables are rich in phenolic constituents. In several fruits, the TPC has been found to be higher in the discarded portions than the edible fleshy part. The peels of peaches, apples, and pears were observed to contain twice the number of polyphenols than the peeled fruits [22]. The total phenolics in seeds of several fruits like mango, avocado, jackfruit, and longan were comparatively higher than the edible flesh, making these a valuable source of antioxidant phenolics. In the case of tomato, peels and seeds have been found to be richer source of polyphenols than the fleshy pulp.

5.2.2 DIETARY FIBERS (DFS)

Dietary fiber constitutes the sum of the nondigestible components in food derived from plants. The insoluble dietary fiber includes cellulose, hemicellulose, and lignin whereas soluble dietary fiber consists of pectin, fructan, and arabinoxylan (AX) components [23]. Dietary fiber extracts from vegetable by-products can be rich in polyphenols, carotenoids, and flavonoids (e.g., hydrolysable tannins, proanthocyanidins) constituting a single matrix often referred to as antioxidant dietary fiber. The by-products of asparagus constitute a good dietary fiber source as the powder of *A. officinalis* showed a high amount (62–77%) of total dietary fiber content [24]. The yogurt enriched using fiber obtained from nonedible parts of asparagus showed increased viscosity with neither enhanced nor worsened organoleptic characteristics. Pomegranate peels fiber hold promising functional properties due to its excellent insoluble to soluble dietary fiber ratio, oil and water holding capacity as well as solubility [25]. Moreover, its high glucose retention capacity and antioxidant activity makes this fiber of high value. Significant concentration of inulin has been reported in artichoke waste. An inulin content of 6.2 and 7.5% (dry matter) has been found in bracts and stems of artichoke, respectively [26]. Crude polysaccharides extracted from mesocarp of pomegranate fruit exhibited *in vitro* prebiotic properties with Lactobacillus and Bifidobacterium strains [27]. Addition of raspberry pomace partially with the flour in cookies is a possible way to enhance dietary fiber content in the diet. Brown skin and top-bottom onions had potential as functional food ingredient due to its rich composition in dietary fiber as well as total phenolics and flavonoids of high antioxidant activity. The impact of many digestive disorders, diabetes mellitus, obesity, and cardiovascular problems can be minimized to a large extent with the consumption of foods rich in dietary fiber [28].

5.2.3 EDIBLE OILS

An important constituent present in fruit or vegetable by-product is the lipid or oil. The residues of fruits are composed of lipid content up to 10 times higher than the edible portion. Banana peels have high content of linoleic and α-linolenic acid contributing to prevention of cancer, atherosclerosis, diabetes, etc. [29]. The oil obtained from mango seed kernel is edible and its triglycerides profile resemble to cocoa butter. Dry tomato seeds have 17% oil with fatty acid composition similar to soyabean and sunflower oils. This oil is characterized by high thermal stability, efficient physicochemical profile after heating and antioxidant activity higher than the commercial edible oils like soyabean or olive oil [30]. The passion fruit seed oil possesses about 87.6% of unsaturated fatty acids and show high antioxidant activity [31]. Citrus peels possess significant quantity of essential oils which could be safely utilized as food additive due to its antimicrobial properties and flavors [32]. Several *in vivo* and *in vitro* studies on grape seed oil revealed its cardio-protective, antimicrobial, anticancer, and anti-inflammatory activity [33]. Essential oils derived from carrot by-products has been utilized in flavor industry and applications of food preservation.

5.2.4 PECTIN

A wide distribution of pectin can be found in fruits and vegetables (10–30%). Pectin has gained focus among researchers for its potential role in prevention and reduction of carcinogenesis [34]. Citrus peels contain 20–30% pectin while apple pomace and mango peels possess 10–15 and 10–20% pectin content, respectively [35, 36]. Mango peel possess appreciable quantity of pectin with potential to compete with apple pomace and is less likely to degrade in presence of enzyme. It could possibly be employed in food, cosmetic, and pharmaceutical industries [37]. Citrus pectin impart thickening, emulsifying, and stabilizing effect in food products like jams, jellies, and marmalades [38]. The pH and temperature-modified citrus pectin may be used to target tumors as it shows anti-adhesive and apoptosis-inducing properties. Enzymatically extracted pectin from beetroot by-products can be effectively used as a food additive. On incorporation of beetroot powder in pectin-based edible films, the optical and antioxidant properties were enhanced [39]. The wastewater generated from olive mills is a dark brown colored fluid and was found to be highly rich in pectin. Pectin isolation from tomato by-products has been successfully done using solid state fermentation [40].

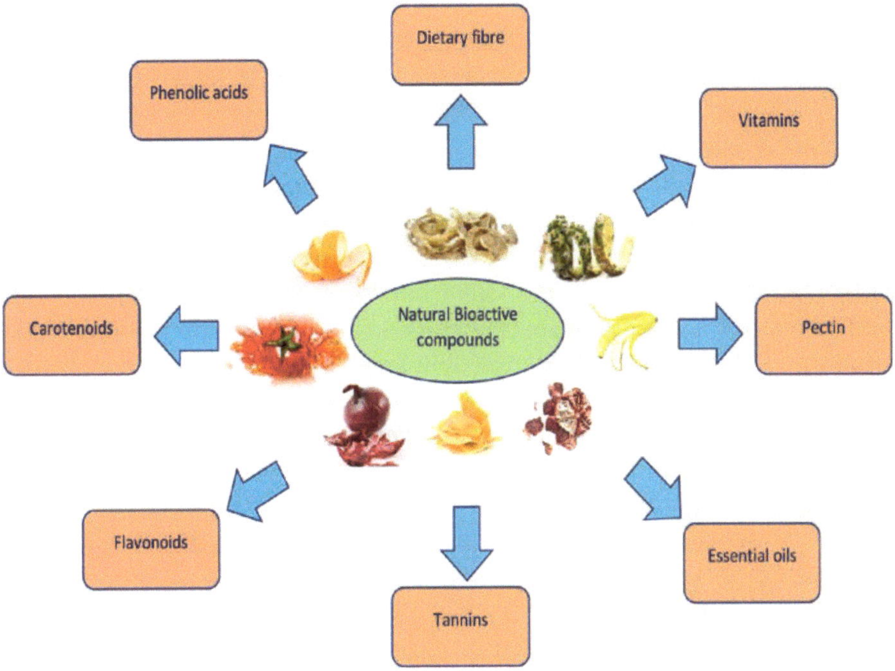

FIGURE 5.1 Major bioactive compounds present in fruits and vegetables processing by-products.

5.3 MAJOR FRUIT PROCESSING BY-PRODUCTS

5.3.1 CITRUS BY-PRODUCTS

The major by-products obtained from citrus waste include the pith, albedo, seed, and flavedo. The annual citrus production in tropical and subtropical regions of the planet exceeds 12 million tons [41]. The global production of oranges, tangerines/mandarins and lemons/limes during the year 2020–2021 were estimated to be 48.6 million tons, 33.3 million tons and 8.4 million tons, respectively. A significant portion of this fruit (40–60%) is left unconsumed and discarded as waste. A huge quantity of citrus by-products (including seeds, peels, pomace, and wastewater) is generated after industrial processing accounting for 55–60% of the total fruit weight. These organic by-products are biologically unstable and poses a threat to environment [42]. Therefore, the rich organic composition of citrus by-products should be known in-depth for developing effective solutions for their disposal/valorization [43]. Citrus peels and seeds are rich source of phenolic compounds including both phenolic acids and flavonoids with the content being higher in peels than in the seeds (Table 5.1). The main flavonoid compounds found abundantly in citrus species are naringin, hesperidin, narirutin, and eriocitrin [44]. The peels of citrus fruits comprise of four main phenolic acids, namely ferulic, caffeic, sinapinic, and p-coumaric acids [45]. The cuticles and peels of citrus fruit contains substantial amount of essential oil, the major component of which is D-limonene. Essential oils from citrus possess antimicrobial, anti-inflammatory, and antioxidant properties with potential applications in food industry. Citrus waste has been widely recognized as a source of polyphenols, animal feed, dietary fiber, and biofuels. Albedo portion of citrus waste specifically is a rich source of fiber, i.e., pectin [46]. Pectin has proven useful for improvement in cooking yield, enhancement of texture in food products, and reduction in formulation cost [47]. Citrus fruit is considered an important source of dietary nutrients due to its abundance in carotenoids. More than 115 carotenoids can be found in citrus which imparts yellow to orange color to the fruit [48]. The dietary fiber and antioxidants extracted from citrus wastes are used as health promoting ingredients in food processing for the development of functional foods.

TABLE 5.1 Chemical Composition and Bioactive Components Present in Citrus Peels

Component	Content	References
Moisture*	75.3 ± 10.2	[49]
Crude protein*	10.2 ± 3.70	
Crude fiber*	57.0 ± 10.0	
Crude fat*	2.22 ± 6.10	
Total ash*	3.33 ± 0.50	
Total phenolics #	147.6 ± 6.3	[50]
Total flavonoids $	49.2 ± 1.33	[51]
Hesperidin ‖	29.5 ± 0.32	
Naringin ‖	0.54 ± 0.02	
Neohesperidin ‖	0.11 ± 0.00	
Total carotenoids ^	2.04 ± 0.0	
Antioxidant activity **	656.18 ± 10.91	[52]

Note: *g/100 g DM; # mg GAE/g DM; $ mg QE/g DM; ‖ mg/g DM; ^ μg/g DW; and ** μmol Fe^{2+}/g DM.

5.3.2 BANANA BY-PRODUCTS

Banana is one of the widely cultivated tropical fruit widely consumed for its fruit pulp. Approximately 115 million tons of banana production were recorded during the year 2017–2019 [53]. The processing of banana fruit by food processing industries generates a large amount of waste in the form of stalk, stem, and peels. Among these, peels constitute major portion (30–40% w/w) of the fruit [54]. Banana peels are generated in large volumes by food-processing industries; however, its disposal is of significant concern. Researchers have suggested banana peels as a valuable source of bioactive compounds [54]. The chemical and elemental analysis of banana peels revealed the presence of lipids, proteins, crude fibers, carbohydrates, water-soluble sugars and minerals (Table 5.2). It is considered a rich source of phenolic compounds as it possesses TPC ranging from 4.95 to 47 mg GAE/g dry matter [55]. It possesses high reducing activity and radical scavenging activity in comparison to other fruit peels [56]. More than 40 phenolic compounds have been identified in banana peels broadly classified as flavonols, flavan-3-ols, catecholamines, and hydroxycinnamic acids. Among these, flavan-3-ols constitute the largest group of phenolics in banana peels [57] with rutin and its conjugates being the most dominant components [58]. Extractable pectin in banana peels is found in larger quantity as compared to other fruit peels. Extraction of aromatic compounds in banana peels is widely recognized in which isoamyl acetate the main aromatic compound is applied to foods as banana flavor [59]. Banana peels has been used to make flour as a source of carbohydrates (mainly dietary fiber) and thickening agent in food industries.

TABLE 5.2 Chemical and Bioactive Components Present in Banana Peels

Component	Content	References
Moisture*	20.87 ± 0.19	[60]
Carbohydrate*	67.29 ± 0.19	
Crude protein*	2.48 ± 0.00	
Crude fat*	1.24 ± 0.00	
Ash content*	3.95 ± 0.00	
Total flavonoids#	21.04 ± 0.28	[61]
Total phenolics$	17.89 ± 0.16	
Total tannins@	24.21 ± 0.17	
Carotenoids^	13.61–20.63	[62]

Note: * g/100 g dry basis; # mg QAE/g dw; $ mg GAE/g dw; @ mg TA/g dw; and ^ μg/g dw.

5.3.3 PINEAPPLE BY-PRODUCTS

Pineapple fruit of the *Bromeliaceae* family is marketed around the world. In 2019, production of pineapple around the world amounted to approximately 28.18 million metric tons [63]. The increase in production of pineapple processed items leads to massive waste generation which comprise of residual pulp, peels, stem, and leaves. Waste production from pineapple processing accounts 25–35% of the total weight of fruit [64] depending on the variety and type of processing of pineapple. Pineapple waste is an enriched raw material, composed of phenolic compound, pectin, sugars, protein, vitamins, minerals, and insoluble fibers [65]. Chemical and bioactive components present in pineapple peels reported by previous studies were described in Table 5.3. The waste generated from pineapple processing industries have been utilized as a substrate for bromelain, ethanol, organic acids, etc., as it is potential source of vitamins, sugars, and growth factors. Polyphenols from pineapple residues have been less studied, although some of the phenolic compounds like salicylic acid, tannic acid, trans-cinnamic acid, p-coumaric acid and myricetin have been identified in a high dietary fiber powder obtained from pineapple shell. These compounds have been reported as potent antioxidants [66]. One of the most valuable and studied component in pineapple waste is bromelain. It is a proteolytic enzyme naturally found in shell, core, crown, and leaves with numerous advantages in cardiovascular and digestive systems. It also serves as an anticancer, anti-inflammatory, and anti-microbial agent [67]. For this reason, bromelain has been utilized in applications of food, cosmetics, pharmaceuticals, and other industries.

TABLE 5.3 Chemical and Bioactive Components Present in Pineapple Peels

Component	Content	References
Moisture	82.7	[68]
Fat	1.57	
Protein	9.13	
Carbohydrates	42.3	
Ash	4.81	
Total dietary fiber	42.2	
Soluble dietary fiber	5.90	
Insoluble dietary fiber	36.3	
Total phenolics*	80.42	[67]
Total flavonoids$	18.8	[69]

Note: *mg CAE/g dry matter; and $ mg/100g all other values are reported in g/100 g dry basis.

5.3.4 MANGO BY-PRODUCTS

The processing of mango generates two by-products-peels and kernels which collectively represent 35–60% of the fruit weight. In 2019, global mango production accounted to 55.85 million metric tons [63]. Mango by-products possess high amount of various health-enhancing compounds like carotenoids, polyphenols, vitamin C and dietary fiber [70]. Mango peels have been reported to contain a variety of macronutrients like carbohydrates, proteins, enzymes as well as dietary fiber (Table 5.4). Polyphenol content of mango peels have been reported to be more than that of mango pulp. The main polyphenolic constituents of mango peel include mangiferin, rhamnetin, quercetin, kaempferol, ellagic acid, and their related conjugates [71]. Like mango peels, mango seeds are also discarded as by-products after industrial processing of mango pulp. Mango seed kernel contains high amount of carbohydrates, lipids, proteins, and several minerals. The fat present in mango kernel has gained considerable attention in research field due to its unique physico-chemical characteristics which are similar to those of cocoa butter, shea, kokum, illipe, and Sal butter [72]. Moreover, it is a promising and safe, naturally occurring source of edible fats as it is free of any trans fatty acid. Mango seed kernel is rich in protein, carbohydrates, lipids, and several minerals. The valuable components of mango by-products have potential in various application of food industry.

TABLE 5.4 Chemical and Bioactive Components Present in Mango Peels and Kernels

Component	Mango Peel	Mango Seed Kernel	References
Moisture*	4.92	6.57	[73]
Crude protein*	3.6	7.76	
Crude fat*	1.23	8.15	
Carbohydrates*	77.04	–	
Ash*	3.88	1.46	
Total phenolics#	2,032–3,185	11,228–20,034	[70]
Ascorbic acid#	68.49–84.74	61.22–74.48	
Carotenoids%	1,880–4,050	370–790	

Note: * g/100 g dry basis; # mg/100 g dry basis; and % μg/100 g dry basis.

5.3.5 POMEGRANATE BY-PRODUCTS

Globally recognized as a 'super-food,' pomegranate market value was recorded as USD 8.2 billion in 2018 with expected value of USD 23.14 billion by the year 2026 [74]. The consumption of pomegranate in the form of juices, jams, dietary supplements, etc., has increased in the last decade due to its numerous therapeutic properties. The industrial processing of pomegranate generates substantial quantity of by-products in the form of peels and seeds. These by-products usually discarded as waste, are a rich source of number of high value compounds like phenolic compounds, polysaccharides, proteins, lipids, sugars, etc. (Table 5.5). Depending on the cultivar, peels constitute about 49–55% of the total fruit weight whereas seeds account for 12–20% of the total weight of fruit [75]. The pomegranate peels is a potential source of nutrients and a number of secondary metabolites such as phenolics and flavonoids. The main flavonoids reported in the pomegranate peels include flavonoles (quercetin and kemferol), flavones (luteolin) and glycosylated flavonoles (rutin, kemferol-3-O-glycoside). The other constituents responsible for color of the fruit include anthocyanins (cyanidine, delfinidine, pelargonidine as well as their glycosylated forms) and naringin [76]. Other phenolic compounds present in the peels reported are hydroxycinnamic acids (chlorogenic acid, caffeic acid, p-cumaric acid), hydroxybenzoic acids (ellagic acid and gallic acid), catechins, and epicatechins [77]. Pomegranate peel extracts have been documented to show various *in vitro* bioactivities, such as anti-inflammatory, antioxidant, antitumor, and anti-proliferative activities. Due to the high antioxidant and antimicrobial activities of peels, it may be used as potential natural additives in elongation of shelf life and quality enhancement of food products. The seeds of pomegranate fruit also possess therapeutic properties. It possesses considerable amounts of proteins, lipids, essential minerals and sugars [78]. The oil present in pomegranate seeds have properties which makes it an attractive nutraceutical ingredient. The pomegranate seed oil constitutes 10–20% of the total seed weight with an ideally high content of unsaturated fatty acid with punicic acid (rare conjugate of linoleic acid) as the main component. The lipid fraction of seed is also composed of phenolic compounds (hydroxybenzoic acid, gallic acid and ellagic acid) which imparts antioxidant properties [79]. The other compounds of importance reported in pomegranate seeds include tocopherols (γ-tocopherols), sterols (stigmasterol, beta-sitosterol, camesterol, cholesterol, and daucosterol) and ursolic acid (UA) [80].

TABLE 5.5 Chemical and Bioactive Components Present in Pomegranate Peels and Seeds

Component	Pomegranate Peel	Pomegranate Seed	References
Moisture*	69.7 ± 0.35	20.8 ± 0.45	[81]
Crude protein*	7.8 ± 0.16	9.2 ± 0.35	
Crude fat*	1.2 ± 0.29	4.8 ± 0.29	
Carbohydrate*	64.84 ± 0.53	64.85 ± 0.07	
Ash*	5.6 ± 0.05	5.3 ± 0.19	
Total phenolics#	64.2 ± 0.50	4.96 ± 0.23	[82, 83]
Total flavonoid#	1.4 ± 0.03	0.68 ± 0.08	

Note: *g/100 g dry basis, and # mg/g dry weight.

5.4 MAJOR VEGETABLE PROCESSING BY-PRODUCTS

5.4.1 TOMATO BY-PRODUCTS

Tomato is one of the widely cultivated vegetable crop with global production of 41.52 million tons reported in 2020 [84]. The waste generated after commercial processing of tomato includes tomato peels, seeds, fibrous part and a small fraction of the pulp (Table 5.6). The rich composition of tomato by-products in high value compounds like soluble dietary fiber, natural antioxidants and vitamins [85] suggests its possible utilization in functional food development. The antioxidants present abundantly in tomato pomace include phenolic compounds, carotenoids, and ascorbic acid with strong radical scavenging properties. The main carotenoid present in tomato pomace is lycopene constituting 80–90% of total carotenoids associated with reduced incidence of cardiovascular diseases and prostate cancer [86]. The skin and seed fraction of tomato are comparatively richer in phenolic compounds than pulp and hence possess higher antioxidant activity. Tomato skin possess high content of dietary fiber [87] and seeds are considered rich source of oil [88]. The protein content of tomato seed is more than double the amount of protein present in most varieties of wheat [89].

TABLE 5.6 Chemical and Bioactive Components Present in Tomato Peels, Seeds, and Pomace

Component	Tomato Peel	Tomato Seed	Tomato Pomace
Protein*	10.50$	32.0#	16.0#
Oil*	4.04$	22.0#	2.0#
Ash*	5.90$	5.0#	4.0#
Fiber*	78.56$	16.0#	46.0#
Total phenolics**	208.2&	95.2&	10.08–22.75@

Note: * g/100 g dm; ** mg GAE/100 g dry weight; $ [90]; # [91]; @ [92]; and & [93].

5.4.2 POTATO BY-PRODUCTS

Potato is one of the chief crops consumed worldwide with global production of about 370 million metric tons [53]. The processing of potato involves peeling of potato with production loss of about 15 to 40% depending on the peeling method. The potato by-products contain large quantity of polyphenols, protein, non-starch polysaccharides, and small portion of lipids. The rich composition of potato by-products in such biomolecules makes it a valuable base material for extraction of high value substances like antioxidants, dietary fiber, biopolymers, etc. [94]. Chemical and bioactive components present in potato peels were described in Table 5.7. The potato peels have caught increasing attention as a new source of dietary fiber since it represents 40–45% of its dry weight. The dietary fiber components include cellulose, hemicellulose, pectin, gums, lignin, etc. [94]. The phenolic content in potato peels is almost 10 times higher than that found in potato flesh [95]. The main phenolic compounds in potatoes are the phenolic acids of which the major part is composed of chlorogenic, neochlorogenic, and caffeic acids [96]. The flavonoid component of potato peels includes flavonols, flavones, flavanones, flavan-3-ols and anthocyanins. The waste slurry generated after potato starch manufacturing process contains considerable amount of starch and various bioactive substances including proteins, minerals, etc. It has applications as functional additives, thickeners, gluing agents, or stabilizers in a wide range of food and non-food products [97].

TABLE 5.7 Chemical and Bioactive Components Present in Potato Peels

Component	Content	References
Moisture	83.8–85.1	[98]
Protein	1.2–2.3	
Carbohydrate	8.7–12.4	
Lipids	0.1–0.4	
Total dietary fiber	2.5	
Ash	0.9–1.6	
Total phenolics*	3.78	[99]
Total flavonoids#	0.13	

Note: * mg GAE/g dw; # mg rutin/g dw. All other values are expressed in g/100 g.

5.4.3 CAULIFLOWER BY-PRODUCTS

Cauliflower is one of the nutritionally rich cruciferous vegetable with recorded global production of 25.97 million tons in 2020 [100]. Cauliflower has a high waste index generating a huge amount of solid organic waste which creates foul odor on decomposition. The non-edible portion of cauliflower constitute 45–60% of the total vegetable weight [101]. Chemical and bioactive components present in cauliflower peels were described in Table 5.8. It is a rich source of dietary fiber, protein, cellulose, hemicellulose, and possess antioxidant and anticarcinogenic activity. The phenolic compounds, vitamin C, carotenoids, and vitamin E contribute to the high antioxidant activity of this vegetable [102]. The dried leaves of cauliflower are reported to be good source of minerals including iron, copper, zinc, and manganese [103]. The phenolic compounds as well as the total flavonoids concentration present in cauliflower by-products has been found to be comparatively higher than in the edible portion [104]. The presence of compounds like isothiocyanates specifically in cauliflower is associated with prevention of numerous chronic degenerative diseases, cardiovascular diseases, cancer, and diabetes [105].

TABLE 5.8 Chemical and Bioactive Components Present in Cauliflower Peels

Component	Content	References
Moisture^	81.1	[106]
Fat^	0.49	
Crude fiber^	34.58	
Protein^	13.8	
Ash^	4.32	
Total phenolics$	3.45	[107]
Total flavonoids&	0.35	

Note: ^ All values expressed in percentage on dry basis (except moisture on wet basis); $ mg GAE/g; and & mg CE/g.

5.4.4 CARROT BY-PRODUCTS

Carrot is one of the top six vegetable crops grown around the world with global production accounting to 44.76 million metric tons in 2019 [53]. The processing of carrot results in large quantities of carrot waste which is an excellent source of bioactive substances and can be utilized for human and animal consumption. The pomace and mash constitute the main by-products obtained after processing of carrots. Some of the compounds of interest present in this vegetable include carotenoids, sugars, pectin, and other fibers [108]. One of the important nutrients with diverse functions and effects in food is the carotenoids which imparts the attractive color to carrots. The main pigments present in this class include alpha-carotene, beta-carotene, lutein, and lycopene. Their intake has several useful effects on human body such as reduction in occurrence of diseases like cardiovascular diseases, cancer, age related muscular degeneration, and enhancement of immune system. Polyphenols are another class of bioactive substances abundantly present in carrots, especially in the peels which contribute to 11% of fresh weight of carrot [108]. The main phenolic compound reported in carrots is the chlorogenic acid contributing to the organoleptic characteristics of fresh and processed carrots. A total of 11 phenolic acids, mainly hydroxybenzoic acid have been found in orange, yellow, white, and purple carrots. The vegetable is considered an excellent source of dietary fiber, mainly the pomace fraction of carrot which account for 63.6% of total dietary fiber on dry basis [109]. The nutritive aspects and quality characteristics of carrot by-products makes it suitable for value addition of food products (Table 5.9).

TABLE 5.9 Chemical and Bioactive Components Present in Carrot Peels

Component	Content	References
Carbohydrates	67.867	[110]
Protein	6.428	
Fats	0.714	
Ash	6.144	
Fiber	22.785	
Total phenolics	106.36$^{\$}$	[111]
Total flavonoids	27.83$^{\#}$	
Total carotenoids	2.75^	[112]

Note: $^{\$}$ µg/g GAE; $^{\#}$ µg/g QE; and ^ mg/g dw. All other values reported in percentage dry matter.

5.4.5 *ONION BY-PRODUCTS*

Owing to its nutritional and medicinal value, production of onions has increased significantly in the past decade with global production of 4.49 million metric tons reported in 2020 [113]. Waste generated after onion processing includes the outer layers, brown skin, stalks, and roots as well as the onions not large enough for commercial use. The brown kin and external layers of onion are particularly rich in phenolic compounds like quercetin, flavonoids, and fiber [114]. Table 5.10 shows the chemical and bioactive components present in onion peels. The phytochemical composition of onion waste includes phenolic compounds, flavonols, flavonoids, and antioxidants. Two major flavonoid subgroups present in onion includes flavonols and anthocyanins. Flavonols such as quercetin and its derivatives are responsible for production of yellow and brown compounds in many varieties of onion. Anthocyanins contribute to the red/purple color in some varieties [114]. Quercetin is one of the important compounds present chiefly in onion. It possesses antioxidant activity along with anticancer, antiviral, and anti-inflammatory activities [115]. The major components of onion also include sulfur and carbohydrates. S-alk (en) yl-L-cysteine sulfoxide (ACSOs) is the compound responsible for the characteristic flavor of onions. Variations in the flavor of different species of onions are mainly due to the variable composition and concentration of ACSOs [116]. Onion waste obtained from top and bottom part comprise of outer, inner, and skin section and therefore possess high amount of dietary fiber [117]. A considerably high amount of magnesium, zinc, iron, and manganese are present in the top-bottom section of onion whereas high concentration of selenium and potassium can be found in the inner scaly portion. The brown skin of onion is specifically rich in calcium [118]. Apart from this, high quantities of unsaturated fatty acids are present in outer scale of onion while saturated fatty acids are predominantly found in the top-bottom waste. The outer scales of onion possess linoleic acid as the major fatty acid which has potential as dietary supplement for polyunsaturated fatty acids [119].

TABLE 5.10 Chemical and Bioactive Components Present in Onion Peels

Component	Content	References
Moisture*	9.98	[120]
Carbohydrate*	88.56	
Crude fiber*	0.15	
Fat*	0.04	
Protein*	0.88	
Ash*	0.39	
Total phenolics#	289.04	[4]
Total flavonoids$	168.77	

Note: * percent dry matter; # mg GAE/g dw; and $ mg QE/g dw.

Source: Adapted from Ref. [120].

5.5 EXTRACTION OF FUNCTIONAL COMPOUNDS FROM FRUITS AND VEGETABLE BY-PRODUCTS

5.5.1 CONVENTIONAL EXTRACTION TECHNIQUES

The conventional or traditional extraction method is based on the solvent extraction power and application of heat or used in combination. The commonly used conventional techniques include maceration or solvent extraction, Soxhlet extraction and hydro-distillation [121]. Different approaches for the extraction of bioactive compounds from fruit and vegetable processing by-products described in Figure 5.2. Solvent extraction is generally utilized for isolation of bioactives using a solvent of appropriate polarity [9]. This extraction approach involves contact of suitably sized raw material with different organic solvents, which takes up soluble components of interest like coloring and flavoring substances (e.g., lycopene and anthocyanins) with antioxidant properties [122]. The type of solvent is an important factor in extraction of functional compounds from fruit and vegetable by-products [123]. Solvents like acetonitrile, acetone, ethyl acetate, methanol, ethanol, and water are generally employed for extracting functional compounds from fruit and vegetable by-products. Water soluble compounds are extracted using solvents like water or dilute alcohol whereas nonpolar solvents are utilized for extraction of lipophilic components present in plant matrix [124]. Non-polar solvents like hexane, acetone, methanol has been used for the extraction of carotenoids, lycopene, and anthocyanins from gac fruit peel, tomato pulp and eggplant, respectively [125–127]. The beneficial aspect of solvent extraction includes its low processing cost and ease of operation as compared to other methods. However, involvement of large quantity of toxic solvents, extended period of time and an additional concentration/evaporation step for recovery make its use limited. Moreover, susceptibility of natural bioactive compounds to thermal degradation because of high temperature of the solvents and long exposure time is an additional concern in the use of this method [128].

Hydro distillation is one more conventional method employed for the extraction of bioactive compounds and essential oils from plant matrix. It does not include use of organic solvents and can be useful before the dehydration of plant material. Hydro distillation can be of three types – direct steam distillation, steam and water distillation, and water distillation [129]. The three main processes involved in this method include hydro diffusion, hydrolysis, and decomposition by heat. The main drawback of this method is that it cannot be used for extraction of thermo labile compounds because at high temperature, some of the volatile components of the plant sample may be lost [130]. Soxhlet extraction is the classical technique originally for the extraction of lipids, however, its use has been widened for recovery of functional compounds from natural sources. In this technique, a minor quantity of dry sample is placed in the equipment and the solvent is allowed to pass through until the extraction process gets completed. Extraction of several bioactive compounds such as polyphenols from grape peels and carotenoids from orange peels has been successfully done using Soxhlet technique [131, 132]. However, the technique is associated with major drawback of high extraction time and large volume of solvent [133].

5.5.2 GREEN EXTRACTION TECHNIQUES

The limitations of conventional techniques make their application uneconomical and unsustainable. This increased the interest in search of cost-effective green techniques for the recovery of functional compounds from fruit and vegetable waste with a wider scope of applications in food and pharmaceutical sectors [134]. Table 5.11 shows the green technologies used for the extraction of functional compounds from fruit and vegetable processing by-products. Currently, novel techniques like ultrasound, microwave, pulse electric field (PEF), supercritical fluid extraction (SCFE), enzyme assisted extraction (EAE), etc., has been gaining increasing importance owing to several advantages associated with it. The main aim of using these techniques lies behind several reasons such as reduced processing steps, efficient energy use, enhanced mass and heat transfer, etc. [130, 135]. Some of the commonly utilized novel techniques for bioactive compounds extraction from fruit and vegetable peels have been discussed in subsections.

5.5.2.1 ULTRASOUND ASSISTED EXTRACTION

Ultrasound assisted extraction (UAE) is a novel technique which involves propagation of ultrasonic waves through the sample matrix because of which, compression, and expansion cycles are originated creating a phenomenon of cavitation. This process implies the simultaneous production, growth, and collapse of bubbles which drives the solvent penetration within the cell and facilitates the extraction process [136]. Ultrasound energy used in this technique is beneficial in terms of effective mixing, faster distribution of energy, reduced thermal gradients and extraction temperature with a high overall yield of target compound [137]. The precise control of UAE parameters such as power, frequency, temperature, time, duty cycle, solvent type, liquid to solid ratio is needed for effective extraction. UAE technique could be implemented at industrial level for the extraction of variety of biomolecules like carotenoids, polyphenols, essential oils, polysaccharides, proteins, minerals, vitamins, and organic acids [11, 138] from fruits and vegetable by-products. Pectin from orange, mango, and banana fruit by-products [139–141] as well as jackfruit, eggplant, and tomato vegetable by-products [142–144] have been successfully extracted using UAE. A new process involving ultrasound has been utilized in combination with other green techniques for the recovery of carotenoids from pomegranate peels by vegetable oils as green solvent

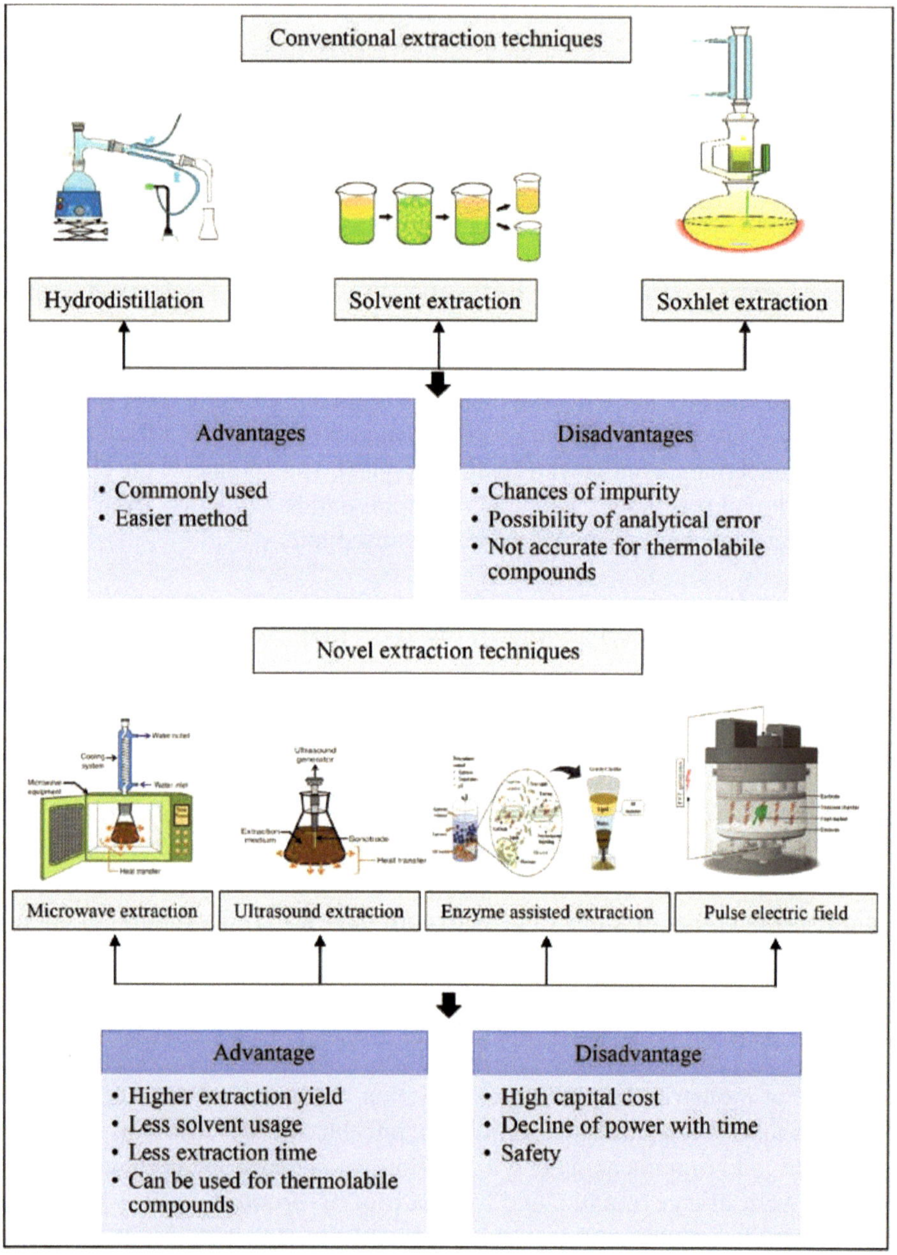

FIGURE 5.2 Different approaches for the extraction of bioactive compounds from fruit and vegetable processing by-products.

[145]. Plant pigments like betalains, carotenoids, lycopene, anthocyanins, etc., have been extracted from fruit and vegetable by-products with the use of ultrasound technique [146]. UAE has been effective in total phenolic compounds extraction from fruits and vegetable by-products [147, 148] as well as for the extraction of specific phenolic components like mangiferin from mango peels and hesperidin from citrus by-products [149, 150].

5.5.2.2 MICROWAVE ASSISTED EXTRACTION

Microwave-assisted extraction (MAE) technique is based on the generation of electromagnetic wave that causes the cell disruption of plant material and facilitates the dissolution of functional compounds in solvent [151]. Recovery of functional compounds from fruits and vegetable by-products using MAE technique has received much attention, as this can efficiently reduce the energy consumption, solvent utilization, and time of extraction. Citrus processing by-product like peel was found to be rich in essential oil, terpenoids, and pectin. Extraction of these compounds by microwave and conventional methods suggested less treatment time and higher extraction yield in the MAE method than conventional extraction method [152]. Phenolic compounds were found to be highly heat-sensitive and extraction of these at the alleviated temperature may reduce the yield. Therefore, the MAE technique was found to be an effective method for the extraction of polyphenolic compounds from fruit processing byproducts viz. seeds, skin, pulp, and pomace [151]. Extraction of phenolic compounds from olive leaves indicated higher yield and lesser treatment time in MAE method when compared with maceration method [153]. Bioactive compounds like polyphenols and other phenolic subclasses of apple pomace were recovered more with MAE method than conventional methods [154]. Extraction yield of phenolic compounds viz. chlorogenic acid, gallic acid, rutin, triterpene, etc., of apple pomace was found to be higher in the extract obtained with MAE than maceration, pressurized liquid, and ultrasound assisted extraction method [155]. Broccoli by-products like stalks, leaves, and inflorescences suggested that the microwave hydro-diffusion gravity technique can be a promising method to valorize these by-products [156]. They have reported that extract obtained with microwave hydro-diffusion retains a high quantity of phenolic compounds, amino acids, glucosinolates, pectic polysaccharides, fructose, and glucose.

5.5.2.3 PULSED ELECTRIC FIELD EXTRACTION

The mechanism of extracting bioactive compounds by the pulsed electric field is based on the cell membrane electroporation and thereby increasing the inside materials diffuse out. In this technique, electric energy permits through the cell membrane followed by the separation of the molecules based on their charge, and form pores that increase permeability [157]. Extraction yield of polyphenol and flavonoids and the antioxidant activity of obtained extract from orange peel by pulsed electric field treatment was significantly higher than those obtained by the conventional method [158]. Cherry processing by-product like press cake was found to be an excellent source of anthocyanin and phenolic compounds. It was observed that the extraction yield of these compounds obtained through PEF was higher than those obtained through the conventional and freeze-thaw methods. In addition to extraction yield, antioxidant activity was also higher for the extract obtained by PEF than that obtained by conventional and freeze-thaw method [159]. Extract of blueberry by-product obtained through PEF showed higher anthocyanin content and antioxidant activity than that obtained through conventional method [160]. PEF treatment was found to be necessary to recover some of the phenolic compounds from grape processing by-product (pomace and peel), those were not possible to recover by other extraction methods [161]. Tomato processing by-products like peel showed a higher extraction yield of lycopene and antioxidant activity after PEF treatment as compared to untreated samples [162].

5.5.2.4 SUPERCRITICAL FLUID EXTRACTION (SCFE)

Above the critical temperature and pressure boundary between the liquid and gas phase, disappear, and fluids behave like a single-phase having characteristics of both liquid and gas phase. This phase of fluid is utilized in the SCFE which diffuses into the cell membrane as gas and dissolves target compounds as a liquid. Supercritical carbon dioxide has been used for the extraction of a wide range of bioactive compounds like phenolic compounds, flavonoids, and essential oils [163]. To valorize grape seed and skin, hazelnut waste and coffee spent SCFE can be an efficient technique for the recovery of phenolic compounds and triglycerides from these by-products [164]. Mango processing by-products like peel constitute a large number of carotenoids, flavonoids, and phenolic compounds. It was observed that these compounds can be efficiently extracted from mango peel using the SCFE technique. The carotenoid content of mango peel extract obtained by SCFE was found to be higher than that obtained with the pressurized ethanol extraction method [147]. Mandarin peel, a popular citrus processing by-product comprise a high quantity of total phenolic compounds. Recovery of phenolic compounds of the mandarin peel by supercritical carbon dioxide with 10% ethanol resulted in the highest yield than that with other concentrations of ethanol. It was also observed that obtained extract with supercritical fluid (SF) with 10% ethanol had high antioxidant activity [165]. Tomato peel as a major vegetable processing by-product contains a high quantity of lycopene. A study demonstrated the efficient extraction of lycopene with oleoresin from tomato peel by SCFE technique [166].

5.5.2.5 ENZYME ASSISTED EXTRACTION (EAE)

Extraction of bioactive functional compounds by enzyme-assisted extraction is one of the best techniques among the available green techniques. In this extraction, method enzymes are added to the extraction medium to break or soften the cell structure of plant material and enhance the recovery of active compounds [163]. The major benefits of this method are shorter treatment time, reduce the utilization of solvent, and increases the purity of functional compounds. The enzyme extraction method has been used as a suitable method for extracting polyphenols and their subclass like pro-anthocyanidins which are usually difficult to extract [167]. Grape pomace polyphenols were found to be successfully extracted by enzyme-assisted extraction method using cellulose, pectinase, and tannase. These enzymes were found to release phenolic acids like gallic acid and *p*-coumaric acid, and anthocyanin compounds like malvidin-3-*O*-glucoside from grape pomace [168]. Optimizing various treatment parameters viz. type of enzyme, hydrolysis time, temperature for hydrolysis, and particle size of the substrate, it was possible to extract antioxidant polyphenols from black current pomace [169]. Commercial enzymes such as Novoferm, celluclast, and Pectinex Ultra at 40°C and treatment time 48 h were successfully used to recover the antioxidant polyphenols from grape waste [170]. Enzyme-assisted extraction with pancreatin enzyme enhanced the recovery of lycopene from vegetable processing by-products like tomato skin [171].

5.5.2.6 EXTRACTION USING IONIC LIQUIDS AND DEEP EUTECTIC SOLVENTS

Extraction of functional compounds using solvent extraction method with unsustainable solvent, volatile solvent, a mixture of these solvents, and highly acidic or alkaline aqueous solution have been practiced for a long time. But this extraction method is not only exceptionally energy exhaustive, requiring long treatment time or high temperatures but also possesses low selectivity to the active compounds. Two types of solvents viz. ionic liquid and deep eutectic solvent have received much attention to competing with the conventional solvents for extracting functional compounds [172]. These solvents are capable of extracting a wide variety of polar and non-polar compounds from plant materials. Efficient extraction of polyphenolic compounds, lignin, and cellulose was reported from chestnut shell waste using choline-chloride based deep eutectic solvent and bio-based ionic liquids [173]. Magnesium chloride hexahydrate and urea-based deep eutectic solvent showed remarkable extractability of betalain from beetroot waste. The extraction ability of this solvent was attributed to the low melting point, good thermal stability, low conductivity, and variation in viscosity [174]. The deep eutectic solvent in combination with microwave and ultrasound treatment has been used for the extraction of many functional compounds like phenolic acids, flavonoids, anthocyanins, and proteins from olive leaves, gape peels, onion, and tomato by-products [175].

TABLE 5.11 Green Technologies for the Extraction of Functional Compounds from Fruit and Vegetable Processing By-Products

Technique	Fruit and Vegetable Processing By-Products	Extracted Compound	References
Ultrasound assisted extraction	Pomegranate peel	Carotenoids	[50]
	Avocado seed	Polyphenols	[176]
	Jackfruit peel	Pectin	[177]
	Cauliflower peel	Fiber, polyphenols, and flavonoids	[107]
	Citrus waste	Organic acids, and enzymes	[41]
	Mango peel	Polyphenols	[178]
Microwave assisted extraction	Citrus peel	Terpenoids, essential oil, and pectin	[152]
	Gac peel	Carotenoid	[179]
	Broccoli stalk, and leaves	Phenolic compounds	[156]

TABLE 5.11 *(Continued)*

Technique	Fruit and Vegetable Processing By-Products	Extracted Compound	References
	Eggplant peel	Phenolics, flavonoids, and anthocyanins	[180]
	Grape pomace	Phenolics, and anthocyanins	[181]
Pulse electric field	Grape seed	Polyphenols	[182]
	Mango peel	Phenolic compounds	[183]
	Blueberry waste	Anthocyanin	[160]
	Tomato peel	Lycopene	[161]
Supercritical fluid extraction	Grape seed, and skin	Polyphenol	[164]
	Mango peels	Phenolic acids, carot-enoids, and flavonoids	[147]
	Apple pomace	Phenolic compounds	[184]
	Tomato peel	Lycopene	[166]
Enzyme assisted extraction	Tomato skin	Lycopene	[171]
	Kiwifruit pomace	Pectin, and non-starch polysaccharides	[185]
	Grape pomace	Phenolic acids, and anthocyanin	[168]
	Mango peel	Dietary fiber	[186]
Ionic liquids and deep eutectic solvents	Orange peel	Ferulic acid	[187]
	Potato peel	Xylan, glucan, arabinan	[188]
	Apple pomace	Xylan, glucan, arabinan	
	Onion skin	Polyphenol, flavonoid, and anthocyanin	[175]

5.6 ANALYTICAL TECHNIQUES USED FOR QUANTIFICATION AND CHARACTERIZATION OF BIOACTIVE COMPOUNDS EXTRACTED FROM FRUITS AND VEGETABLE BY-PRODUCTS

5.6.1 NUCLEAR MAGNETIC RESONANCE (NMR) SPECTROSCOPY

Nuclear magnetic resonance (NMR) spectroscopy is a well-established technique for identification of individual bioactive components in a complex mixture (Figure 5.3). The usefulness of this tool lies in its non-invasiveness, rapid speed and sensitivity for a wide variation of compounds in a single measurement without involving sample pretreatment. It is an increasingly recognized technique as it offers several advantages such as simplicity of the sample preparation and measurement procedures, the instrumental stability and the ease of spectral interpretation. Standard H, C, and high-resolution magic angle spinning (HR/MAS) NMR spectra can contribute to a wide array of information relating to the chemical information on liquid and semisolid food material. The NMR spectra of the plant material sample (fruit and vegetable by-product) can act as "fingerprints" for comparison, discrimination or classification of samples. The specific peak heights or integrals in the spectra that particularly characterize the sample are also used instead of whole spectra. It is often assisted with chemometric techniques for data analysis as the spectral information is of highly complex. Solid samples of fruit and vegetable by-products are generally freeze-dried, grounded, and then extraction in a deuterated solvent is performed. Liquid samples, such as oils, are simply dissolved at a desired concentration in an appropriate deuterated solvent. In NMR spectroscopy, the intensity of peak is directly proportional to the number of nuclei. In turn, the signal produced is linearly associated to the absolute concentration of the analyte in solution [189]. Therefore, NMR spectroscopy has a similar response factor for all analytes and signals of interest. NMR spectroscopic analysis has been applied for identification of bioactive constituents in grape by-products generated in the process of winemaking [190]. The characterization of bioactive molecules using NMR spectroscopy has been performed in apple pomace [191]. The chemical composition of pectin obtained from tomato peel waste was confirmed using H1 NMR [192].

5.6.2 FOURIER TRANSFORM INFRARED (FTIR) SPECTROSCOPY

Infrared spectroscopy is one of the most utilized analytical technique for analyzing any sample of any state. Fourier transform infrared (FTIR) spectroscopy is an effective instrumental technique based on measurement of molecular vibration caused by infrared radiation excitation at a specific wavenumber range. The sample molecule to be investigated using IR spectroscopy should be polar with a dipole moment. FTIR is a time and cost effective nondestructive analytical tool that furnishes fingerprints to provide structural details of extracted biomolecules. The correlation between spectral features and phenolic extractability of the F&Vs by-products can be found using FTIR. This method is considered an excellent alternative over traditional method as it is cheap, offers faster response and simple to handle.

Moreover, it involves use of small sample without significant pretreatment [193]. The by-products of pineapple were characterized for total phenolic content (TPC) and antioxidant capacity using this technique [194]. Quantification of bioactive compounds of grape seeds were successfully done using FTIR [195].

5.6.3 CHROMATOGRAPHIC TECHNIQUES

Chromatographic methods can be employed for identification and quantification of high value compounds from fruit and vegetable by-products. The development of stationary-phase and high-pressure module technologies has given rise to new possibilities for the separation of components with similar structure and properties from a complex system. Individual ingredients can be detected and identified based on their electrical properties, molecular ionization characteristics and ability to emit or absorb electromagnetic waves. The modern chromatographic techniques allow isolation of mixture ingredients for further analysis or constituting mixtures with predesigned properties. The two most commonly used methods for the analysis and quantification of biologically active substances from fruit and vegetables by-products are high performance liquid chromatography (HPLC) and GC.

5.6.3.1 HIGH PERFORMANCE LIQUID CHROMATOGRAPHY (HPLC)

High performance liquid chromatography (HPLC) is the most frequently used analytical technique for separation and characterization of bioactive substances from F&V by-products. HPLC is a versatile adaptable technique that offers high selectivity and sensitivity with increased resolution and precision. Compounds from complex mixtures are separated on the basis of solubility, polarity, and size properties of individual compound. A photodiode-array detector (HPLC-DAD) coupling with HPLC can provide extensive information on complex structure of bioactive molecules. The identity of a particular compound can be obtained by comparison of the retention time (RT) of the compound with standard. For a quantitative evaluation, a series of standards of desired concentration range are analyzed, and the sample analytes are quantified as per the calibration curve of each standard. Despite being one of the most used techniques, HPLC present some limitations in detection and quantification limit. HPLC technique has been applied for characterization and quantification of bioactive compounds in by-products of many fruits like mango, papaya, pineapple, dragon fruit, etc. [196]. Determination of phenolics and amino acids from tomato processing waste was done using HPLC technique [85]. Resveratrol and coumarin content were determined in by-products of tropical fruits including guava, cherry, passion fruit using HPLC technique [197].

5.6.3.2 GAS CHROMATOGRAPHY-MASS SPECTROSCOPY (GC-MS)

Gas chromatography (GC) has been identified as one of the most successful technique for identification of bioactive components of long chain hydrocarbons, acids, alcohols, esters, steroids, alkaloids, amino acids and nitro compounds [198–200]. GC can be used for quantitative determination of compounds present in very low concentrations with the use of flame ionization detector and electron capture detector (with high precision). It is a simple, sensitive, and effective technique for separating component of mixtures and therefore, is one of the most important tools in the field of chemistry and food. Higher plants as a source of bioactive compounds continues to gain attention in the research field due to its dominant role in maintaining health by prevention of a number of diseases. Profiling of bioactive constituents derived from plant sources provides useful insights of chemical diversities, therapeutic potential and toxicity concerns and therefore provides opportunity for new drug discoveries [201, 202]. When compared to LC and high-performance liquid chromatography, GC-MS provides more accuracy. However, poor compound identification with several unidentified peaks is a major challenge [203]. Limitations in MS data acquisition speeds due to incomplete metabolomes causes difficulty in identification of polyphenolic compounds [203].

5.6.4 CALORIMETRY

Thermal analysis (TA) in food system is a convenient, reproducible, and a significant method for characterization of heterogenous organic compounds. One of the most widely recognized thermal technique for bioactive compound characterization is differential scanning calorimetry (DSC). It is a fast, cheap, sensitive technique which do not consume much time in processing of test samples. It helps determine energetics of phase transitions and conformational changes for temperature depending quantification of

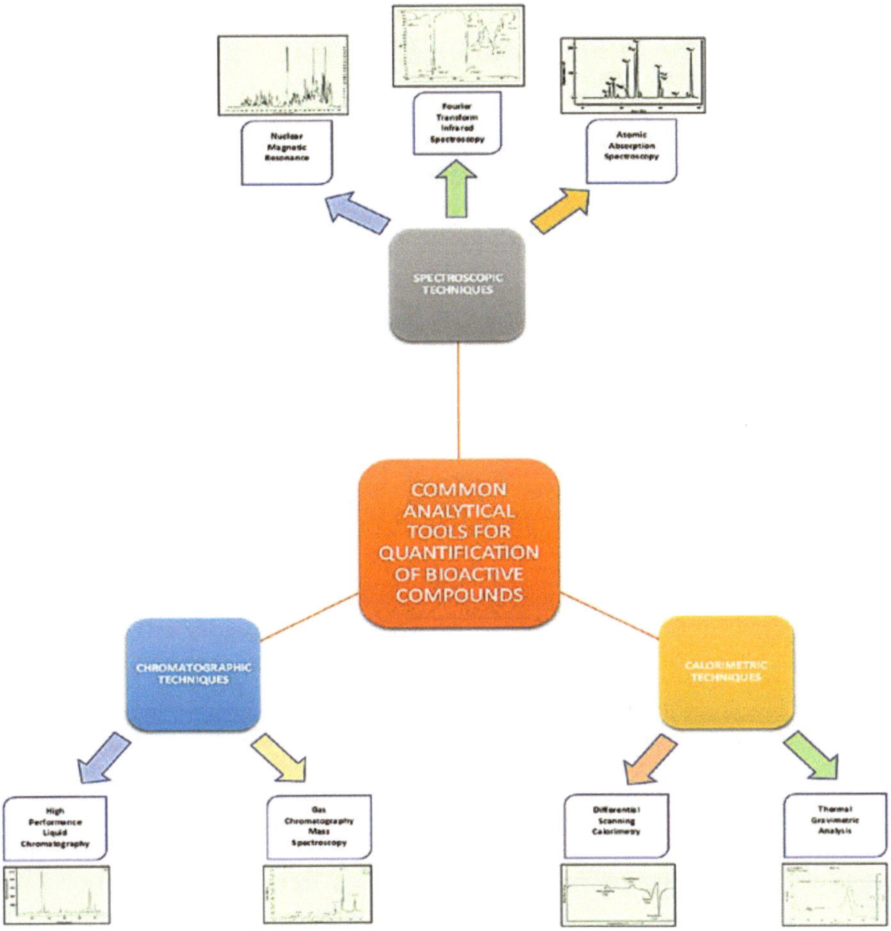

FIGURE 5.3 Common analytical tools used for quantification of bioactive compounds from fruit and vegetable processing by-products.

food materials. In TA, measurement of thermal properties is done based on relationships between sample properties and temperature of sample heating or cooling [204]. The TA based on DSC curves gives endo- and exo-thermic peaks. Higher stability of material is reflected by the higher temperature of peaks. Thermally stable regions and mass losses are indicated by thermo-gravimetric analysis (TGA) curve.

5.7 HEALTH BENEFITS OF BIOACTIVE COMPOUNDS IN FRUIT AND VEGETABLE PROCESSING BY-PRODUCTS

The observed biological effects after identification, extraction, quantification, and evaluation of bioactive constituents from fruit and vegetable by-products have been of major interest in health and medical-related research. This might be due to reduced risk of several cardiometabolic diseases and cancer [17]. Bioactive compounds due to their health enhancing effect have been observed to lower the risk of diseases like Alzheimer, cancer, Parkinson, cataract, etc. These beneficial effects might be due to their antioxidant properties and radical scavenging activities which can inhibit or delay oxidation of proteins, DNA, and lipids. The antimicrobial properties exhibited by these compounds play an important role in protection of fruit and vegetables against pathogenic agents which can cause lysis of microorganism's cell membrane [205]. Compounds like flavonoids, anthocyanins, vitamin E, vitamin C, fibers, phenolic compounds, carotenoids, and other antioxidants are key bioactive components present in fruit and vegetable by-products. These can be extracted using green techniques and reused as functional food ingredients in food supplements and products to provide required antioxidants and health benefits [146].

The characteristic polyphenol named phloridzin is a chalcone derivative present in apple, has been observed to possess antidiabetic effect as it interferes with glucose absorption causing inhibition of sodium-linked glucose transporters [206]. The bioactive composition in banana peels extract plays a significant role in lowering of cholesterol level, reduction of hyperglycemia, and stimulation of gastric mucosa growth. Moreover, it acts as efficient neurotransmitter as well as against aging and free radical formation [207, 208]. Grape seed derived oil possesses antimicrobial, cardioprotective, anti-inflammatory, and anti-cancer effects [33]. Similarly, mango peel extracts have been observed to give promising results in diabetes and obesity, neuroprotection, cardiovascular disease, intestinal health, and skin cancers [209]. By-products of vegetables like beetroot have been reported to impart several health benefits like antioxidant, anticarcinogenic, anti-inflammatory, antibacterial, anti-hypertensive, detoxifying, and hepato-protective effects [210]. Dietary fiber in carrot peels helps prevent diseases like constipation, cancer, heart disease, and control blood sugar levels whereas polyphenols help fight free radicals and possess antitumor and antimutagenic effects [108]. Quercetin present in onion by-products was considered to possess pronounced effects on asthma, arthritis, cancer, diabetes, coronary artery disease, osteoporosis, and neurodegenerative disorders [211–213]. Cauliflower by-products have been found to exhibit beneficial effect against diabetes, and hypertension as well as glucose lowering effects [214]. The oil found in tomato seeds is highly thermostable and possess excellent physiochemical profile after heating with high antioxidant activity than common edible oils like soyabean and olive oil [30]. The role of these natural compounds has been gaining importance owing to the increasing demand of consumer for reducing the utilization of synthetic preservatives in food products [3].

5.8 CONCLUSIONS

Fruits and vegetable processing industry generates large quantity of by-products such as peel, skin, seeds, pomace, stalks, etc., of citrus, banana, mango, pineapple, tomato, and carrot processing section. These by-products contain many functional compounds like phenolic acids, flavonoids, anthocyanins, essential oils, and pectin. Techniques used for the extraction of functional compounds play an important role on their extractability because no single extraction method is appropriate for all by-products. For extracting the bioactive compounds from fruits and vegetable processing by-products, green technologies viz. ultrasound, microwave, SCFE, EAE, etc., can play significant role. Green techniques are far better than conventional techniques as they are environmental friendly, improve the extraction yields, reduce treatment time and the usage of environmentally harmful solvents. Various innovative analytical techniques such as FTIR, HPLC, GC-MS, DSC, etc., can help to generate additional information with respect to quantification and characterization of bioactive compounds. Functional compounds of fruits and vegetable processing by-products possess various health benefits and can be used as an ingredient for the development of value added food products.

KEYWORDS

- **bioactive**
- **by-products**
- **Fourier transform infrared**
- **gas chromatography-mass spectroscopy**
- **green techniques**
- **quantification**
- **supercritical fluid extraction**

REFERENCES

1. Esparza, I., Jiménez-Moreno, N., Bimbela, F., Ancín-Azpilicueta, C., & Gandía, L., (2020). Fruit and vegetable waste management: Conventional and emerging approaches. *J Environ. Management, 265*, 110510.
2. Ajila, C. M., Aalami, M., Leelavathi, K., & Rao, U. P., (2010). Mango peel powder: A potential source of antioxidant and dietary fiber in macaroni preparations. *Innov. Food Sci. Emerging Technol., 11*(1), 219–224.
3. Jiménez-Moreno, N., Esparza, I., Bimbela, F., Gandía, L. M., & Ancín-Azpilicueta, C., (2020). Valorization of selected fruit and vegetable wastes as bioactive compounds: Opportunities and challenges. *Crit. Rev. Environ. Sci. Technol., 50*(20), 2061–2108.
4. Sagar, N. A., Pareek, S., Sharma, S., Yahia, E. M., & Lobo, M., (2018). Fruit and vegetable waste: Bioactive compounds, their extraction, and possible utilization. *Comprehen. Rev. Food Sci. Food Safety, 17*(3), 512–531.
5. Ajikumar, P. K., Tyo, K., Carlsen, S., Mucha, O., Phon, T. H., & Stephanopoulos, G., (2018). Terpenoids: Opportunities for biosynthesis of natural product drugs using engineered microorganisms. *Mol. Pharm., 5*, 167–190.
6. Kosseva, M., (2013). Sources, characterization, and composition of food industry wastes. *Food Industry Wastes*. 10.1016/B978–0-12–391921–2.00003–2.
7. Rudra, S. G., Nishad, J., Jakhar, N., & Kaur, C., (2015). Food industry waste: Mine of nutraceuticals. *Int. J. Sci. Environ. Technol, 4*(1), 205–229.
8. Banerjee, J., Vijayaraghavan, R., Arora, A., MacFarlane, D. R., & Patti, A., (2016). Lemon juice based extraction of pectin from mango peels: Waste to wealth by sustainable approaches. *ACS Sustain. Chem. Eng., 4*(11), 5915–5920.
9. Galanakis, C. M., (2012). Recovery of high added-value components from food wastes: Conventional, emerging technologies and commercialized applications. *Trends Food Sci. Technol., 26*(2), 68–87.
10. Saini, A., Panesar, P. S., & Bera, M., (2019). Valorization of fruits and vegetables waste through green extraction of bioactive compounds and their nanoemulsions-based delivery system. *Biores. Bioproces., 6*(1), 1–12.
11. Gil-Chávez, G. J., Villa, J. A., Ayala-Zavala, J. F., Basilio, H. J., Sepulveda, D., Yahia, E. M., & González-Aguilar, G. A., (2013). Technologies for extraction and production of bioactive compounds to be used as nutraceuticals and food ingredients: An overview. *Compr. Rev. Food Sci. Food Saf., 12*, 5–23.
12. Day, L., Seymour, R. B., Pitts, K. F., Konczak, I., & Lundin, L., (2009). Incorporation of functional ingredients into foods. *Trends Food Sci. Technol., 20*(9), 388–395.
13. Lemes, A. C., Sala, L., Ores, J. D. C., Braga, A. R. C., Egea, M. B., & Fernandes, K., (2016). A review of the latest advances in encrypted bioactive peptides from protein-rich waste. *Int. J Mol. Sci., 17*(6), 950.
14. Baiano, A., (2014). Recovery of biomolecules from food wastes—A review. *Molecules, 19*(9), 14821–14842.
15. Renard, C. M., (2018). Extraction of bioactives from fruit and vegetables: State of the art and perspectives. *LWT, 93*, 390–395.
16. Ayala-Zavala, J. F., Rosas-Domínguez, C., Vega-Vega, V., & González-Aguilar, G. A., (2010). Antioxidant enrichment and antimicrobial protection of fresh-cut fruits using their own byproducts: Looking for integral exploitation. *J Food Sci., 75*(8), R175–R181.

17. Pandey, K. B., & Rizvi, S., (2009). Plant polyphenols as dietary antioxidants in human health and disease. *Oxid. Med. Cell. Long.*, *2*(5), 270–278.
18. Ran, X. L., Zhang, M., Wang, Y., & Adhikari, B., (2019). Novel technologies applied for recovery and value addition of high value compounds from plant byproducts: A review. *Crit. Rev. Food Sci. Nutrition*, *59*(3), 450–461.
19. Sharif, M. K., & Khalid, R., (2018). Nutraceuticals: Myths versus realities. In: *Therapeutic Foods* (pp. 3–21). Academic Press.
20. Ignat, I., Volf, I., & Popa, V., (2011). A critical review of methods for characterization of polyphenolic compounds in fruits and vegetables. *Food Chem.*, *126*(4), 1821–1835.
21. Puri, B., & Hall, A., (1998). *Phytochemical Dictionary: A Handbook of Bioactive Compounds from Plants*. CRC press.
22. Gorinstein, S., Martin-Belloso, O., Lojek, A., Číž, M., Soliva-Fortuny, R., Park, Y. S., Caspi, A., et al., (2002). Comparative content of some phytochemicals in Spanish apples, peaches and pears. *J. Sci. Food Agri.*, *82*(10), 1166–1170.
23. Dhingra, D., Michael, M., Rajput, H., & Patil, R., (2012). Dietary fiber in foods: A review. *J. Food Sci. Technol.*, *49*(3), 255–266.
24. Fuentes-Alventosa, J. M., Rodríguez-Gutiérrez, G., Jaramillo-Carmona, S., Espejo-Calvo, J. A., Rodríguez-Arcos, R., Fernández-Bolaños, J., Guillén-Bejarano, R., & Jiménez-Araujo, A., (2009). Effect of extraction method on chemical composition and functional characteristics of high dietary fiber powder obtained from asparagus by-products. *Food Chem.*, *113*, 665–671.
25. Viuda-Martos, M., Ruiz-Navajas, Y., Martin-Sánchez, A., Sánchez-Zapata, E., Fernández-López, J., Sendra, E., Sayas-Barberá, E., et al., (2012). Chemical, physico-chemical and functional properties of pomegranate (*Punica granatum L.*) bagasses powder co-product. *J. Food Eng.*, *110*, 220–224.
26. Domingo, C. S., Soria, M., Rojas, A. M., Fissore, E. N., & Gerschenson, L., (2015). Protease and hemicellulase assisted extraction of dietary fiber from wastes of *Cynara cardunculus*. *Int. J Mol. Sci.*, *16*(3), 6057–6075.
27. Khatib, M., Giuliani, C., Rossi, F., Adessi, A., Al-Tamimi, A., Mazzola, G., Di Gioia, D., et al., (2017). Polysaccharides from by-products of the wonderful and Laffan pomegranate varieties: New insight into extraction and characterization. *Food Chem.*, *235*, 58–66.
28. Hussain, S., Jõudu, I., & Bhat, R., (2020). Dietary fiber from underutilized plant resources—A positive approach for valorization of fruit and vegetable wastes. *Sustainability*, *12*(13), 5401.
29. Emaga, T. H., Andrianaivo, R. H., Wathelet, B., Tchango, J. T., & Paquot, M., (2007). Effects of the stage of maturation and varieties on the chemical composition of banana and plantain peels. *Food Chem.*, *103*(2), 590–600.
30. Shao, D., Venkitasamy, C., Li, X., Pan, Z., Shi, J., Wang, B., & McHugh, T., (2015). Thermal and storage characteristics of tomato seed oil. *LWT-Food Sci. Technol.*, *63*(1), 191–197.
31. Malacrida, C. R., & Jorge, N., (2012). Yellow passion fruit seed oil (*Passiflora edulis* f. flavicarpa): Physical and chemical characteristics. *Brazilian Arch. Biol. Technol.*, *55*(1), 127–134.
32. Lang, G., & Buchbauer, G., (2012). A review on recent research results (2008–2010) on essential oils as antimicrobials and antifungals: A review. *Flav. Frag. J.*, *27*(1), 13–39.

33. Garavaglia, J., Markoski, M. M., Oliveira, A., & Marcadenti, A., (2016). Grape seed oil compounds: Biological and chemical actions for health. *Nutr. Metabol. Insights*, *9*, NMI-S32910.
34. Maxwell, E. G., Belshaw, N. J., Waldron, K. W., & Morris, V., (2012). Pectin–an emerging new bioactive food polysaccharide. *Trends Food Sci. Technol.*, *24*(2), 64–73.
35. Koubala, B. B., Kansci, G., Garnier, C., Mbome, I. L., Durand, S., Thibault, J. F., & Ralet, M., (2009). Rheological and high gelling properties of mango (*Mangifera indica*) and ambarella (*Spondias cytherea*) peel pectins. *Int. J. Food Sci. Technol.*, *44*, 1809–1817.
36. Srivastava, P., & Malviya, R., (2011). Sources of pectin, extraction and its applications in pharmaceutical industry: An overview. *Indian J. Nat. Prod. Res.*, *2*, 10–18.
37. Geerkens, C. H., Nagel, A., Just, K. M., Miller-Rostek, P., Kammerer, D. R., Schweiggert, R. M., & Carle, R., (2015). Mango pectin quality as influenced by cultivar, ripeness, peel particle size, blanching, drying, and irradiation. *Food Hydrocol.*, *51*, 241–251.
38. Satari, B., & Karimi, K., (2018). Citrus processing wastes: Environmental impacts, recent advances, and future perspectives in total valorization. *Res, Conserv. Recycle*, *129*, 153–167.
39. Chaturvedi, K., Sharma, N., & Yadav, S., (2019). Composite edible coatings from commercial pectin, corn flour and beetroot powder minimize post-harvest decay, reduces ripening and improves sensory liking of tomatoes. *Int. J. Biol. Macromol.*, *133*, 284–293.
40. Antunes, S., Freitas, F., Sevrin, C., Grandfils, C., & Reis, M., (2017). Production of FucoPol by *Enterobacter* A47 using waste tomato paste by-product as sole carbon source. *Biores. Technol.*, *227*, 66–73.
41. Mahato, N., Sharma, K., Sinha, M., Dhyani, A., Pathak, B., Jang, H., Park, S., et al., (2021). Biotransformation of citrus waste-I: Production of biofuel and valuable compounds by fermentation. *Processes*, *9*(2), 220.
42. Panwar, D., Saini, A., Panesar, P. S., & Chopra, H., (2021). Unraveling the scientific perspectives of citrus by-products utilization: Progress towards circular economy, *Trends Food Sci. Technol.*, *111*, 549–562.
43. Panwar, D., Panesar, P. S., & Chopra, H., (2021). Recent trends on the valorization strategies for the management of citrus by-products, *Food Rev. Int.*, *37*(1), 91–120.
44. Schieber, A., Stintzing, F. C., & Carle, R., (2001). By-products of plant food processing as a source of functional compounds. *Trends Food Sci. Tech.*, *12*, 401–413.
45. Mahato, N., Sinha, M., Sharma, K., Koteswararao, R., & Cho, M., (2019). Modern extraction and purification techniques for obtaining high purity food-grade bioactive compounds and value-added co-products from citrus wastes. *Foods (Basel, Switzerland)*, *8*(11), 523.
46. Chavan, P., Singh, A. K., & Kaur, G., (2018). Recent progress in the utilization of industrial waste and by-products of citrus fruits: A review. *J. Food Proc. Eng.*, *41*(8).
47. Jimenez-Colmenero, F., (1996). Technologies for developing low-fat meat products. *Trends Food Sci. Tech.*, *7*, 41–48.
48. Dugo, P., Škeříková, V., Kumm, T., Trozzi, A., Jandera, P., & Mondello, L., (2006). Elucidation of carotenoid patterns in citrus products by means of comprehensive normal-phase reversed-phase liquid chromatography. *Anal. Chem.*, *78*, 7743–7750.

49. Chau, C. F., & Huan, Y., (2003). Comparison of the chemical composition and physicochemical properties of different fibers prepared from the peel of *Citrus sinensis* L. Cv. Liucheng. *J. Agri. Food Chem., 51*(9), 2615–2618.
50. Goulas, V., & Manganaris, G., (2012). Exploring the phytochemical content and the antioxidant potential of *Citrus* fruits grown in Cyprus. *Food Chem., 131*(1), 9–47.
51. Wang, Y. C., Chuang, Y. C., & Hsu, H., (2008). The flavonoid, carotenoid and pectin content in peels of citrus cultivated in Taiwan. *Food Chem., 106*(1), 277–284.
52. De Moraes, B. H. R., De Castro, F. T. A. P., & Genovese, M., (2012). Antioxidant capacity and mineral content of pulp and peel from commercial cultivars of citrus from Brazil. *Food Chem., 134*(4), 1892–1898.
53. FAO, *Banana Facts and Figures.* [Online] Available at: http://www.fao.org/economic/est/estcommodities/bananas/bananafacts/en/#.XS8Id-gzaUk (accessed on 26 September 2023).
54. Pathak, P. D., Mandavgane, S. A., & Kulkarni, B., (2016). Valorization of banana peel: A biorefinery approach. *Rev. Chem. Eng., 32*(6), 651–666.
55. Hernández-Carranza, P., Ávila-Sosa, R., Guerrero-Beltrán, J. A., Navarro-Cruz, A. R., Corona-Jiménez, E., & Ochoa-Velasco, C. E., (2016). Optimization of antioxidant compounds extraction from fruit by-products: Apple pomace, orange and banana peel. *J. Food Proces. Preserv., 40*(1), 103–115.
56. Morais, D. R., Rotta, E. M., Sargi, S. C., Schmidt, E. M., Bonafe, E. G., Eberlin, M. N., Sawaya, A. C., & Visentainer, J., (2015). Antioxidant activity, phenolics and UPLC–ESI (–)–MS of extracts from different tropical fruits parts and processed peels. *Food Res. Int., 77*, 392–399.
57. Rebello, L. P. G., Ramos, A. M., Pertuzatti, P. B., Barcia, M. T., Castillo-Muñoz, N., & Hermosin-Gutierrez, I., (2014). Flour of banana (Musa AAA) peel as a source of antioxidant phenolic compounds. *Food Res. Int., 55*, 397–403.
58. Tsamo, C. V. P., Herent, M. F., Tomekpe, K., Emaga, T. H., Quetin-Leclercq, J., Rogez, H., Larondelle, Y., & Andre, C., (2015). Phenolic profiling in the pulp and peel of nine plantain cultivars (Musa sp.). *Food Chem., 167*, 197–204.
59. Ji, L., & Srzednicki, G., (2013). Extraction of aromatic compounds from banana peels. *Acta Horticult., 1088*, 541–546.
60. Oyeyinka, B. O., & Afolayan, A., (2019). Comparative evaluation of the nutritive, mineral, and antinutritive composition of *Musa sinensis* L. (banana) and *Musa paradisiaca* L. (plantain) fruit compartments. *Plants, 8*(12), 598.
61. Aboul-Enein, A. M., Salama, Z. A., Gaafar, A. A., Aly, H. F., Abou-Elella, F., & Ahmed, H., (2016). Identification of phenolic compounds from banana peel (*Musa paradaisica L.*) as antioxidant and antimicrobial agents. *J. Chem. Pharm. Res., 8*(4), 46–55.
62. Septiany, G. J., Putri, W. D. R., Panca, I. N., Heriyanto, H., & Limantara, L., (2019). Carotenoid analysis of ripe banana flesh and peel from three cultivars of banana. *Indonesian J. Nat. Pig., 1*(2), 60.
63. Statista, (2021). *Global Pineapple Production.* [Online] Available at https://www.statista.com/statistics/298505/global-pineapple-production/ (accessed on 26 September 2023).
64. Selani, M. M., Brazaca, S. G. C., Dos Santos, D. C. T., Ratnayake, W. S., Flores, R. A., & Bianchini, A., (2014). Characterization and potential application of pineapple pomace in an extruded product for fiber enhancement. *Food Chem., 163*, 23–30.

65. Diaz-Vela, J., Totosaus, A., Cruz-Guerrero, A. E., & De Lourdes Pérez-Chabela, M., (2013). In vitro evaluation of the fermentation of added-value agro-industrial by-products: Cactus pear (*Opuntia Ficus-indica* L.) peel and pineapple (*Ananas comosus*) peel as functional ingredients. *Int. J. Food Sci. Technol.*, *48*(7), 1460–1467.
66. Larrauri, J. A., Rupérez, P., & Calixto, F., (1997). Pineapple shell as a source of dietary fiber with associated polyphenols. *J. Agricult. Food Chem.*, *45*(10), 4028–4031.
67. Santos, D. I., Martins, C. F., Amaral, R. A., Brito, L., Saraiva, J. A., Vicente, A. A., & Moldão-Martins, M., (2021). Pineapple (*Ananas comosus* L.) By-products valorization: Novel bio ingredients for functional foods. *Molecules*, *26*(11).
68. Huang, Y. L., Chow, C. J., & Fang, Y., (2011). Preparation and physicochemical properties of fiber -rich fraction from pineapple peels as a potential ingredient. *J. Food Drug. Anal.*, *19*, 318–323.
69. Saleh, M., Amro, L., Barakat, H., Baker, R., Reyash, A. A., Amro, R., & Qasem, J., (2021). Fruit by-product processing and bioactive compounds. *J. Food Quality*.
70. Sogi, D. S., Siddiq, M., Greiby, I., & Dolan, K., (2013). Total phenolics, antioxidant activity, and functional properties of 'Tommy Atkins' mango peel and kernel as affected by drying methods. *Food Chem.*, *141*(3), 2649–2655.
71. Masibo, M., & He, Q., (2008). Major mango polyphenols and their potential significance to human health. *Comprehen. Rev. Food Sci. Food Safety*, *7*(4), 309–319.
72. Jahurul, M. H. A., Zaidul, I. S. M., Ghafoor, K., Al-Juhaimi, F. Y., Nyam, K. L., Norulaini, N. A. N., Sahena, F., & Omar, A., (2015). Mango (*Mangifera indica* L.) by-products and their valuable components: A review. *Food Chem.*, *183*, 173–180.
73. Ashoush, I. S., & Gadallah, M. G., (2011). Utilization of mango peels and seed kernels powders as sources of phytochemicals in biscuit. *World J. Dairy Food Sci.*, *6*(1), 35–42.
74. APEDA, (2020). *Market Intelligence Report for Pomegranates*. [Online] Available at: Agriexchange.apeda.gov.in (accessed on 26 September 2023).
75. Viuda-Martos, M., Fernández-López, J., & Pérez-Álvarez, J. A., (2010). Pomegranate and its many functional components as related to human health: A review. *Compr. Rev. Food Sci. Food Saf.*, *9*, 635–654.
76. Hernandez, F., Melgarejo, P., Tomas-Barberan, F. A., & Artes, F., (1999). Evolution of juice anthocyanins during ripening of new selected pomegranate (*Punica granatum*) clones. *Euro. Food Res. Technol.*, *210*(1), 39–42.
77. De Pascual-Teresa, S., Santos-Buelga, C., & Rivas-Gonzalo, J., (2000). Quantitative analysis of flavan-3-ols in Spanish foodstuffs and beverages. *J. Agri. Food Chem.*, *48*(11), 5331–5337.
78. El-Nemr, S. E., Ismail, I. A., & Ragab, M., (1990). Chemical composition of juice and seed of pomegranate fruit. *Die Nahrung, 34*, 601–606.
79. Abbasi, H., Rezaei, K., Emamdjomeh, Z., & Mousavi, S. M., (2008). Effect of various extraction conditions on the phenolic contents of pomegranate seed oil. *European J. Lipid Sci. Technol.*, *110*, 435–440.
80. Coronado-Reyes, J. A., Cortés-Penagos, C. D. J., & González-Hernández, J., (2021). Chemical composition and great applications to the fruit of the pomegranate (*Punica granatum*): A review. *Food Sci. Technol.*
81. Sharma, K., & Akansha, C. E., (2018). Comparative studies of proximate, mineral and phytochemical compositions of pomegranate (*Punica granatum*) in peel, seed and whole fruit powder. *Int. J. Food Sci. Nutr.*, *3*(2), 192–196.

82. Al-Rawahi, A. S., Edwards, G., Al-Sibani, M., Al-Thani, G., Al-Harrasi, A. S., & Rahman, M., (2014). Phenolic constituents of pomegranate peels (*Punica granatum* L.) cultivated in Oman. *European Journal of Medicinal Plants*, 315–331.
83. Peng, Y., (2019). Comparative analysis of the biological components of pomegranate seed from different cultivars. *Int. J. Food Pro.*, *22*(1), 784–794.
84. FAO, (2020). *World Food and Agriculture-Statistical Yearbook 2020.* Rome. https://doi.org/10.4060/cb1329en.
85. Nour, V., Panaite, T. D., Ropota, M., Turcu, R., Trandafir, I., & Corbu, A., (2018). Nutritional and bioactive compounds in dried tomato processing waste. *CyTA-J Food, 16*(1), 222–229.
86. Fărcaş, A. C., Socaci, S. A., Michiu, D., Biriş, S., & Tofană, M., (2019). Tomato waste as a source of biologically active compounds. *Bull. UASVM Food Sci. Technol, 76*, 85–88.
87. Navarro-González, I., García-Valverde, V., García-Alonso, J., & Periago, M., (2011). Chemical profile, functional and antioxidant properties of tomato peel fiber. *Food Res. Int.*, *44*(5), 1528–1535.
88. Botineştean, C., Gruia, A. T., & Jianu, I., (2015). Utilization of seeds from tomato processing wastes as raw material for oil production. *J. Mat. Cycles Waste Manage*, *17*(1), 118–124.
89. Al-Wandawi, H., Abdul-Rahman, M., & Al-Shaikhly, K., (1985). Tomato processing wastes as essential raw materials source. *J. Agri. Food Chem., 33*(5), 804–807.
90. Elbadrawy, E., & Sello, A., (2016). Evaluation of nutritional value and antioxidant activity of tomato peel extracts. *Arabian J. Chem.*, *9*, S1010–S1018.
91. Fuentes, E., Carle, R., Astudillo, L., Guzmán, L., Gutiérrez, M., Carrasco, G., & Palomo, I., (2013). Antioxidant and antiplatelet activities in extracts from green and fully ripe tomato fruits (*Solanum lycopersicum*) and pomace from industrial tomato processing. *Evidence-Based Complement. Altern. Med.*
92. Silva, Y. P. A., Borba, B. C., Pereira, V. A., Reis, M. G., Caliari, M., Brooks, M. S. L., & Ferreira, T., (2019). Characterization of tomato processing by-product for use as a potential functional food ingredient: Nutritional composition, antioxidant activity and bioactive compounds. *Int. J. Food Sci. Nutr.*
93. Valdez-Morales, M., Espinosa-Alonso, L. G., Espinoza-Torres, L. C., Delgado-Vargas, F., & Medina-Godoy, S., (2014). Phenolic content and antioxidant and antimutagenic activities in tomato peel, seeds, and by products. *J Agric Food Chem.*, *62*(23), 5281–5289.
94. Al-Weshahy, A., & Rao, V., (2012). *Potato Peel as a Source of Important Phytochemical Antioxidant Nutraceuticals and Their Role in Human Health: A Review*. IntechOpen.
95. Wu, T., Yan, J., Liu, R., Marcone, M. F., Aisa, H. A., & Tsao, R., (2012). Optimization of microwave-assisted extraction of phenolics from potato and its downstream waste using orthogonal array design. *Food Chem.*, *133*(4), 1292–1298.
96. Maldonado, A. F. S., Mudge, E., Gänzle, M. G., & Schieber, A., (2014). Extraction and fractionation of phenolic acids and glycoalkaloids from potato peels using acidified water/ethanol-based solvents. *Food Res. Int.*, *65*, 27–34.
97. Zhang, M., & Mu, T., (2017). Identification and characterization of antioxidant peptides from sweet potato protein hydrolysates by Alcalase under high hydrostatic pressure. *Innovative Food Sci. Emerg. Technol*, *43*, 92–101.

98. Javed, A., Ahmad, A., Tahir, A., Shabbir, U., Nouman, M., & Hameed, A., (2019). Potato peel waste-its nutraceutical, industrial and biotechnological applications. *AIMS Agri. Food, 4*(3), 807–823.
99. Helal, M. M., Adawy, T. E., Beltagy, A. E., Bedawey, A. E., & Youssef, S., (2020). Evaluation of potato peel extract as a source of antioxidant and antimicrobial substances. *Menoufia J. Food Dairy Sci.*, *5*(6), 79–90.
100. Statista, (2021). *Vegetable Production Worldwide by Type*. [Online] Available at: https://www.statista.com/statistics/264065/global-production-of-vegetables-by-type/ (accessed on 26 September 2023).
101. Oberoi, H. S., Kalra, K. L., Uppal, D. S., & Tyagi, S., (2007). Effects of different drying methods of cauliflower waste on drying time, color retention and glucoamylase production by *Aspergillus Niger* NCIM 1054. *Int. J. Food Sci. Technol.*, *42*(2), 228–234.
102. Podsędek, A., (2007). Natural antioxidants and antioxidant capacity of brassica vegetables: A review. *LWT-Food Sci. Technol.*, *40*(1), 1–11.
103. Wani, T. A., Sood, M., & Kaul, R., (2011). Nutritional and sensory properties of roasted wheat noodles supplemented with cauliflower leaf powder. Annals. *Food Sci. Technol.*, *12*(2), 102–107.
104. Llorach, R., Espín, J. C., Tomás-Barberán, F. A., & Ferreres, F., (2003). Valorization of cauliflower (*Brassica oleracea* L. var. botrytis) by-products as a source of antioxidant phenolics. *J. Agri. Food Chem.*, *51*(8), 2181–2187.
105. Fimognari, C., Turrini, E., Ferruzzi, L., Lenzi, M., & Hrelia, P., (2012). Natural isothiocyanates: Genotoxic potential versus chemoprevention. *Muta. Res. Rev. Muta. Res.*, *750*(2), 107–131.
106. Khedkar, M. A., Nimbalkar, P. R., Chavan, P. V., Chendake, Y. J., & Bankar, S., (2017). Cauliflower waste utilization for sustainable biobutanol production: Revelation of drying kinetics and bioprocess development. *Biop. Biosys. Eng.*, *40*(10), 1493–1506.
107. Munir, A., Sultana, B., Bashir, A., Ghaffar, A., Munir, B., Shar, G. A., Nazir, A., & Iqbal, M., (2018). Evaluation of antioxidant potential of vegetables waste. *Polish J. Environ. Studies*, *27*(2).
108. Sharma, K. D., Karki, S., Thakur, N. S., & Attri, S., (2012). Chemical composition, functional properties, and processing of carrot—A review. *J. Food Sci. Technol.*, *49*(1), 22–32.
109. Chau, C. F., Chen, C. H., & Lee, M. H., (2004). Comparison of the characteristics, functional properties, and in vitro hypoglycemic effects of various carrot insoluble fiber-rich fractions. *LWT-Food Sci. Technol.*, *37*(2), 155–160.
110. Clementz, A., Torresi, P. A., Molli, J. S., Cardell, D., Mammarella, E., & Yori, J. C., (2019). Novel method for valorization of by-products from carrot discards. *LWT*, *100*, 374–380.
111. John, S., Priyadarshini, S., Monica, S. J., & Arumugam, P., (2017). Phytochemical profile and thin layer chromatographic studies of *Daucus carota* peel extracts. *Int. J. Food Sci. Nutr*, *2*, 23–26.
112. Lau, W. K., Van, C. H., & Vuong, Q., (2018). Physical properties, carotenoids and antioxidant capacity of carrot (*Daucus carota* L.) peel as influenced by different drying treatments. *Int. J. Food Eng.*, *14*(3).
113. FAO, (2019). *World Food and Agriculture-Statistical Yearbook 2020*. Rome. https://doi.org/10.4060/cb1329en (accessed on 26 September 2023).

114. Downes, K., Chope, G. A., & Terry, L., (2009). Effect of curing at different temperatures on biochemical composition of onion (*Allium cepa* L.) skin from three freshly cured and cold stored UK-grown onion cultivars. *Post. Biol. Technol.*, *54*(2), 80–86.
115. Bonaccorsi, P., Caristi, C., Gargiulli, C., & Leuzzi, U., (2008). Flavonol glucosides in Allium species: A comparative study by means of HPLC–DAD–ESI-MS–MS. *Food Chem.*, *107*(4), 1668–1673.
116. Randle, W. M., Lancaster, J. E., Shaw, M. L., Sutton, K. H., Hay, R. L., & Bussard, M., (1995). Quantifying onion flavor compounds responding to sulfur fertility-sulfur increases levels of alk(en)yl cysteine sulfoxides and biosynthetic intermediates. *J. American Society Horti. Sci.*, *120*(6), 1075–1081.
117. Benítez, V., Mollá, E., Martín-Cabrejas, M. A., Aguilera, Y., López-Andréu, F. J., Cools, K., Terry, L. A., & Esteban, R., (2011). Characterization of industrial onion wastes (*Allium cepa L.*) dietary fibre and bioactive compounds. *Plant Foods Human Nutr.*, *66*(1), 48–57.
118. Bello, M. O., Olabanji, I. O., Abdul-Hammed, M., & Okunade, T., (2013). Characterization of domestic onion wastes and bulb (*Allium cepa L.*) fatty acids and metal contents. *Int. Food Res. J.*, *20*(5), 2153.
119. Boelhouwer, C., (1983). Trends in chemistry and technology of lipids. *J. American Oil Chem. Society*, *60*(2), 457–462.
120. Ifesan, B. O. T., (2017). Chemical composition of onion peel (*Allium cepa*) and its ability to serve as a preservative in cooked beef. *Int. J. Sci. Res. Methodol.*, *7*(4).
121. Khoddami, A., Wilkes, M. A., & Roberts, T., (2013). Techniques for analysis of plant phenolic compounds. *Molecules*, *18*(2), 2328–2375.
122. Vyas, M., (2019). Physicochemical analysis of leaves of *Eriobotrya japonica* and antioxidant and antidiabetic evaluation of its methanolic extract. *Int. J. Green Pharm. (IJGP)*, *13*(3).
123. Dailey, A., & Vuong, Q., (2015). Effect of extraction solvents on recovery of bioactive compounds and antioxidant properties from macadamia (*Macadamia tetraphylla*) skin waste. *Cogent Food Agri.*, *1*(1), 1115646.
124. Aberoumand, A., (2011). A review article on edible pigments properties and sources as natural biocolorants in foodstuff and food industry. *WJDFS*, *6*(1), 71–78.
125. Chuyen, H. V., Roach, P. D., Golding, J. B., Parks, S. E., & Nguyen, M., (2017). Optimization of extraction conditions for recovering carotenoids and antioxidant capacity from Gac peel using response surface methodology. *Int. J. Food Sci. Technol.*, *52*(4), 972–980.
126. Poojary, M. M., & Passamonti, P., (2015). Optimization of extraction of high purity all-trans-lycopene from tomato pulp waste. *Food Chem.*, *188*, 84–91.
127. Boulekbache-Makhlouf, L., Slimani, S., & Madani, K., (2013). Total phenolic content, antioxidant, and antibacterial activities of fruits of *Eucalyptus globulus* cultivated in Algeria. *Indus. Crops Prod.*, *41*, 85–89.
128. Kumar, K., Yadav, A. N., Kumar, V., Vyas, P., & Dhaliwal, H., (2017). Food waste: A potential bioresource for extraction of nutraceuticals and bioactive compounds. *Biores. Biop.*, *4*(1), 1–14.
129. Vankar, P. S., (2004). Essential oils and fragrances from natural sources. *Resonance*, *9*(4), 30–41.

130. Azmir, J., Zaidul, I. S. M., Rahman, M. M., Sharif, K. M., Mohamed, A., Sahena, F., & Omar, A. K., (2013). Techniques for extraction of bioactive compounds from plant materials: A review. *J. Food Eng.*, *117*(4), 426–436.
131. Caldas, T. W., Mazza, K. E., Teles, A. S., Mattos, G. N., Brígida, A. I. S., Conte-Junior, C. A., & Tonon, R., (2018). Phenolic compounds recovery from grape skin using conventional and non-conventional extraction methods. *Indus. Crops Prod.*, *111*, 86–91.
132. Kodal, S. Ü. H., & Aksu, Z. Ü. M., (2001). Optimization of carotene pigment production by Soxhlet extraction from waste orange peels. *Food Chem.*, *72*(2), 145–171.
133. Heleno, S. A., Diz, P., Prieto, M. A., Barros, L., Rodrigues, A., Barreiro, M. F., & Ferreira, I., (2016). Optimization of ultrasound-assisted extraction to obtain mycosterols from *Agaricus bisporus* L. by response surface methodology and comparison with conventional Soxhlet extraction. *Food Chem.*, *197*, 1054–1063.
134. Gorinstein, S., Martín-Belloso, O., Park, Y. S., Haruenkit, R., Lojek, A., Ĉíž, M., & Trakhtenberg, S., (2001). Comparison of some biochemical characteristics of different citrus fruits. *Food Chem.*, *74*(3), 309–315.
135. Chemat, F., Vian, M. A., Fabiano-Tixier, A. S., Nutrizio, M., Jambrak, A. R., Munekata, P. E., & Cravotto, G., (2020). A review of sustainable and intensified techniques for extraction of food and natural products. *Green Chem.*, *22*(8), 2325–2353.
136. Barba, F. J., Galanakis, C. M., Esteve, M. J., Frigola, A., & Vorobiev, E., (2015). Potential use of pulsed electric technologies and ultrasounds to improve the recovery of high-added value compounds from blackberries. *J. Food Eng.*, *167*, 38–44.
137. Chemat, F., Tomao, V., & Virot, M., (2008). Ultrasound-assisted extraction in food analysis. *Handbook Food Anal. Instr.*, 85–103.
138. Le, H. V., & Le, V. V., (2012). Comparison of enzyme-assisted and ultrasound-assisted extraction of vitamin C and phenolic compounds from acerola (*Malpighia emarginata* DC.) fruit. *Int. J. Food Sci. Technol.*, *47*(6), 1206–1214.
139. Hosseini, S. S., Khodaiyan, F., Kazemi, M., & Najari, Z., (2019). Optimization and characterization of pectin extracted from sour orange peel by ultrasound assisted method. *Int. J. Biol. Macromol.*, *125*, 621–629.
140. Guandalini, B. B. V., Rodrigues, N. P., & Marczak, L. D., (2019). Sequential extraction of phenolics and pectin from mango peel assisted by ultrasound. *Food Res. Int.*, *119*, 455–461.
141. Maran, J. P., Priya, B., Al-Dhabi, N. A., Ponmurugan, K., Moorthy, I. G., & Sivarajasekar, N., (2017). Ultrasound assisted citric acid mediated pectin extraction from industrial waste of *Musa balbisiana, Ultrason. Sonochem.*, *35*, 204–209.
142. Moorthy, I. G., Maran, J. P., Ilakya, S., Anitha, S. L., Sabarima, S. P., & Priya, B., (2017). Ultrasound assisted extraction of pectin from waste *Artocarpus heterophyllus* fruit peel, *Ultrason. Sonochem.*, *34*, 525–530.
143. Kazemi, M., Khodaiyan, F., & Hosseini, S., (2019). Eggplant peel as a high potential source of high methylated pectin: Ultrasonic extraction optimization and characterization. *LWT*, *105*, 182–189.
144. Grassino, A. N., Brnčić, M., Vikić-Topić, D., Roca, S., Dent, M., & Brnčić, S., (2016). Ultrasound assisted extraction and characterization of pectin from tomato waste. *Food Chem.*, *198*, 93–100.
145. Goula, A. M., Ververi, M., Adamopoulou, A., et al., (2017). Green ultrasound-assisted extraction of carotenoids from pomegranate wastes using vegetable oils. *Ultrason Sonochem.*, *34*, 821–830.

146. Sharma, M., Usmani, Z., Gupta, V. K., & Bhat, R., (2021). Valorization of fruits and vegetable wastes and by-products to produce natural pigments. *Crit. Rev. Biotechnol.*, *41*(4), 535–563.
147. Garcia-Mendoza, M. P., Paula, J. T., Paviani, L. C., Cabral, F. A., & Martinez-Correa, H., (2015). Extracts from mango peel by-product obtained by supercritical CO_2 and pressurized solvent processes. *LWT-Food Sci. Technol.*, *62*(1), 131–137.
148. Ferarsa, S., Zhang, W., Moulai-Mostefa, N., Ding, L., Jaffrin, M. Y., & Grimi, N., (2018). Recovery of anthocyanins and other phenolic compounds from purple eggplant peels and pulps using ultrasonic assisted extraction. *Food Bioprod. Process*, *109*, 19–28.
149. Ruiz-Montanez, G., Ragazzo-Sanchez, J. A., Calderon-Santoyo, M., Velazquez-De La, C. G., De Leon, J. A., & Navarro-Ocana, A., (2014). Evaluation of extraction methods for preparative scale obtention of mangiferin and lupeol from mango peels (*Mangifera indica* L.). *Food Chem.*, *159*, 267–272.
150. Ma, Y., Ye, X., Wu, H., Wang, H., Sun, Z., Zhu, P., & Han, Z., (2015). Evaluation of the effect of ultrasonic variables at locally ultrasonic field on yield of hesperidin from penggan (*Citrus reticulata*) peels. *LWT*, *60*(2), 1088–1094.
151. Fierascu, R. C., Sieniawska, E., Ortan, A., Fierascu, I., & Xiao, J., (2020). Fruits by-products–A source of valuable active principles. A short review. *Front. Bioeng. Biotechnol.*, *8*, 319.
152. Chen, Q., Hu, Z., Yao, F. Y. D., & Liang, H., (2016). Study of two-stage microwave extraction of essential oil and pectin from pomelo peels. *LWT-Food Sci. Technol.*, *66*, 538–545.
153. Rafiee, Z., Jafari, S. M., Alami, M., & Khomeiri, M., (2011). Microwave-assisted extraction of phenolic compounds from olive leaves; a comparison with maceration. *J. Anim. Plant Sci.*, *21*(4), 738–745.
154. Chandrasekar, V., Martín-González, M. S., Hirst, P., & Ballard, T., (2015). Optimizing microwave-assisted extraction of phenolic antioxidants from red delicious and Jonathan apple pomace. *J. Food Proces. Eng.*, *38*(6), 571–582.
155. Cristina-Gabriela, G., Emilie, D., Gabriel, L., & Claire, E., (2012). Bioactive compounds extraction from pomace of four apple varieties. *J. Eng. Stud. Res.*, *18*(1), 96.
156. Ferreira, S. S., Passos, C. P., Cardoso, S. M., Wessel, D. F., & Coimbra, M., (2018). Microwave assisted dehydration of broccoli by-products and simultaneous extraction of bioactive compounds. *Food Chem.*, *246*, 386–393.
157. Rajha, H. N., Boussetta, N., Louka, N., Maroun, R. G., & Vorobiev, E., (2015). Effect of alternative physical pretreatments (pulsed electric field, high voltage electrical discharges and ultrasound) on the dead-end ultrafiltration of vine-shoot extracts. *Sep. Purific. Technol.*, *146*, 243–251.
158. Luengo, E., Álvarez, I., & Raso, J., (2013). Improving the pressing extraction of polyphenols of orange peel by pulsed electric fields. *Innov. Food Sci. Emerg. Technol.*, *17*, 79–84.
159. Bobinaitė, R., Pataro, G., Visockis, M., Bobinas, Č., Ferrari, G., & Viškelis, P., (2017). Potential application of pulsed electric fields to improve the recovery of bioactive compounds from sour cherries and their by-products. In: *11th Baltic Conf. on Food Science and Technology "Food Science and Technology in A Changing World" FOODBALT* (pp. 70–74).
160. Pataro, G., Bobinaitė, R., Bobinas, Č., Šatkauskas, S., Raudonis, R., Visockis, M., & Viškelis, P., (2017). Improving the extraction of juice and anthocyanins from blueberry

fruits and their by-products by application of pulsed electric fields. *Food Biop. Technol.*, *10*(9), 1595–1605.

161. Khalil, J., (2011). *Pulsed Electric Field (PEF) and Pectinase for the Extraction of Polyphenols from Grape Pomace and Peel.*
162. Pataro, G., Carullo, D., Falcone, M., & Ferrari, G., (2020). Recovery of lycopene from industrially derived tomato processing by-products by pulsed electric fields-assisted extraction. *Innov. Food Sci. Emerg. Technol.*, *63*, 102369.
163. Panja, P., (2018). Green extraction methods of food polyphenols from vegetable materials. *Cur. Opinion Food Sci.*, *23*, 173–182.
164. Manna, L., Bugnone, C. A., & Banchero, M., (2015). Valorization of hazelnut, coffee and grape wastes through supercritical fluid extraction of triglycerides and polyphenols. *J. Supercrit. Fluids*, *104*, 204–211.
165. Quispe-Fuentes, I., Uribe, E., López, J., Contreras, D., & Poblete, J., (2021). A study of dried mandarin (*Clementina orogrande*) peel applying supercritical carbon dioxide using co-solvent: Influence on oil extraction, phenolic compounds and antioxidant activity. *J. Food Proces. Preserv.*, e16116.
166. Mihalcea, L., Crăciunescu, O., Prelipcean, A. M., Enachi, E., Barbu, V., Bahrim, G. E., & Stănciuc, N., (2021). Supercritical CO_2 extraction and microencapsulation of lycopene-enriched oleoresins from tomato peels: Evid. Antiprolif. Cytocomp. Act. *Antioxid.*, *10*(2), 222.
167. Radenkovs, V., Juhnevica-Radenkova, K., Górnaś, P., & Seglina, D., (2018). Non-waste technology through the enzymatic hydrolysis of agro-industrial by-products. *Trends Food Sci. Technol.*, *77*, 64–76.
168. Meini, M. R., Cabezudo, I., Boschetti, C. E., & Romanini, D., (2019). Recovery of phenolic antioxidants from Syrah grape pomace through the optimization of an enzymatic extraction process. *Food Chem.*, *283*, 257–264.
169. Landbo, A. K., & Meyer, A., (2001). Enzyme-assisted extraction of antioxidative phenols from black currant juice press residues (Ribes nigrum). *J. Agri. Food Chem.*, *49*(7), 3169–3177.
170. Gómez-García, R., Martínez-Ávila, G. C., & Aguilar, C., (2012). Enzyme-assisted extraction of antioxidative phenolics from grape (*Vitis vinifera* L.) residues. *Biotech.*, *2*(4), 297–300.
171. Dehghan-Shoar, Z., Hardacre, A. K., Meerdink, G., & Brennan, C., (2011). Lycopene extraction from extruded products containing tomato skin. *Int. J. Food Sci. Technol.*, *46*(2), 365–371.
172. Morais, E. S., Lopes, A. M., Freire, M. G., Freire, C. S., Coutinho, J. A., & Silvestre, A., (2020). Use of ionic liquids and deep eutectic solvents in polysaccharides dissolution and extraction processes towards sustainable biomass valorization. Molecules, *25*(16), 3652.
173. Husanu, E., Mero, A., Rivera, J. G., Mezzetta, A., Ruiz, J. C., D'Andrea, F., & Guazzelli, L., (2020). Exploiting deep eutectic solvents and ionic liquids for the valorization of chestnut shell waste. *ACS Sust. Chem. Eng.*, *8*(50), 18386–18399.
174. Hernández-Aguirre, O. A., Muro, C., Hernández-Acosta, E., Alvarado, Y., & Díaz-Nava, M. D., (2021). Extraction and stabilization of betalains from beetroot (*Beta vulgaris*) wastes using deep eutectic solvents. *Molecules, 26*(21), 6342.

175. Socas-Rodríguez, B., Torres-Cornejo, M. V., Álvarez-Rivera, G., & Mendiola, J., (2021). Deep eutectic solvents for the extraction of bioactive compounds from natural sources and agricultural by-products. *App. Sci., 11*(11), 4897.
176. Segovia, F. J., Corral-Pérez, J. J., & Almajano, M., (2016). Avocado seed: Modeling extraction of bioactive compounds. *Indus. Crops Prod., 85*, 213–220.
177. Xu, S. Y., Liu, J. P., Huang, X., Du, L. P., Shi, F. L., Dong, R., & Cheong, K., (2018). Ultrasonic-microwave assisted extraction, characterization, and biological activity of pectin from jackfruit peel. *LWT, 90*, 577–582.
178. Kaur, B., Panesar, P. S., & Anal, A., (2021). Standardization of ultrasound assisted extraction for the recovery of phenolic compounds from mango peels. *J. Food Sci. Technol.,* 1–8.
179. Chuyen, H. V., Nguyen, M. H., Roach, P. D., Golding, J. B., & Parks, S., (2018). Microwave-assisted extraction and ultrasound-assisted extraction for recovering carotenoids from Gac peel and their effects on antioxidant capacity of the extracts. *Food Sci. Nutr., 6*(1), 189–196.
180. Doulabi, M., Golmakani, M. T., & Ansari, S., (2020). Evaluation and optimization of microwave-assisted extraction of bioactive compounds from eggplant peel by-product. *J. Food Proces. Preserv., 44*(11), e14853.
181. Chimshirova, R., Karsheva, M., Diankov, S., & Hinkov, I., (2021). Conventional ultrasound and microwave-assisted extraction of phenolic compounds from mavrud grape, pomace and seeds. *J. Chem. Technol. Metal., 56*(1).
182. Boussetta, N., Vorobiev, E., Le, L. H., Cordin-Falcimaigne, A., & Lanoisellé, J., (2012). Application of electrical treatments in alcoholic solvent for polyphenols extraction from grape seeds. *LWT-Food Sci. Technol., 46*(1), 127–134.
183. Parniakov, O., Barba, F. J., Grimi, N., Lebovka, N., & Vorobiev, E., (2016). Extraction assisted by pulsed electric energy as a potential tool for green and sustainable recovery of nutritionally valuable compounds from mango peels. *Food Chem., 192*, 842–848.
184. Ferrentino, G., Morozova, K., Mosibo, O. K., Ramezani, M., & Scampicchio, M., (2018). Biorecovery of antioxidants from apple pomace by supercritical fluid extraction. *J. Clean. Prod., 186*, 253–261.
185. Yuliarti, O., Goh, K. K., Matia-Merino, L., Mawson, J., & Brennan, C., (2015). Extraction and characterization of pomace pectin from gold kiwifruit (*Actinidia chinensis*). *Food Chem., 187*, 290–296.
186. Kaur, B., Panesar, P. S., & Thakur, A., (2021). Extraction and evaluation of structural and physicochemical properties of dietary fiber concentrate from mango peels by using green approach. *Biom. Conv. Bioref.*, 1–10.
187. Ozturk, B., Parkinson, C., & Gonzalez-Miquel, M., (2018). Extraction of polyphenolic antioxidants from orange peel waste using deep eutectic solvents. *Sep. Puri. Technol., 206*, 1–13.
188. Procentese, A., Raganati, F., Olivieri, G., Russo, M. E., Rehmann, L., & Marzocchella, A., (2018). Deep eutectic solvents pretreatment of agro-industrial food waste. *Biotechnol. Biof., 11*(1), 1–12.
189. Bharti, S. K., & Roy, R., (2012). Quantitative 1H NMR spectroscopy. *TrAC Trends Anal. Chem., 35*, 5–26.
190. Spatafora, C., & Tringali, C., (2012). Valorization of vegetable waste: Identification of bioactive compounds and their chemo-enzymatic optimization. *Open Agri. J., 6*(1).

191. Bhushan, S., Kalia, K., & Sharma, M., (2008). Processing of apple pomace for bioactive molecules. *Crit. Rev. Biotechnol.*, *28*(4), 285–296.
192. Grassino, A. N., Ostojić, J., & Miletić, V., (2020). Application of high hydrostatic pressure and ultrasound-assisted extractions as a novel approach for pectin and polyphenols recovery from tomato peel waste. *Innov. Food Sci. Emerg. Technol.*, *64*, 102424.
193. Dion, M. A. M., & Saskia, M., (2008). An overview of analytical methods for determining the geographical origin of food products. *Food Chem.*, *107*, 897–911.
194. Santos, D. I., Neiva, C. M. J., & Mateus, M., (2019). Fourier transform infrared (FT-IR) spectroscopy as a possible rapid tool to evaluate abiotic stress effects on pineapple by-products. *App. Sci.*, *9*(19), 4141.
195. Ismail, E. H., Khalil, M. M. H., Al Seif, F. A., & El-Magdoub, F., (2014). Biosynthesis of gold nanoparticles using extract of grape (*Vitis vinifera*) leaves and seeds. *Prog. Nanotechnol. Nanomater.*, *3*, 1–12.
196. Cheok, C. Y., Mohd, A. N., Abdul, R. R., Zainal, A. N. H., Hussain, N., Sulaiman, R., & Chong, G. H., (2018). Current trends of tropical fruit waste utilization. *Crit. Rev. Food Sci. Nutr.*, *58*(3), 335–361.
197. Ribeiro Da, S. L. M., Teixeira De, F. E. A. T., & Silva, R. N. M., (2014). Quantification of bioactive compounds in pulps and by-products of tropical fruits from Brazil. *Food Chem.*, *143*, 398–404.
198. Subramanian, S., & Ramakrishnan, N., (2011). Chromatographic fingerprint analysis of *Naringi crenulata* by HPTLC technique. *Asian Pac. J. Trop. Biomed.*, *1*, S195–S198.
199. Muthulakshmi, A., Joshibhi-Margret, R., & Mohan, V., (2012). GC-MS analysis of bioactive components of *Feronia elephantum* correa (Rutaceae). *J. Appl. Pharm Sci, 2*, 69–74.
200. Yamunadevi, M., Wesely, E. G., & Johnson, M., (2011). Chromatographic finger print analysis of steroids in *Aerva Lanasa* L. by HPTLC technique. *Asian Pac. J. Trop. Biomed.*, *1*(6), 428–433.
201. Chikezie, P. C., Ibegbulem, C. O., & Mbagwu, F., (2015). Bioactive principles from medicinal plants. *Res. J. Phytochem.*, *9*, 88–115.
202. Fadeyi, S. A., Fadeyi, O. O., Adejumo, A. A., Okoro, C., & Myles, E., (2013). In vitro anticancer screening of 24 locally used Nigerian medicinal plants. *BMC Compl. Altern. Med.*, *13*, 79.
203. Cui, L., Lu, H., & Lee, Y., (2018). Challenges and emergent solutions for LC-MS/MS based untargeted metabolomics in diseases. *Mass Spectrom. Rev.*, *37*(6), 772–792.
204. Lever, T., Haines, P., & Rouquerol, J., (2014). ICTAC nomenclature of thermal analysis (IUPAC Recommendations 2014). *Pure App. Chem.*, *86*(4), 545–553.
205. Ayala-Zavala, J. F., & González-Aguilar, G., (2011). *Use of Additives to Preserve the Quality of Fresh-Cut Fruits and Vegetables* (pp. 231–254). CRC Press, Boca Raton.
206. Manzano, S., & Williamson, G., (2010). Polyphenols and phenolic acids from strawberry and apple decrease glucose uptake and transport by human intestinal Caco-2 cells. *Mol. Nutr. Food Res.*, *54*(12), 1773–1780.
207. Pudakalkatti, P. S., & Shivanaikar, S., (2015). Detection of antimicrobial activity of banana peel (*Musa paradisiaca* L.) on *Porphyromonas gingivalis* and *Aggregatibacter actinomycetemcomitans*: An *in vitro* study. *Contemp. Clin. Dent.*, *6*(4), 496–499.
208. Vu, H. T., Scarlett, C. J., & Vuong, Q. V., (2018). Phenolic compounds within banana peel and their potential uses: A review. *J. Funct. Foods, 40*, 238–248.

209. Burton-Freeman, B. M., Sandhu, A. K., & Edirisinghe, I., (2017). Mangos and their bioactive components: Adding variety to the fruit plate for health. *Food Func.*, *8*(9), 3010–3032.
210. Chhikara, N., Kushwaha, K., Sharma, P., Gat, Y., & Panghal, A., (2019). Bioactive compounds of beetroot and utilization in food processing industry: A critical review. *Food Chem.*, *272*, 192–200.
211. Ko, E. Y., Sharma, K., & Nile, S. H., (2016). Effect of harvesting practices, lifting time, curing methods, and irrigation on quercetin content in onion (*Allium cepa* L.) cultivars. *Emirates J. Sci. Food Agric, 28*(9), 594–600.
212. Curti, V., Di Lorenzo, A., Dacrema, M., Xiao, J., Nabavi, S. M., & Daglia, M., (2017). In vitro polyphenol effects on apoptosis: An update of literature data. *Seminars Cancer Biol.*, 30209–30212.
213. Nile, S. H., Nile, A. S., Keum, Y. S., & Sharma, K., (2017). Utilization of quercetin and quercetin glycosides from onion (*Allium cepa* L.) solid waste as an antioxidant, urease and xanthine oxidase inhibitors. *Food Chem.*, *235*, 119–126.
214. Xu, Y., Li, Y., Bao, T., Zheng, X., Chen, W., & Wang, J., (2017). A recyclable protein resource derived from cauliflower by-products: Potential biological activities of protein hydrolysates. *Food Chem.*, *221*, 114–122.

PART II

Technology of Bioactive Compounds and Nutraceuticals from Food Grain Sources

CHAPTER 6

Technological and Analytical Aspects of Bioactive Compounds and Nutraceuticals from Major Cereal Grain Sources

FARHAN MOHIUDDIN BHAT,[1] RANJANA VERMA,[1] and CHARANJIT S. RIAR[2]

[1]*CSK Himachal Pradesh Agricultural University, Palampur, Himachal Pradesh, India*

[2]*Sant Longowal Institute of Engineering and Technology, Sangrur, Punjab, India*

ABSTRACT

Cereals are the staple food item of the Indian diet, having a richness of calories, protein, micronutrients, dietary fiber, and resistant starch, coupled with a low glycemic index. Due to all these properties, they have an essential role in managing almost all metabolic diseases. Whole grain cereals are associated with several health benefits than their processed or refined fractions. Bioactive substances, dietary fibers (DFs), omega-3 fatty acids, and resistant starches abound in cereals. Because of the antioxidative, antimutagenic, and anticarcinogenic properties of phytochemicals found in cereals, these are excellent sources for creation of functional foods. In terms of macro- and micro-nutrients, millets are three to five times more nutritious than widely marketed rice and wheat and are the least allergenic crops and are rich sources of proteins, calcium, phosphorus, iron. A cereal-based formulation has formed an important part of folk medicine since prehistoric times in China and India due to its nutritive and pharmacological value and is used by Ayurvedic and Unani practitioners for the treatment of numerous ailments.

Bioactive Compounds and Nutraceuticals from Plant Sources: Extraction Technology, Analytical Techniques, and Potential Health Prospects. Charanjit Singh Riar, Parmjit S. Panesar (Eds.)

Barley is a functional food because it contains numerous health-promoting components, such as β-glucan, that have been found to have favorable health effects. Because of its involvement in lowering hike in blood glucose level and risk of coronary heart disease through lowering bad cholesterol, beta-glucans have been approved as a health-beneficial soluble fiber in many countries. The bioactive compounds with widely variations in their chemical structures, properties, and functions are often produced in small quantities either as primary metabolites or secondary metabolites such as terpenoids, polyphenols, vitamins, and alkaloid. These compounds are used for the formulation and fortification of functional foods and pharmaceuticals.

6.1 INTRODUCTION

Cereal grains are staple foods sources of the world population and are important sources of nutrients and energy worldwide than any other crops [1]. Cereals are usually consumed as a whole or as an ingredient in food stuffs as a valuable nutritional source along with phytochemicals including phenolic acids, flavones, organic acids, polyphenols, anthocyanins, sterols, tocopherols, tocotrienols, and phytosterols. The amount of starch in cereal grains gives a lot of energy, but the fat and protein level also helps with energy production. Cereals' carbohydrate content ranges from 65% to 75%. Proteins have been found ranging from 6% to 12% and fat ranging from 1% to 5%, as well as moisture content, cellulose, traces of minerals, and vitamins. Bioactive substances are all abundant in cereals and due to their antioxidative properties, phytochemicals found in cereals make them excellent for the creation of functional foods and nutraceuticals [2]. Cereals are far more widely produced than other crops around the world, and are consequently regarded as a staple food crop [1]. Cereals are consumed by as a whole or as an ingredient in different formulation of food choices. Cereals are also fed to livestock and poultry that are usually consumed by human population in the form of meat, dairy, and poultry products. Besides these, cereals are having widespread applications for industrial manufacturing of glucose, oils, adhesives, thickeners, and alcohols. Bioactive substances are a combination of nutritional ingredients and chemical components found in modest amounts in the bran and germ parts of cereal grains. These provide numerous health benefits like antioxidant, anti-carcinogen, anti-diabetic, and anti-inflammation properties, although they are not considered essential for human health [3]. Wheat, rice, oats, corn, rye, barley, buckwheat, sorghum, and millet as minor grains are the major cereals consumed globally, out of which wheat, corn, and rice are cultivated and produced much larger than other types of cereals [4].

6.1.1 BIOACTIVE COMPOUNDS IN CEREAL AND THEIR FUNCTIONAL IMPORTANCE

The bioactive compounds are having differences in their chemical compositions, and activities are often produced in small quantities either as primary metabolites or secondary metabolites such as terpenoids, polyphenols, vitamins, alkaloids [5]. These compounds (Table 6.1) are used for the formulation and fortification of functional foods and pharmaceuticals. Bioactive chemicals are created in specialized cell types under precise conditions during a specific growth stage, making extraction and purification of these compounds extremely complex [6]. Phenolic compounds have been discovered to have high antioxidant effects, protecting the body against diseases such as cancer and coronary artery disease [7]. Phenolic acids represent hydroxylated derivatives of benzoic acid that includes both free and bound forms like p-hydroxybenzoic, protocatechuic, vanillic, tannic, syringic, caffeic, chlorogenic, and ferulic acid [8]. Anthocyanins, flavonols, flavanones, flavones, and flavonols belonging to flavonoids possess two aromatic rings linked together through three-carbon linkages. Carotenoids are pigments with more than 600 naturally occurring forms and gives yellow, orange, and red colors to different cereals. The carotenoids are focusing attention nowadays in the formulation of nutraceuticals owing to their pro-vitamin and antioxidant activity. The carotenoids consist of two groups with α-carotene, β-carotene, γ-carotene, and β-cryptoxanthin included in the carotene class, and the other one includes xanthophylls with lutein, astaxanthin, and zeaxanthin as main compounds. Carbon and hydrogen atoms make up the majority of B-carotene, -carotene, and lycopene, whereas oxygen atoms make up xanthophylls. Xanthophylls consisting of lutein, zeaxanthin, and astaxanthin contain hydroxyl and keto groups in their structural conformations [9]. Carotenoids are prized for their capacity to scavenge free radicals by quenching singlet oxygen reactive oxygen species, hydroxyl radical and superoxide anion. Among carotenoids, lycopene has been found to retain higher antioxidant capability than β-carotene and lutein. The *in-vivo* antioxidant capability of carotenoids is mainly attributed to their ability to inhibit lipid peroxidation by extinguishing singlet oxygen. β-Carotene has stronger antioxidative activity than vitamin E and vitamin C against singlet oxygen radicals [10]. B-carotene is regarded as a potent antioxidant for its role in preventing liver damage from toxic effects of drugs, photoprotective effects, and anticarcinogenic activity, whereas α-carotene has been found best in lowering chances of heart diseases and cancers. *B*-Cryptoxanthin is more bio-efficacious than α-carotene due to its bipolar nature along with its ability to reduce the risk of lung cancer. Lutein and zeaxanthin are carotenoids that are stored in the human retina and eyes. A higher intake of these carotenoids lowers the risk of cataract by more than 20% and age-related macular degeneration by 40% [11]. Astaxanthin retains 10-fold higher antioxidant activity than β-carotene and 500 times the antioxidant activity of vitamin E [12]. It also prevents the body from various toxic radicals, Ultraviolet effects, LDL-cholesterol damage and tumors.

6.1.2 FUNCTIONAL AND BIOACTIVE POTENTIAL OF MAJORLY GROWN GLOBAL CEREALS

The bioactive compounds that occur in cereal grains and are mostly confined to bran and aleurone layers can be used for various functional purposes (Table 6.2). Maize contains antioxidant rich carotene, xanthophylls, vanillic acid, ferulic acid, anthocyanins, phytoesterols, pantothenic acid and vitamins E, C, and K [13]. These antioxidant-rich compounds in maize have been to possess notable nutraceutical properties and Anti-obesity, anti-hypertensive, anti-prostatic, anti-fatigue, and anti-inflammatory health benefits. It also leads to prevention of neurodegenerative disorders, anti-diabetic, and anti-carcinogenic activities [14]. The carotenoids prevalent in most of the yellow maize kernels have been found to be lutein, zeaxanthin, α and β cryptoxanthin while as pelargonidin, and peonidin-3-glucoside are the anthocyanins found in pigmented blue, red, and purple colored maize kernels in higher concentration [15]. Flavonoids, being an important class of polyphenols, comprise more than 4,000 compounds in different types of corn varieties. The major class of flavonoids in corn have been found to be kaempferol, morin, rutin, and quercetin. Corn has been found to contain thiamine that has been found to boost memory and neuronal health. The flavonoids in corn possess antibacterial, antiviral, antimicrobial, diuretic, and hypertensive properties [16]. Xanthophylls in maize play an essential role in optical muscle health is improved by lowering macular degeneration [17]. The inclusion of carotenoids and xanthophylls in maize cultivars has been shown to help reduce deficiency of vitamin A and oxidative stresses.

TABLE 6.1 Important Classes of Bioactive Compounds in Cereals

Compound	Compound Class	Sources
Luteolin-7-*O*-rutinoside	Flavonoid glycoside	Wheat, rice
Ferulic acid hexose derivative	Hydroxycinnamic acid derivative	Rice, wheat, oats
Proanthocyanidin trimer	Flavonoid	Rice, maize
Epigallocatechin-3-O-gallate	Flavonoid	Rice, oats, maize
(epi)-catechin	Flavonoid	Rice, wheat, oats
Ellagic acid	Phenol	Rice, wheat
Luteolin-7-*O*-glucoside	Flavonoid glycoside	Rice, wheat, maize
Caffeic acid	Hydroxycinnamic acid	Rice, wheat
Cyanidin-3-*O*-galactoside	Anthocyanin	Rice, wheat, maize
Myrecitin	Flavonoid	Rice, wheat
Dicaffeoyl-protocatechuic acid diglucoside	Hydroxybenzoic acid derivatives	Rice, maize
Cyanidin-3-O-rutinoside	Anthocyanin	Rice, wheat, maize
Procyanidin B1	Flavonoid	Rice, wheat
Phloretic acid	Phenol	Rice, wheat, oats
Pelargonidin-3-O-diglucoside	Anthocyanin	Rice, wheat,
Luteolin	Flavonoid	Rice, wheat, oats, maize
Thymol	Phenol	Rice, wheat, maize
Quercetin-3-O-rhamnoside	Flavonoid	Rice, wheat, oats
Apigenin	Flavonoid	Rice, wheat, oats, maize
Apigenin-6,8-di-C-glucoside	Flavonoid	Rice, wheat, oats
(E)-Coniferaldehyde	Phenol	Rice, wheat, maize
Chlorogenic acid	Phenol	Rice, wheat, oats
Quercetin-3-O-rutinoside	Flavonoid	Rice, wheat, oats
p-Coumaric acid	Hydroxycinnamic acid	Rice, oats, maize
Dicaffeoylquinic acid	Hydroxycinnamic acid	Rice, wheat, oats
Quercetin-3-glucoside	Flavonoid glucoside	Rice, oats, wheat
Apigenin-7-O-glucoside	glycosyloxyflavone	Rice, maize
Hyperoside	Flavonol glycoside	Rice, oats
3-succinylarabinoside.	Anthocyanin	Rice, wheat, maize
Tricaffeoyl-hydroxyferulic acid	Phenolic acid	Rice, oats, wheat
Ellagic acid deoxyhexoside	Phenolic acid	Rice

Wheat contains phenolic acids, tocopherols, carotenoids, phytosterols, alkyl resorcinols, benzoxazinoids, and lignans, among other bioactive phytochemicals. That has been associated with reduction of chronic health complications like type 2 diabetes, neurological disorders, cataracts, cardiovascular diseases, atherosclerosis, and cancer. Germ and bran parts of grain contain the highest concentration of bioactive phytochemicals. Wheat bran alkyl resorcinols have been demonstrated to impede platelet binding to fibrinogen, causing thromboxane generation to increase and triglyceride accumulation to decrease [18]. Ferulic acid among phenolic acid is majorly present in wheat, with minor amounts of p-hydroxybenzoic, salicylic, and sinapic acids [19]. The carotenoids Lutein and zeaxanthin are mostly predominant in whole wheat, with concentrations ranging from 0.5–1.44 and 0.2–0.39 μg/g of grain along with lower concentrations of β-cryptoxanthin. All of these components have potential health benefits by inhibiting cholesterol biosynthesis, thereby lowering LDL cholesterol levels along with increasing the absorption of iron in the human body [20]. The carotenoids are present in the highest amount in the germ of grain, while the bran and endosperm possess minimum content. Wheat bran contains bioactive Ferulic acid, which donate electrons to free radicals and transferring hydrogen to them. In addition to ferulic acid, wheat grains also contain hydroxycinnamic acids coumaric, syringic, sinapic, salicylic, and caffeic acid that are also having antioxidant activity [21]. The bran and germ of wheat grains exhibit beneficial positive effects on gut microbiota by providing indigestible fibers that are metabolized by the residing microbiota and thus have positive health benefits, particularly on gastrointestinal tract, such as diverticular disease, constipation, and irritable bowel syndrome [22]. Phenolic chemicals, beta-glucan, inositols, and betaine are all important bioactive substances found in wheat grains that promotes its formulation for the development of functional foods. All these components are associated with prevention of Several dietary and age-related disorders, such as elevated glucose content, bad cholesterol, colon, and breast cancers [23]. Majority of the bioactive compounds having antioxidant properties remain in bound form in the wheat grains that survive gastrointestinal digestion. Only 0.5% to 5% of ferulic acid has the ability of absorption by small intestines, and this typical phenolic acid in wheat plays an essential role in prevention of colon cancer [24]. The bran and germ of wheat grain were found to retain phytosterols, while assteryl ferulates are mainly found to be present in the bran [25]. The higher concentration of pelargonidin and peonidin derivatives in pigmented wheat genotypes, coupled with their excellent antioxidant properties, makes them suitable for the development of functional foods. The phenolic compounds feruloyl oligosaccharides in wheat bran protects human erythrocytes from oxidative damage caused by free radicals [26].

Traditional cereal-based formulations form an important part of folk medicine since prehistoric times in China and India. These Traditional land-races possess nutritive and pharmalogical value and could be used for the treatment of numerous disorders, it is utilized in the Ayurvedic and Unani systems of medicine. The traditional cereals have become a part of China's cultural heritage and have received widespread popularity throughout the world [27]. An increased interest and attention are being paid to by both government and scientific bodies to reveal the various health-promoting benefits of these traditional cereals that could prove beneficial in relieving several chronic diseases. The cereals have been found to be rich source of phenolic and flavonoids [28]. These polyphenols have antioxidant properties by reducing the formation of free radicals and preventing the body from various types of ailments, including cancers and heart disease [29].

Rice ranks second after maize in global production and is the staple diet of the majority of the world's people [30]. Rice is mainly consumed in milled form as white rice. However, rice available in the form of pigmented rice have reddish, purple or even blackish color. The main component of the rice is the starch present in the rice endosperm. The functional role of starch is determined by the two main polymers, amylose (AM) and amylopectin (AP) [31]. Nontoxic, anti-allergic, bland odor, and higher acid resistance are some of the basic characteristics of rice starch that make it preferred over other cereal starches [32]. Pigmented rice consumption is rapidly increasing among the consumers because of its diverse health benefits that marked it an ingredient in formulation of functional foods. There are different shades of pigmented rice varieties grown in the world that have been found to possess more proteins, dietary fiber, minerals, and essential amino acid lysine than other non-pigmented rice varieties. The antioxidant properties of rice could be attributed to its phytochemicals content, including carotenoids, catechins, flavonoids, and phenolic acids, which make it valuable in addition to its high nutritional profiles. The polyphenols in rice bran have been found to prevent oxidation of bad cholesterol and their building up in the arteries resulting in reduction of coronary heart diseases besides their ability in fighting against chronic diseases [33]. The presence of polyphenols and anthocyanins in rice bran have been reported in preventing type 2 diabetes by slowing the rate of sugar absorption in human body besides their antioxidant properties. The polyphenols in rice bran have been found to prevent the chances of cancers by preventing carcinogenic cells from multiplying [34].

After wheat, rice, and corn, barley is cultivated throughout the world as hulled and hull-less barley [35]. Around three-fourth of all barley produced is utilized in manufacturing of animal feed. The proportion of barley used as malt and food product formulations account for 20% and 5% [36]. It comes in a variety of kinds, including hulled, hull-less, normal, waxy, low and high β-glucan barley. Barley products have been utilized in Ayurveda medicine to cure a variety of diseases, such as "Sattu," which is ingested in the summer due to its cooling properties [37]. Barley is a functional food in general because it contains so many health-promoting components that have been proved to have favorable health effects, such as -glucan. Barley's unique properties allow it to be used in the creation of a variety of functional meals. Dietary changes are one of the recommended methods for preventing these disorders. Furthermore, research has indicated that including nutraceuticals such as soluble dietary fibers (DFs) in one's diet can assist regulate blood cholesterol and glucose levels while also benefiting gastrointestinal health in recent years. Beta-glucan, a type of dietary fiber found in barley grains, is found in higher concentrations (3%–7%). Because of its involvement in reducing rise in blood glucose level and the cardiovascular disease through lowering bad cholesterol, beta-glucans have been approved as a health-beneficial soluble fiber in many countries [38].

The eating of hull-less barley has been linked to the Tibetan population's low incidence of hyperlipidemia and diabetes. Complex carbohydrates found in barley provide a delayed discharge of glucose into the bloodstream after digestion, decreasing the glycemic index and making it essential for diabetic patients [39]. Cereals have been utilized in traditional health care systems for ages, and many cultures around the world continue to rely on these traditional cereal-based formulations for their basic health care. Due to recent advancements in plant sciences, there has been a great surge in the use of these cereals and products generated from such cereals for the treatment of numerous maladies in both developing and developed countries [40].

6.2 EXTRACTION OF BIOACTIVE COMPOUNDS FROM CEREAL GRAINS

Bioactive compounds are gaining more popularity due to their antioxidant and medicinal properties. Due to this, there is tremendous exploitation of bioactive compounds in the formulation of different functional foods, chemicals, nutraceuticals, and pharmaceutical industries. The phenolic compounds in

TABLE 6.2 Bioactive Compounds and Their Functional Attributes in Treatment of Ailments

Bioactive Compounds	Chemical Structures	Mode of Actions
Cyanidin	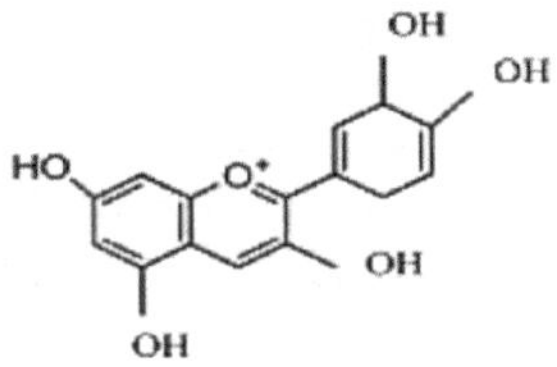	An anthocyanin subcomponent having anti-diabetic, anti-inflammatory, antioxidant properties, anticarcinogenic activity, vasoprotective, and anti-diabetes effects.
Anthocyanins	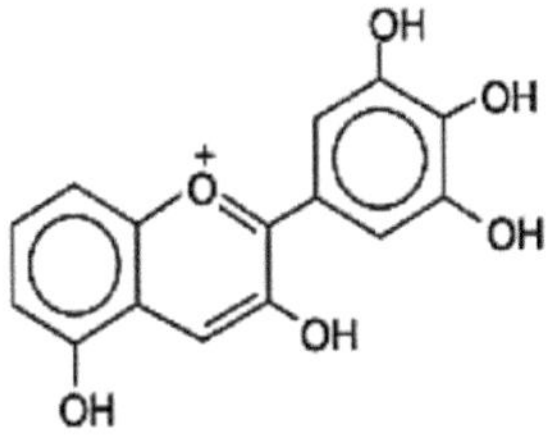	Anthocyanins, which belong to the flavonoid family of phytochemicals, can protect humans from a variety of diseases. Anthocyanins have estrogenic action, enzyme inhibition, cytokine production stimulation, anti-inflammatory activity, lipid peroxidation, capillary permeability and fragility reduction, and membrane strengthening properties.
Gallic acid	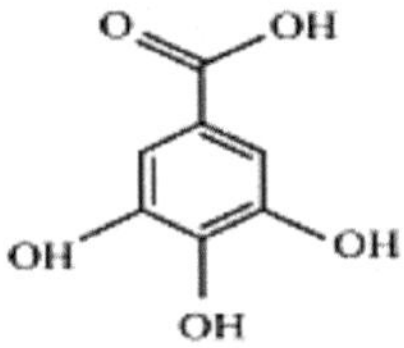	Gallic acid is the most common trihydroxybenzoic acid and is classed as a phenolic acid. Gallic acid has multiple biological functions, such as antiinflammatory, antimicrobial, and anticarcinogenic.
Terpenes	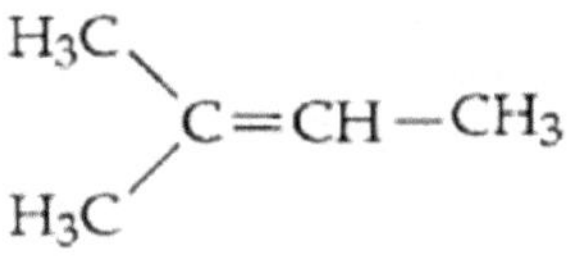	The terpenoids in essential oils have the ability to profoundly influence the immunological response of phagocytes and monocytes. Terpenes can also reduce B cell activation by blocking kappa-light-chain-enhancer, which helps to fight neuroinflammation.
Quercetin	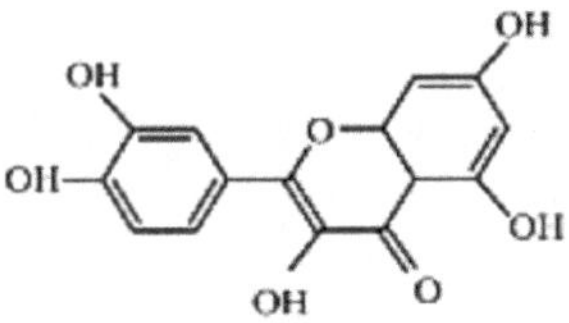	Quercetin is an antioxidant and anti-inflammatory flavonoid that helps to reduce edema, destroy cancer cells, regulate blood sugar, and prevent heart disease.
Flavonoids	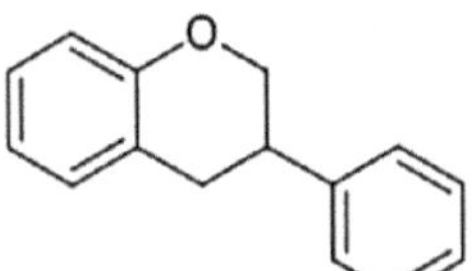	Flavonoids have anti-inflammatory properties and have a good inhibitory effect on edema development and myeloperoxidase activity.

TABLE 6.2 *(Continued)*

Bioactive Compounds	Chemical Structures	Mode of Actions
Phenolics	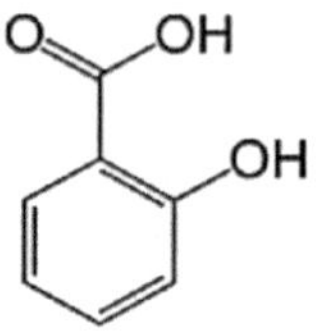	Phenolics can block or slow the breakdown of biopolymers by interfering with the oxidative cycle. Totarol is a topical anti-inflammatory drug that works well in *Podicarpus* spp. and is currently in demand economically.
Alkaloids	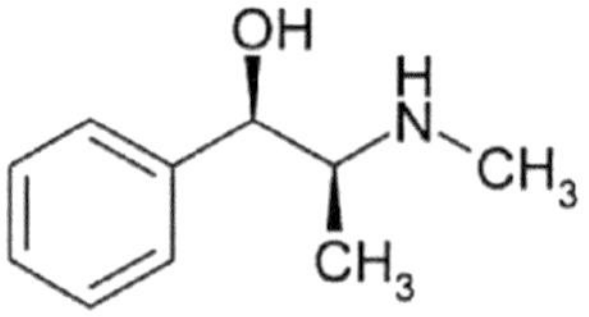	Inflammatory cytokines and inflammatory mediators were shown to be inhibited by alkaloids.
Saponins	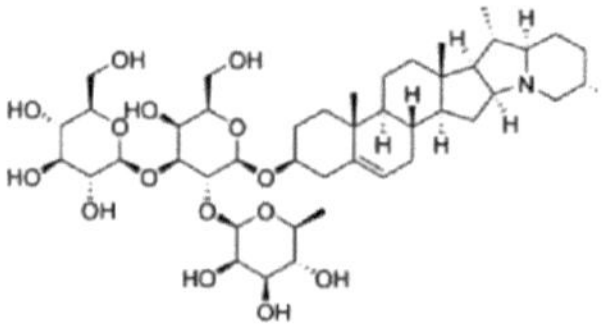	Saponins have been demonstrated to have antiviral effects, which are important in preventing fever.
Amino acids	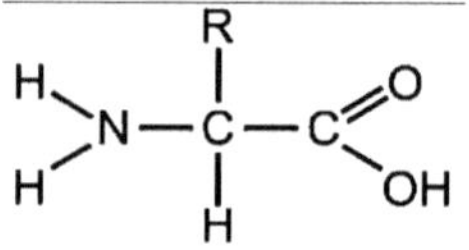	Enzymatic action links the ordered amino acids to make a protein, which regulates the expression of anti-inflammatory cytokines.
Hydroxycinnamic acids	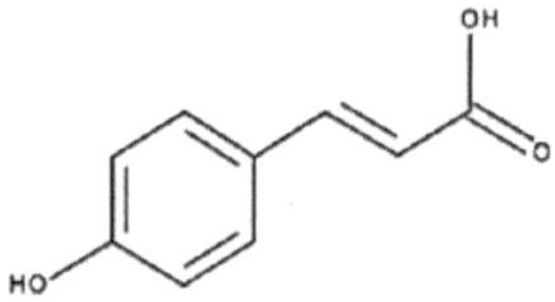	hydroxycinnamic acid are all significant polyphenolic chemicals having powerful antioxidant and anti-inflammatory capabilities, and they may be useful in diabetic and hyperlipidemia patients.
Hydroxybenzoic acid	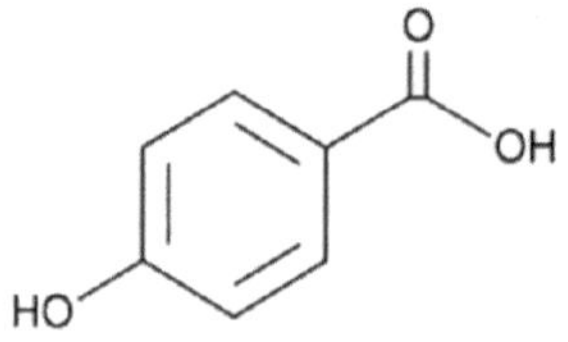	Hydroxybenzoic acids suppress the enzymes -amylase and -glucosidase, which are responsible for breaking down complex carbs, lowering the glycemic index.
B-Carotene	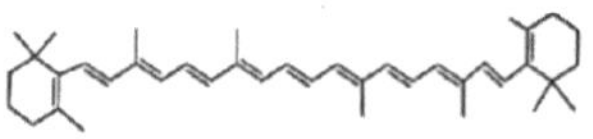	B-Carotene is a nutrient that transforms Beta Carotene into vitamin A, which is essential for eyesight. It also aids in cell growth and the maintenance of healthy organs such as the heart, lungs, and kidneys.

cereals are generally derivatives of benzoic acid or cinnamic acid, with ferulic acid being the most abundant found mostly in the outer layers of cereals [41]. The presence of these bioactive compounds like phenolics compound, tannins, and carotenoids in cereals contributed in antioxidant activities of cereals by different mechanisms [42]. Keeping into consideration the presence of phytochemicals and their antioxidant activities, different solvents have been used for the isolation of bioactive compounds. Free phenolic are present mainly in the outer layers of the pericarp while bound phenolic acids are ester-linked to cell walls, solvent extraction requires an acid, base, or enzymatic hydrolysis [43]. Different solvent mixes have been reported to improve extraction rates and yields of free phenols as compared to pressured liquid extraction procedures as validated by Bonoli et al. [44]. The extraction efficiency of any organic solvent used in extraction of bioactive compounds; the polarity of certain phenolic compounds has an impact. As a result of the dilution of solvents, a medium polarity is generated, allowing for the extraction of additional components that are not soluble in organic solvents [45]. The polarity of different compounds found in the tested grains could explain the variability in total phenolic content (TPC) extracted using varied solvent concentrations. The alcohols ethanol and methanol form hydrogen bonds between the molecules' hydroxyl groups and oxygen atoms, resulting in higher extraction efficiency than acetone [46]. The addition of water to an organic solvent improves its polarity, allowing it to extract different polarity molecules in excess concentrations. The stereochemistry of phenolic compounds and their intermolecular hydrogen bonding interactions and the solvents have been linked to the solubility tendency. For the determination and characterization of bioactive chemicals, several sensitive and selective analytical techniques have been developed. The most effective methods include spectrophotometric, chromatographic, fluorometric, enzymatic, and electrophoretic.

6.2.1 ULTRASOUND-ASSISTED EXTRACTION

Ultrasound-assisted extraction is an advanced, inexpensive, and highly efficient extraction technique for extraction of bioactive compounds from cereal grains. This technique (Figure 6.1) is utilizes the application of high-frequency sound waves in a limited amount of solvent for an effective extraction of the compounds from the solid matrix. In ultrasound-assisted extraction, sound waves are applied for generation of mechanical energy in the form of cavitations through the sample. The cavitations generated from sonications resulted in formation and collapse of Small vacuum bubbles or gaps in the liquid rupture plant cells and tissues, resulting at high temperatures of 4,500C and pressures of roughly 50 Mpa, causing disturbance inside the matrix. The application of sonic waves through the sample aids in better solvent-solute interactions, penetration of the solvent into the interior of cells, resulting in sonolysis, destruction of biological membranes thereby accelerates the extraction of intracellular material. The application of high-power, low-frequency ultrasound waves into slurry consisting of cereals in organic solvent creating alternating high-pressure, low-pressure cycles that causes acoustic cavitation. This phenomenon results in the generation of high temperatures, pressures, pressure gradients, and high shear forces that lead to decreased extraction time and accelerate extraction and release of phenolic compounds due to the compression and rarefaction waves induced in the molecules that cause perforation of cell walls and cell membranes (sonoporation) and disruption of the cell wall. The cavitation of bubbles in the sonication process also creates macro-turbulences and micro-mixing that further accelerates the extraction and isolation of bioactive compounds. The amount of phenolic chemicals extracted from any solid matrix is determined by extraction duration, temperature, solvent composition.

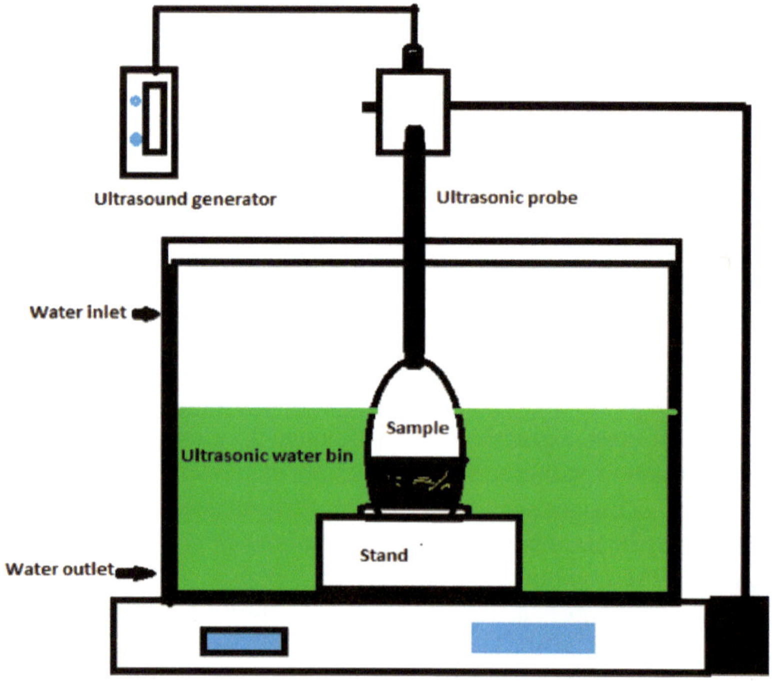

FIGURE 6.1 Schematic representation of ultrasound assisted extraction.

6.2.2 PRESSURIZED LIQUID EXTRACTION (PLE)

Pressurized liquid extraction (PLE) (Figure 6.2) is an advanced extraction technique that is also called accelerated solvent extraction (ASE) or pressurized fluid extraction (PFE) involves the use of organic solvents under high temperature and pressure for extracting phenolic compounds from a sample. The application of high temperature (40–200°C) enhances the mass transfer rate due to its reduction in viscosity and surface tension, due to which the solvent interacts more easily with the sample, increasing solvation power, thereby accelerating the extraction rate. While as lower boiling point is maintained for solvent under high pressure (3.3–20.3 Mpa) that ensures higher penetration of the solvent in the solute matrix resulting in higher diffusion rate and solubility.

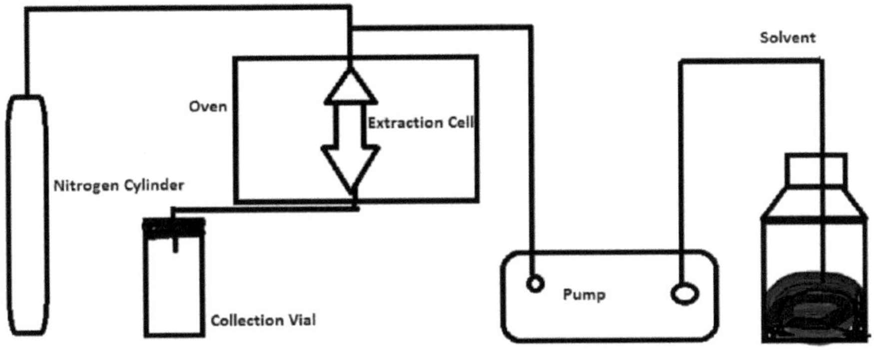

FIGURE 6.2 Schematic diagram of pressurized liquid extraction system.

6.2.3 SOLID PHASE EXTRACTION

Solid-phase extraction is a chromatographic isolation technique that employs a solid adsorbent to absorb dissolved chemicals in a liquid mixture. SPE is used to isolate desired analytes from a liquid matrix by utilizing the differences in affinity between an analyte and interferents for a sorbent. This technique enables the separation of desired solutes from the liquid or mobile phase by passing it through the active sites of stationary (solid) phase that retains either the desired or undesired components. The desired analytes from the stationary phase can be isolated by rinsing it with a suitable eluent. Solid-phase extraction uses a solid adsorbent (sorbent) packed most in a cartridge device that has a space for entrance of liquid mixture and is connected to vacuum to elute the applied solvents. The two principal mechanisms for interaction of solute with sorbent in most of the solid phase extraction processes are polarity and ion exchange. The polarity mode that can be used for separation of analytes in solid phase extraction process is either normal or reversed. A pairing of relatively polar solid phase extracting sorbent (e.g., bare silica, alumina) with a normal mode containing nonpolar mobile phase is considered best for retaining polar constituents of a sample while eluting nonpolar components. Similarly, a reversed mode involving coupling of nonpolar solid phase extracting sorbent (e.g., silica with a C18 or C8 bonded phase) with a relatively polar mobile phase will retain nonpolar constituents and elutes polar constituents. Some sorbents (e.g., CarboPrep Plus, Diol) can operate in either normal or reversed modes depending on the choice of solvent used and the sample. Ion exchange interaction follows the rule of electrostatic attraction. Unlike like-dissolves-like model of interactions in polarity based solid phase extraction, ion exchange operates on the rule of opposites attract. If a solid phase has a positive charge on the surface, it is balanced by an anion bound to it. Upon introducing a negatively charged analyte into the system, it will displace the initially bound anion and couple with the positively charged surface of solid phase extracting medium resulting in retention of the analyte.

6.2.4 *MICROWAVE ASSISTED EXTRACTION*

Microwave Assisted Extraction (Figure 6.3) is an advanced technique that combines the use of microwaves in the conventional solvent extraction system. The microwaves induce dipole rotations and ionic polarization in samples, resulting in the generation of heat and destruction of hydrogen bonding that increases the diffusion rate of solvent in the cell matrices. The heating of free water molecules in the cells of the samples due to microwaves leads to rupture of cell membranes/cell walls resulting in increasing extraction of phytochemicals in the solvent. The extraction in Microwave Assisted processes is due to changes in the cell structure due to application of electromagnetic waves.

6.2.5 *ENZYME ASSISTED EXTRACTION (EAE)*

Enzyme-assisted extraction (Figure 6.4) is an environmentally friendly 'green' extraction technique having more efficiency than conventional extraction technology. This technique uses specific enzymes that are having

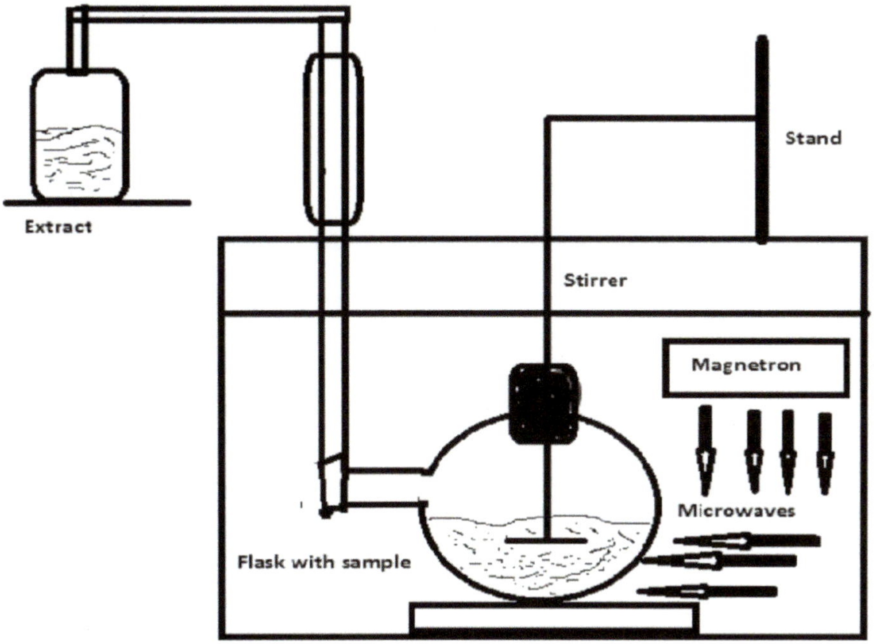

FIGURE 6.3 Schematic representation of microwave-assisted extraction.

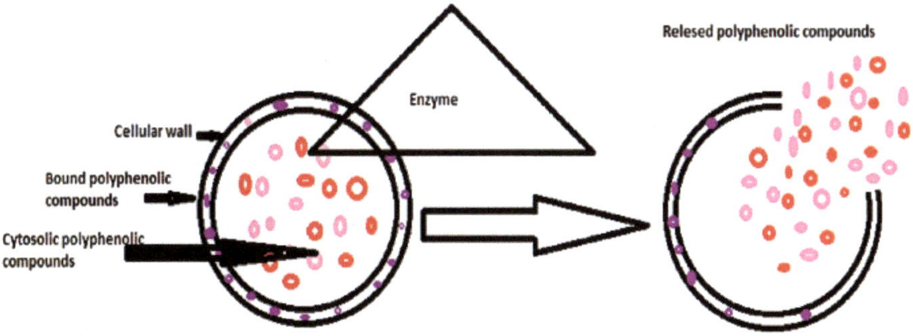

FIGURE 6.4 Schematic representation of enzyme-assisted extraction.

the ability to hydrolyze the cell wall for efficient extraction from cells. The bioactive compounds are mixed with proteins, pectin, lipid, starch, and cellulose in cells due to which there extraction becomes difficult with conventional extraction techniques. The addition of enzymes hydrolyzes the complex bonding between chemical compositional molecules, resulting in the release of the bioactive polyphenolic compounds. Enzyme-assisted extraction technique can be used in combination with various other techniques for increasing discharge rate of bioactive compounds.

6.3 ANALYTICAL TECHNIQUES

6.3.1 DETERMINATION OF TOTAL PHENOLIC CONTENT (TPC)

The Folin–Ciocalteu reagent was used to determine the TPC of cereal grains. About 9 ml of distilled water and 1 ml of the Folin–Ciocalteu reagent are added to 1 ml of the cereal grain extract. The mixture is then diluted with 10 mL of a 7% (w/v) Na_2CO_3 solution and 25 mL of purified water. The material is forcefully mixed before being incubated at 30°C for 90 minutes. A spectrophotometer is used to measure the absorbance against a reagent blank at 750 nm. The TPC is measured in milligrams of gallic acid equivalent (GAE) per gram of weight.

6.3.2 DETERMINATION OF TOTAL FLAVONOID CONTENT (TFC)

The colorimetric approach of Abu Bakar et al. [47] can be used to measure total flavonoid content (TFC). In a test tube, combine 0.5 mL of cereal grains extract with 2.25 mL distilled water, then add 0.15 mL of $NaNO_2$ solution (5% w/v). After a 6-minute rest period, the resultant solution is combined with 0.3 mL of a 10% (w/v) $AlCl_3.6H_2O$ solution. In addition, 1.0 mL of 1 M NaOH is added to the solution and forcefully shaken. The absorbance is measured at 510 nm against a blank sample immediately after a 5-minute rest period. The results are given in mg rutin equivalents per gram of dried material (mg RE/g).

6.3.3 SCAVENGING EFFECT ON DPPH RADICAL

The DPPH radical scavenging method, as described by Sasidharan et al. [48], could be used to assess the free radical scavenging activities of cereal grains extracts. About 3.9 ml of DPPH solution (2.3 mg in 100 ml methanol) was added to 0.1 ml of the prepared cereal grains extracted sample and stirred. The absorbance at 515 nm of the resultant solution is measured after a

30-minute incubation period in the dark at room temperature. The following equation is used to calculate scavenging activity:

$$DPPH\ radical\ scavenging\ percentage = [1/((A515\ nm, sample - A515\ nm, control))] \times 100$$

6.3.4 DETERMINATION OF REDUCING POWER

The modified approach of Yen & Duh [49] was used to determine the reducing power. About 2.5 ml of prepared cereal grains extract, 2.5 ml of phosphate buffer (2.0 M, pH 6.6), and 2.5 ml of 1% potassium ferricyanide are added to 2.5 ml of the prepared cereal grains extract. The resulting mixture was incubated for 20 minutes at 50°C. The mixture was then centrifuged at 700 g for 10 minutes with a 10% solution of tri-chloroacetate (2.5 ml). The solution's upper layer (2.5 ml) was mixed with distilled water (2.5 ml) and 0.1% ferric chloride (0.5 ml), and the absorbance was measured at 700 nm using a UV-vis spectrophotometer. The experiment was carried out three times, and the findings were averaged.

6.3.5 DETERMINATION OF TOTAL ANTHOCYANIN PIGMENT CONTENT

The pH differential approach, as recommended by Hosseinian et al. [50], can be used to determine the total anthocyanin concentration. A 0.5 mL sample of cereal grains is added to 3.5 mL of 0.025 M potassium chloride buffer (pH 1.0). Before measuring absorbance at 515 and 700 nm against distilled water as a blank in a spectrophotometer, the mixture is thoroughly stirred and allowed to stand for 15 minutes. After a 15-minute incubation time, the extract was combined with sodium acetate buffer (0.025 M, pH 4.5) in the same way as it was with KCl buffer, and the absorbance was measured at the same wavelengths. The results were given in milligrams of Cyanidin-3-glucoside equivalents per gram of material.

$$\text{Total anthocyanin content} = \left(\frac{A \times MW \times DF \times 1000}{\in \times 1}\right)$$

where; A (absorbance) = [(A515–A700) pH1 – (A515–A700) pH 4.5]; MW is the molecular weight of cyanidin 3-glucoside (449.2 g/mol); DF is the dilution factor of the sample [8]; and ϵ is molecular absorptive ($L \times cm^{-1} \times mol^{-1}$) of cyaniding 3-glucoside, equal to 26,900; here, L is the path length in cm.

6.3.6 TOTAL ANTIOXIDANT CAPACITY BY PHOSPHOMOLYBDINUM REDUCTION ASSAY

The total antioxidant activity of cereal grains was determined using the phosphomolybdate assay technique, which was slightly modified from the method described by Khan et al. [51]. About 3 ml of reagent solution made with sulfuric acid (0.6 M), sodium phosphate (28 mM), and ammonium molybdate was mixed with 0.3 ml of the produced cereal grains extracted sample (4 mM). For 90 minutes, the mixture was incubated at 95°C in a water bath. After cooling to room temperature, absorbance was measured at 695 nm against a reagent blank that contained 0.3 ml of methanol instead of extract. Ascorbic acid equivalents were used to calculate total antioxidant capacity. The mathematical link developed between ascorbic acid concentration and phosphomolybdic reduction assay was used to assess total antioxidant capacity in terms of phosphomolybdic reduction assay.

6.3.7 INHIBITION OF LIPID PEROXIDATION IN EGG YOLK HOMOGENATE

Badmus et al. [52] used a modified thiobarbituric acid-reactive species (TBARS) assay technique to measure lipid peroxidation inhibition in the egg yolk. In a test tube, 0.5 mL egg yolk homogenate (10% in distilled water, v/v) was carefully mixed with 0.1 mL prepared cereal grains sample, and the volume was increased to 1 mL by adding distilled water. Finally, to promote lipid peroxidation, 0.05 ml $FeSO_4$ (0.07 M) was added to the aforementioned mixture and incubated for 30 minutes. After that, 1.5 mL of 20% acetic acid and 1.5 mL of 0.8% TBA (w/v) in 1.1% sodium dodecyl sulfate (SDS) and 0.05 mL of 20% TCA were added, vortexed, and cooked in a boiling water bath for 15 minutes.

$$\%\ \text{Inhibition} = \left(1 - \frac{\text{A532, Sample}}{\text{A532, control}}\right) \times 100$$

Here, control is tested sample without extract.

6.3.8 HIGH PERFORMANCE-LIQUID CHROMATOGRAPHY (HP-LC)

HPLC is a separation technique, used to separate, identify, and quantities different types of polyphenolic compounds depending on their nature, chemical structure, and molecular weight. HPLC utilizes a high-pressure pump for transporting the mobile phase from a reservoir at a specified flow rate (ml/minute) through an injector. The injector (auto sampler) injects the sample into a reverse-phase C18-packed HPLC column that travels as continuously flowing mobile phase stream for component separation. The column contains the chromatographic packing material or adsorbent typically silica granular particles (3-µm particle size) or polymer that is inert to mobile phase and has the ability to analyze compounds of different chemical natures. The column is called stationary phase and is generally having the 3 mm internal diameter and 100 mm length functionalized with C18 octadecylsilane. The components of a mixture travel at different rates through the stationary column as per their molecular weights and thus are having different retention times (RTs) (s or min). The component is separated from each other in the stationary column because of varying degrees of interaction with the absorbent particles that causes different elution rates for the different components resulting in their separations. Finally, the mobile phase passes through the detector cell, where the absorbance is measured at 220 nm, and can be wasted or collected, as desired.

The detector can be refractive index, ultraviolet (UV), and fluorescence, depending on compound characteristics. Each one of these detectors detects different properties of the molecules coming out of the column and displays a chromatogram. In reverse-phase liquid chromatography (LC), a polar mobile phase is used, and nonpolar molecules are adsorbed on the stationary phase. A chromatogram represents the separation of the compound that has been done chromatically (chemically) in the HPLC system. A series of peaks plotted on time axis rising from a baseline represents the detector response for a different compound.

6.3.9 GEL PERMEATION CHROMATOGRAPHY

Gel permeation is a size exclusion chromatography which separates molecules based on their molecular weight rather than chemical properties. It involves passage of a solution over a stationary phase that contains a column of semipermeable, porous polymer gel beads having a range of pore sizes (Dextran, Agarose gel, Acrylamide gel). The components in the mobile phase travel at different velocities depending on their molecular sizes through the gel beads. Molecules with lower molecular weight pass easily through the pores of the bead and thus possess higher elution time. Whereas components with higher molecular mass excludes from the column beads and elute first. An UV or refractive index detector detects the components eluted during each time interval. The elution time of the components are compared with that of monodisperse samples, whose molecular weight is known for analyzing the entire molecular weight distribution. Gel permeation chromatography does the sequential elution of components in the sample starting from highest molecular weight component.

6.4 CONCLUSION

Cereals are consumed as a whole or as an ingredient in different formulation of food choices. Bioactive substances are a combination of nutritional ingredients and chemical components found in modest amounts in the bran and germ parts of cereal grains. Millets' proteins are gluten-free and non-allergenic, thus eating them lowers triglycerides and C-reactive proteins, which may help prevent cardiovascular disease. Sorghum is beneficial for diabetic patients than other cereals as its starches and sugars are released more slowly in human body and thus are having lower glycemic index. The polarity of some phenolic compounds influences the extraction effectiveness of any organic solvent employed in bioactive compound extraction. As a result, diluting solvents produces a medium polarity, which makes it easier to extract other components that aren't soluble in organic solvents.

KEYWORDS

- **anthocyanin**
- **antioxidant capacity**
- **DPPH radical**
- **gel permeation chromatography**
- **high performance liquid chromatography**
- **lipid peroxidation**
- **phosphomolybdinum reduction assay**
- **total phenolic content**

REFERENCES

1. Sarwar, M. F., Sarwar, M. H., Sarwar, M., Qadri, N. A., & Moghal, S., (2013). The importance of cereals (*Poaceae: Gramineae*) nutrition in human health: A review. *Journal of Cereals and Oil Seeds, 4*(3), 32–35.
2. Serafini, M., Bellocco, R., Wolk, A., & Ekström, A., (2002). Total antioxidant potential of fruit and vegetables and risk of gastric cancer. *Gastroenterology, 123*, 985–991.
3. Kris-Etherton, P. M., Hecker, K. D., Bonanome, A., Coval, S. M., & Binkoski, A. E., (2002). Bioactive compounds in foods: Their role in the prevention of cardiovascular disease and cancer. *American Journal of Medicine, 113*(9B), 71S–88S.
4. *Koehler, P., & Wieser, H., (2013). Chemistry of cereal grains. In: Gobetti, M., & Gaenzle, M., (eds.), Handbook of Sourdough Biotechnology (1st edn., pp. 11–45). Springer, New York.*
5. Yang, B., Jiang, Y., Shi, J., & Ashraf, M., (2010). Extraction and pharmacological properties of bioactive compounds from longan (*Dimocarpus longan* Lour.) fruit: A review. *Food Research International, 44*, 1837–1842.
6. Verpoorte, R., & Memelink, J., (*2012*). Engineering secondary metabolite production in plants. *Current Opinion in Biotechnology, 13*, 181–187.
7. Harborne, J. B., & Williams, C., (2000). Advances in flavonoid research since 1992. *Phytochemistry, 55*, 481–504.
8. Gani, S. M., Wani, F., Masoodi, A., & Hameed, G., (2012). Whole-grain cereals bioactive compounds and their health benefits: A review. *Journal of Food Processing & Technology, 3*, 146–156.
9. Stahl, W., & Sies, H., (2005). Bioactivity and protective effects of natural carotenoids. *Biochim. Biophys Acta, 1740*, 101–107.
10. Sies, H., & Stahl, W., (1996). Vitamins E and C, beta-carotene, and other carotenoids as antioxidants. *American Journal of Clinical Nutrition, 62*(6), 1315S–1321S.
11. Moeller, S. M., Jacques, P. F., & Blumberg, J., (2000). The potential role of dietary xanthophylls in cataract and age-related macular degeneration. *Journal of the American College of Nutrition, 19*(5), 522S–527S.
12. Lopez, M., Arce, L., Garrido, J., Ríos, A., & Valcarcel, M., (2004). Selective extraction of astaxanthin from crustaceans by use of supercritical carbon dioxide. *Talanta, 64*(3), 726–731.
13. *Adom,* K. K.*, & Liu,* R.*, (*2002*)*. Antioxidant activity of grains. *Journal of Agricultural and Food Chemistry, 50*, 6182–6187.
14. Irondi, A. E., Adegoke, B. M., Effion, E. S., Oyewo, S. O., Alamu, E. O., & Boligon, A. A., (2019). Enzymes inhibitory property, antioxidant activity and phenolics profile of raw and roasted red sorghum grains *in vitro*. *Food Sci. Hum. Wellness, 8*, 142–148.
15. Herrera-Sotero, M. Y., Cruz-Hernández, C. D., Trujillo-Carretero, C., Rodríguez-Dorantes, M., GarcíaGalindo, H. S., Chávez-Servia, J. L., Oliart-Ros, R. M., & GuzmánGeronimo, R., (2017). Antioxidant and antiproliferative activity of blue corn and tortilla from native maize. *Chem. Central J., 11*, 110.
16. Cowan, M. M., (1999). Plant products as antimicrobial agents. *Clinical Microbiology Review, 12*, 564–582.
17. Mugode, L., Ha, B., Kaunda, A., Sikombe, T., Phiri, S., & Mutale, R., (2014). Carotenoid retention of biofortified provitamin A maize (*Zea mays* L.) after Zambian traditional

methods of milling, cooking and storage. *Journal of Agricultural and Food Chemistry, 62*, 6317–6125.
18. Ross, S. M., King, J. R., Donovan, J. T., & Spaner, D., (2004). Forage potential on intercropping berseem clover with barley, oat, or triticale. *Agronomy J., 96*(4), 1013–1021.
19. Moore, J., Hao, Z., Zhou, K., Luther, M., Costa, J., & Yu, L., (2005). Carotenoid, tocopherol, phenolic acid, and antioxidant properties of Maryland-grown soft wheat. *Journal of Agricultural and Food Chemistry, 53*, 6649–6657.
20. *Tiwari,* U., *& Cummins,* E., *(*2009*).* Nutritional importance and effect of processing on tocols in cereals. *Trends in Food Science & Technology, 20*, 511–520.
21. Graf, E., (1992). Antioxidant potential of ferulic acid. *Free Radical Bio Med., 13*, 435–448.
22. Gong, L., Cao, W., Chi, H., Wang, J., Zhang, H., Liu, J., & Sun, B., (2018). Whole cereal grains and potential health effects: Involvement of the gut microbiota. *Food Research International, 103*, 84–102.
23. Catoi, A. F., Parvu, A., Muresan, A., Galea, R. F., & Catoi, C, (2006). Evaluation of nitric oxide synthesis in morbid obesity. *Free Radical Research, 40*, S140.
24. Mateo, A. N., Van, D. B. R., Havenor, R., Aalt, B., & Guido, R. M. M., H., (2009). Bioavalibility of ferulic acid is determined by its bioaccesibility. *Journal of Cereal Science, 49*, 296–300.
25. Nurmi, T., Lampi, A. M., Nyström, L., Hemery, Y., Rouau, X., & Piironen, V., (2012). Distribution and composition of phytosterols and sterylferulates in wheat grain and bran fractions. *Journal of Cereal Science, 56*, 379–388.
26. Wang, J., Sun, B., Cao, Y., & Tian, Y., (2009). Protection of wheat bran feruloyl oligosaccharides against free radical-induced oxidative damage in normal human erythrocytes. *Food Chemical Technology, 47*, 1591–1599.
27. Sen, S., & Chakraborty, R., (2017). Revival, modernization, and integration of Indian traditional herbal medicine in clinical practice: Importance, challenges, and future. *Journal of Traditional and Complementary Medicine, 7*(2), 234–244.
28. Mao, P., Hu, Y., ZhiLi, H., Huang Y., JianBo, Y., & TiaoYi, X., (2014). Effects of feeding broad bean and dehulling broad bean diets on growth performance, muscle quality and blood physiological-biochemical indices of grass carp (*Ctenopharyngodon idellus*). *Chinese Journal of Animal Nutrition, 26*(3), 803–811.
29. Kehrer, J. P., (1993). Free radicals as mediators of tissue injury and disease. *Critical Reviews in Toxicology, 23*(1), 21–48.
30. Tyagi, A. K., Khuran, J. P., Khurana, A. P., Raghubanshi, S., Gour, A., Kapur, A., & Sharma, S., (2004). Structural and functional analysis of rice genome. *Journal of Genetics, 83*, 79–99.
31. Singh, N., Kaur, L., Sandhu, K. S., Kaur, J., & Nishinari, K., (2006). Relationships between physicochemical, morphological, thermal, rheological properties of rice starches. *Food Hydrocolloid, 20*, 532–542.
32. Noda, T., Nishiba, Y., Sato, T., & Suda, I., (2003). Properties of starches from several low-amylose rice cultivars. *Cereal Chemistry, 80*(2), 193–197.
33. Min, S. W., Ryu, S. N., & Kim, D., (2010). Anti-inflammatory effects of black rice, cyanidin- 3-O-β-D-glycoside, and its metabolites, cyanidin and protocatechuic acid. *International Immunology, 10*, 959–966.

34. Bhat, F. M., & Riar, C., (2016). Effect of amylose, particle size & morphology on the functionality of starches of traditional rice cultivars. *International Journal of Biological Macromolecules, 92,* 637–644.
35. Mayer, K. X. F., (2012). The international barley genome sequencing consortium. A physical, genetic, and functional sequence assembly of the barley genome. *Nature, 491,* 711–716.
36. Blake, T., Blake, V., Bowman, J., & Abdel-Haleem, H., (2011). *Barley: Production, Improvement and Uses* (pp. 522–531). Wiley-Blackwell, UK.
37. Verma, R. P. S., Kharub, A. S., Kumar, D., Sarkar, B., Selvakumar, R., Singh, R., Malik, R., et al., (2011). *Fifty Years of Coordinated Barley Research in India*. Directorate of Wheat Research, Karnal-132001, Research Bulletin No. 27: 46.
38. Behall, K. M., Scholfield, D. J., & Hallfrisch, J., (2004). Diets containing barley significantly reduce lipids in mildly hypercholesterolemic men and women. *American Journal of Clinical Nutrition, 80,* 1185–1193.
39. Foster-Powell, K., Holt, S. H., & Brand-Miller, J., (2002). International table of glycemic index and glycemic load values. *The American Journal of Clinical Nutrition, 76,* 5–56.
40. Singh, J., Mccarthy, O. J., Singh, H., & Moughan, P., (2008). Low temperature post temperature changes of New Zealand Tewa (Maori potato): Effect on starch physic-chemical and functional characteristics. *Food Chemistry,* 583–596.
41. Abdel-Aal, E. S., Hucl, P., Sosulski, F. W., Graf, R., Gillott, C., & Pietrzak, L., (2001). Screening spring wheat for midge resistance in relation to ferulic acid content. *Journal of Agriculture and Food Chemistry, 49*(8), 3559–3566.
42. Heinio, R. L., Liukkonen, K. H., & Myllymäki, O., (2008). Quantities of phenolic compounds and their impacts on the perceived flavor attributes of rye grain. *Journal of Cereal Science, 47,* 566–575.
43. Sosulski, F., Krygier, K., & Hogge, L., (1982). Free, esterified, and insoluble-bound phenolic acids. 3. Composition of phenolic acids in cereal and potato flours. *Journal of Agriculture and Food Chemistry, 30,* 337–340.
44. Bonoli, M., Marconi, E., & Caboni, M., (2004). Free and bound phenolic compounds in barley (*Hordeum vulgare* L.) flours: Evaluation of the extraction capability of different solvent mixtures and pressurized liquid methods by micellar electrokinetic chromatography and spectrophotometry. *Journal of Chromatography A, 1057,* 1–12.
45. Do, Q., Angkawijaya, A., Tranguyen, P., Huynh, L., Soetaredjo, F., Ismadji, S., & Ju, Y., (2014). Effect of extraction solvent on total phenol content, total flavonoid content, and antioxidant activity of *Limnophila aromatica*. *Journal of Food and Drug Analysis, 22,* 296–302.
46. Galanakis, C. M., Goulas, V., Tsakona, S., Manganaris, G. A., & Gekas, V. A., (2013). Knowledge base for the recovery of natural phenols with different solvents. *International Journal of Food Properties, 16*(2), 382–396.
47. Abu-Bakar, M. F., Mohamed, M., Rahmat, A., & Fry, J., (2009). Phytochemicals and antioxidant activity of different parts of bambangan (*Mangifera pajang*) and tarap (*Artocarpus odoratissimus*). *Food Chemistry, 113,* 479–483.
48. Sasidharan, S., Chen, Y., Saravanan, D., Sundram, K. M., & Latha, L., (2011). Extraction, isolation, and characterization of bioactive compounds from plant extracts. *African Journal of Traditional, Complementary and Alternative Medicines, 8*(1), 1–101.
49. Yen, G. C., & Duh, P., (1993). Antioxidative properties of methanolic extracts from peanut hulls. *Journal of the American Oil Chemists Society, 70*(4), 383–386.

50. Hosseinian, F. S., Li, W., & Beta, T., (2008). Measurement of anthocyanins and other phytochemicals in purple wheat. *Food Chemistry, 109*(4), 916–924.
51. Khan, R. A., Khan, M. R., Sahreen, S., & Ahmed, M., (2012). Assessment of flavonoids contents and *in vitro* antioxidant activity of *Launaea procumbens*. *Chemistry Central Journal, 6*, 1–11.
52. Badmus, J. A., Adedosu, T. O., Fatoki, J. O., Adegbite, V. A., Adaramoye, O. A., & Odunola, O., (2011). Lipid peroxidation inhibition and antiradical activities of some leaf fractions of *Mangifera indica*. *Acta Poloniae Pharmaceutica, 68*(1), 23–29.

CHAPTER 7

Technological and Analytical Aspects of Bioactive Compounds and Nutraceuticals from Minor Cereal Grains (Millets) Sources

SEEMA SHARMA,[1] ROMEE JAN,[2] and SAKSHI SUKHIJA[3]

[1]*Jaipur National University, Jaipur, Rajasthan, India*

[2]*University of Kashmir, Hazratbal, Srinagar, Kashmir, India*

[3]*Amity University, Noida, Uttar Pradesh, India*

ABSTRACT

Millets are low-cost cereal grains that are widely used as a source of food and feed in industries and animal husbandry. Millet grains are high in starch, protein, minerals, vitamins, and specific bioactive compounds with antioxidant properties. There are two types of bioactive phytochemicals that come in soluble, free, and conjugated forms, as well as insoluble, bound forms. Whole millets are high in phenolic acids, carotenoids, tocopherols, alkylresorcinols, benzoxazinoids, phytosterols, and lignans, which are all bioactive phytochemicals. In the last two decades, several classes of compounds that are described in various methods have been isolated from oats and barley, including dietary fiber (mainly glucan), lipid classes (phytosterols), phenolics, tocols, bioactive peptides, and proteins. Many of the compounds are tangled up in the grain's matrix, making extraction challenging. To accurately quantify bioactive phytochemicals, it is critical to understand their number of forms. Because of rapid technological advances, the analytical methods used to quantify bioactive chemicals in whole seeds are constantly changing. This chapter summarized various alternative extraction techniques for analyzing

Bioactive Compounds and Nutraceuticals from Plant Sources: Extraction Technology, Analytical Techniques, and Potential Health Prospects. Charanjit Singh Riar, Parmjit S. Panesar (Eds.)

bioactive phytoconstituents from millets. This is obvious that maximum phenolic occur as a bound categories and their base hydrolysis employing ultrasonication, microwave, and pressurized liquid extraction (PLE) methods results in higher yields. For rapid total quantification, spectrophotometric measurements are utilized. Although, HPLC and GC procedures that use mass spectrometers to separate, accurately quantify, identify, and record metabolic changes have been widely used. The method of analysis could be determined by the resources available and research objectives. Exploring the properties of millet lays a basis for better utilization of millet crops and further development of millet as an industrially important crop.

7.1 INTRODUCTION

Millets are a drought-tolerant, high-yielding grain crop that is necessary for the majority of people in Africa and Asia. Even though millet is found in several forms in various places of the world, five are most commonly grown millets: *Eleusine coracana* (Finger millet), *Panicummilia ceum* (common millet), *Setariaitalica* (Foxtail millet), *Pennisetum typhoideum* (Pearl millet), and *Echinochloa frumentacea* (Barnyard millet) [1]. The common millet (proso millet) is mostly grown in China, middle East, Russia, Central Europe and in India [2]. The proso millet has a 10- to 11-week growing season and can thrive at a variety of heights with less water [3]. Foxtail millet is farmed in 23 countries, with the biggest producers being China, India, and Japan [4]. It's not just a fast-growing crop, but it's also pest and disease resistant. The grains have high nutritional value, health-beneficial effects, and numerous therapeutic characteristics [5]. In Africa's semiarid regions and the Indian subcontinent, finger millet is commonly grown and is critical to the survival of farmers [6]. Millets differ from one another based on their appearance, morphological features, maturity, the type of grain, etc. (Figure 7.1).

As per literature, it has been documented by various researchers that finger millet holds anti-diabetic, cholesterol-lowering, and antiulcer effects [7]. Barnyard millet is a quick-growing cultivar popular in Malaysia, the East Indies, India, China, Africa, Japan, and the United States [8]. The millets contain good quality proteins with all the amino acids (essential) excluding threonine and lysine. Moreover, the seeds also possess higher bioactive compounds and minerals Chethan [9]. Aside from its nutritional content, for its health benefits, millet is used for multiple purposes, including improved healing of wounds, mitigation of cardiovascular disease, and blood glucose level reduction [10]. Prior work suggested that potential triggers like oxidative

FIGURE 7.1 Generally used millets are represented in pictures.

stress could be responsible for the commencement of prolonged health-related problems, e.g., cancer, diabetes, neurodegenerative disorders, cardiovascular disease, and arthritis. Millets are high in antioxidants, which help to prevent oxidation [11]. The common names of millets used in a variety of nations, as well as their taxonomic classification, are summarized in Table 7.1.

TABLE 7.1 The Scientific and Common Names for the Most Common Millets

Scientific Name	Common Name	Taxonomy (Tribe)	Grain Production Zones
Pennisetum glaucum	Munga, cattail, pearl, Sanyo, Seno	Paniceae	Central, Eastern, Western Southern Africa, India
Elusine coracana	African, Rapoko, Finger, bird's foot, ragi	Eragrostideae	India, Sri Lanka, Tanzania, Southern and Eastern Africa, Malaysia, Uganda
Setaria italic	Hungarian, German, Foxtail, Italian	Paniceae	Eastern Europe, China, India
Panicum milliaceum	Broom, hog, common, Proso	Paniceae	India, Japan, China, Russia, Ukraine, Kazakhstan
Panicum sumatrense	Heenmeneri, Little, panic, blue	Paniceae	India
Paspalums crobiculatum	Varagu, Kodo	Paniceae	India
Echinochola crus-galli	Japanese, Sanwa, Barnyard	Paniceae	Japan, Malaysia, China India
Eragro stistef	Abyssinian, Teff	Paniceae	Australia, Eritrea, Ethiopia

Source: Adapted from: Ref. [12].

Minor millets contain a variety of components known as phytochemicals or plant bioactive substances, which have the potential to improve one's health [13]. Nutraceuticals are pharmacologically active substances that have health-enhancing, disease-fighting, or therapeutic properties. Bioactive compounds are nutritional components present in small amounts in meals. These compounds are advantageous to human health, but they are not required for human survival [14]. Millets should be recognized as functional foods and nutraceuticals as they furnish essential nutrients like proteins, vitamins, minerals, dietary fibers (DFs), energy, and antioxidants. The germ/bran fraction of cereal grains contains the majority of bioactive chemicals found in the whole grains of millet. Chandrasekara & Shahidi [15] found a number of flavonoids primarily in soluble fractions from phenolic extracts of millets, including flavan-3-ols (monomers and dimmers), flavanols, their flavones, and glycosides (Table 7.2). The numerous extraction and analytical approaches that have been employed for bioactive chemical identification are discussed in this chapter. Novel processing strategies such as ultrasonic aided extraction and pressurized liquid extraction (PLE) have also been suggested for efficiently extracting various bioactive components from millets and other related cereals.

TABLE 7.2 Important Polyphenols (Phenolic and Flavonoids) Present in Millet Grains

Millet	HBAS	HCAS	FLOS	References
Finger	Gallic, vanillic, protocatechuic, p-hydroxybenzoic	Trans-ferulic, cinnamic, caffeic, p-coumaric, sinapic	Catechin, epicatechin, epigallocatechin, gallocatechin	[17]
Pearl	Vanillic, protocatechuic, p-Hydroxybenzoic	p-coumaric, caffeic, sinapic, cinnamic, trans-ferulic	Myricetin, apigenin	[16]
Proso	p-Hydroxybenzoic, syringic, gallic, protocat-echuic, vanillic	Caffeic, chloro-genic, ferulic	Myricetin Kempherol	[16]
Foxtail	Protocatechuic, gallic, syringic, p-hydroxyben-zoic, vanillic	Transferulic, sinapic, caffeic, p-coumaric	Kempherol, quercetin, catechin	[16]
Kodo	p-hydroxybenzoic, gallic, vanillic, protocatechuic	Caffeic, chlorogenic, trans-ferulic, p-coumaric	Quercetin, kaempferol, luteolin	[16]
Little	Gallic, protocatechuic, vanillic	Caffeic, p-trans-ferulic	Apigenin	[16]
Barnyard	NA	–	N-(p-coumaroyl) serotonin, luteolin	[18]

Note: NA: data not available; HABS: hydroxybenzoic acids and their derivatives; HACS: hydroxycinnamic acids and their derivatives; and FLOS: flavonoids.

7.2 BIOACTIVE COMPOUNDS IN MINOR MILLETS

Bioactive substances such as phenolic, dietary fiber, and vitamins can be found in millet grains, occurring predominantly in external bran layers along with micro-chemicals (tocopherols and carotenoids) furnishing the antioxidant activity to millets.

7.2.1 PHENOLIC COMPOUNDS

The most common phenolic compounds viz. phenolic acid and tannins have been detected in abundance in cereals, whereas flavonoids are in tiny amounts [19]. These have been proven to operate as free radical terminators, reducing agents, metal chelators, and antioxidant agents [20]. Furthermore, the degree of hydroxylation and the position of the phenolic ring affect the phenolic antioxidant activity [21]. The antioxidant activity of millet grains also impacts a variety of other structural characteristics. The phenolic acids of millet are mostly found inbound forms (60%), with the rest existing in free form [22]. The insoluble-bound form of phenolic acids in millet seeds contains the most common phenolics, such as hydroxycinnamic acids and their various forms [23]. The other type of hydroxycinnamic acid, such as ferulic, has anti-inflammatory as well as acts as the strongest antioxidant agent [24]. Besides, the most important part of millet grains are monomeric and dimeric compounds such as ferulate, showing higher antioxidant activity than other compounds [25]. The other form of phenolic compounds with the strongest antioxidant properties are found in bound form. So, these compounds do not need the assistance of microbial activity in digestion [17]. The water-soluble feraxans (feruloyl arabinoxylans) extracted from finger millet have a superior antioxidant activity of about 5,000 times stronger than sulfated polysaccharides [26].

7.2.2 FLAVONOIDS

Flavonoids make up the strongest and most varied category of plant bioactive chemicals and are found in abundance in grains, vegetables, and fruits. Millet contains anthocyanidins, flavanols, aminophenolics, chalcones, flavones, and flavanones in appreciable amount [27] Flavonoids are found as O- or C-form of glycosides in association with sugars and as free aglycones [28]. In the major cereals, the most abundant flavonoids, however, are the C-flavones glycosides [29]. The antioxidant activity of isolated bioactive compounds viz. luteolin, flavones, and tricin, from barnyard millet grains, has been demonstrated [18]. Flavonoids have been found in high concentrations in whole millet grains and pearled fractions and used as functional dietary additives in basic diets.

The color of food such as red, yellow, blue, and purple due to the presence of polyphenolic like flavonoids. They have aromatic rings joined by a heterocyclic ring of three carbon molecules [30]. Several flavonoids from distinct groups of anthocyanins, proanthocyanidins, flavanols, flavones, and flavonols have been discovered in millet grains. Millets have been reported to contain flavonoid concentrations ranging from 25.15 to 1,203 mol CE/g of crude extract, with Kodo millet having the highest value, followed by proso millet and finger millet [16]. Flavonoids are the most common type of polyphenol, with a basic chemical structure containing a heterocyclic $C_6C_3C_6$ skeleton [31]. Some plants have considered being great producers of flavonoids, which can be used in conjunction with food preservatives to add to a good diet. Millets have a large number of flavonoids. Dietary flavonoids are powerful antioxidants than vitamins E and C. Quercetin, catechin, orientin, luteolin, isoorientin, vitexin, myricetin, apigenin, daidzein, violanthin, sponarin, tricin lucenin 1, and isovitexin, are some of the flavonoids found in millets [9, 32, 33]. Millet flavonoids have been shown to have anti-analgesic, antihypertensive, inflammatory, anticancer, and hypolipidemic activities for medical and clinical purposes [34–37].

7.2.3 TOTAL DIETARY FIBER

Plant foods have dietary fiber as an important component. It's a less important nutrient, yet it's crucial for inside upkeep. Dietary fiber is a soluble and insoluble fiber complex carbohydrate that is non-absorbable and found in plant cell walls. Soluble and insoluble fibers are the two forms of fiber that have different nutritional roles. Fiber promotes digestion, lowers cholesterol, and is a low-calorie, animal-fat-free diet [6]. Obesity, heart disease, constipation, colon cancer, and diabetes disease are linked to dietary fiber consumption [38]. Fiber can be utilized for both methodological and nutritional goals, such as a fat replacer. The physiological activities of dietary fiber are the most likely dependent on its physicochemical features, includes water and oil holding capabilities, organic molecule absorption, bacterial deterioration, ion-cation exchange activity, and free radical properties [39]. Millet grain outer cover contains dietary fiber in the form of pectin, cellulose, and hemicelluloses, and all of these forms are resistant to digestive enzyme breakdown [40]. The inclusion of dietary fiber in millet grain seed coats is helpful to human health since it influences various metabolic and digestive processes, including glucose-uptake, and HDL AND LDL levels [41]. The grain outer coat, being a valuable origin of dietary fiber, produces a large number of byproducts, which are mostly used as complex flour in bakery product production.

7.2.4 PHYTOSTEROL COMPONENTS IN MILLETS

Phytosterols are significant biological constituents in plant cells that provide structural integrity to membranes and exert bioactive potential by producing a variety of health-concerned advantages. For the production of other biofunctional substances such as brassino steroids, phytoecdysteroids, glycoalkaloids, steroidal, and saponins, these compounds also act as precursors [42].

The whole phytosterol compounds of finger millet varied from 477 to 568 mg/kg comprising the compounds like campesterol and stigmasterol, out of which the sitosterol is about 85% [43]. These three components in millets had been discovered by Ryan et al. [41] and quantified as campesterol (8.07 milligram/gram), stigmasterol (0.08 milligram/gram) and sitosterol (48.03 milligram/gram) of grains. The phytosterol content of foxtail millet bran oil was determined to be 56.26% sitosterol, 15.91% campesterol, and 15.12% stigmastanol, with low amounts of stigmasterol (5.58%) and fucosterol (4.16%) and ergosterol (2.97%) [44]. Squalene, a long-chain triterpene molecule with a high unsaturation index, acts as a carrier agent in the phytosterol production pathway in millets [45]. While sterols and stanols have same effects on human health, the role of stanols in millet bioactivity needs to be enquired advance [46].

Millet components have free radical scavenging action. Due to an imbalance between greater free-radical synthesis and the weak antioxidant defense chemical mechanism, the degenerative and chronic diseases have been regarded to occur mostly due to oxidative stress [47]. Antioxidant properties in terms of metal complex, reducing power and oxidative agents have been studied in millets [48]. Free radical inhibition activity was found in five decided populations of pearl millet, running from 66.9% to 84.7% [49]. Prolamin peptides from foxtail millet have been used to slow the malondialdehyde and other oxidative reactions while also increasing glutathione levels, resulting in antioxidant effect [46]. Antioxidant activities have been linked to bioactive compounds; hence colored finger millet cultivars have greater oxygen absorption ability and free-radical scavenging activity than in seed cultivars with white seed coat [50]. The other constituents like flavonoids, phenolic acids, carotenoids, vitamin E, tannins, insoluble fiber, xylo oligosaccharides and bioactive peptides are also included.

7.2.5 ANTI-NUTRITIONAL FACTOR

Anti-nutritional components are harmful components found in cereals and legumes that prevent nutrients from being absorbed and lower their bioavailability in the body. Alkaloids, phytic acid, tannins, protease inhibitors and lectins, are among the anti-nutritionals found in cereal grains [50]. Phytic acid and tannin acids, on the other hand, are the most studied anti-nutritional components in millet grains. Proso millet and foxtail millet have phytic acid levels of 7.2 and 9.9 mg/g, respectively. In comparison to polished grains, whole grains have a higher content of phytic acid [51]. Proso and barnyard millet had the greatest inositol hexa-phosphates of all millets, whereas kodo had the highest condensed tannins, accompanied by barnyard and finger millet [52]. The phytic acid concentration of several millet varieties ranged from 3.27–3.60 milligram/gram (barnyard millet) to 5.44–5.78 milligram/gram (finger millet), with trypsin inhibitors of 7.85–8.37% and 8.0%, respectively [53]. Similarly, pearl millet flour contained 9.20 milligrams/gram of phytic acid and 2.20 milligrams/gram of tannin acid, whereas finger millet contained 6.10 milligrams/gram of phytic acid and 3.50 milligrams/gram of tannin acid, according to another study [54]. Kodo millet cultivars have low amount phytic acid and tannin compounds, with phytic acid and tannin content varying from 1.02–1.04 milligram/gram and 1.00–1.02 milligram/gram, respectively [55].

7.3 EXTRACTION METHODS FOR BIOACTIVE COMPOUNDS FROM MINOR MILLET

Optimized extraction of bioactive phytochemicals is critical to accurately evaluate phytochemical content in food and plant products. Accurate quantification enables adequate dietary and safety guidelines to be established. There are two types of bioactive phytochemicals: free and conjugated soluble form and bound insoluble forms. Understanding the various forms of bioactive phytochemicals is critical for precisely quantifying and comprehending their health-promoting and bioavailability characteristics. Peer-reviewed literature has reported on wide-ranging extraction processes and solvents [56, 57]. Sequential extraction techniques employing various solvents were utilized to achieve full and distinctive extraction of the variable phenolic content in millet, and subsequent quantification. Li et al. [58], for example, used solvent for extraction of the total phenolics of pearl millet and barnyard milled flour. The total phenolic content (TPC) (albumins and globulins) have been extracted with 70% ethanol at a temperature below 60°C followed by 2 hours reflux. Similar method has been followed on the total flavonoid content (TFC). Wang et al. [59] extracted the TPC of foxtail millet flour using various solvent. For the isolation of the phenolic components from defatted whole grain foxtail millet, acetone (70%), ethanol (80%), methanol (70%), and a mixture of aqueous/methanol/acetone (6:7:7, v/v/v) have been employed to extract the TPC (albumins and globulins). Table 7.3 shows a full view of the extraction methods of several bioactive compounds of minor millet.

TABLE 7.3 The Broader Classification of Bioactive Compounds in Millets and Their Extraction Methods

Bioactive Compounds	Sources	Solvent Extraction/Techniques Applied	References
Total phenolics (mg GAE/g)	Pearl millet Barnyard millet Italian millet	70% ethanol/the contents were refluxed for 2 h at a temperature not exceeding 60°C.	[58]
	Foxtail millet	To separate phenolic compounds from defatted whole grain foxtail millet, four common solvents were utilized: acetone (70%), ethanol (80%), methanol (70%), and a combination of aqueous/methanol/acetone (6:7:7, v/v/v). Refluxing, homogenization, cold extraction, and sonication were all used.	
Total flavonoids (mgQE/g)	Pearl millet Barnyard millet Italian millet	70% ethanol/the contents were refluxed for 2 hrs. at a temperature not exceeding 600°C.	[60]
Vanillin	Proso millet Barnyard millet	Supercritical fluid extraction	[61]
Proanthocyanidins	Foxtail millet Proso millet	70% (v/v) aqueous acetone; vanillin-HCl.	[62]
	Finger millet	Reagent (2.5 ml of 0.5%), (0.5% vanillin (w/v) in 4% concentrated HCl in methanol)/Rotary evaporator.	
Phytosterol	Foxtail millet Proso millet	(IS) 5α-cholestan-3β-ol (500 μL of 1.0 mg/mL in MTBE) and cholestanyl; LC–GC.	[19]

7.4 ANALYTICAL TECHNIQUES FOR QUALITATIVE AND QUANTITATIVE ANALYSIS

Due to rapid advances in technology, the analytical methods which are used for the quantification of bioactive compounds in whole seeds are continuously changing. The research work and the resources present also affect the methods which are used for analysis (Table 7.4). It has been possible to measure, identify, and separate the bioactive components in cereal extracts by utilizing high-performance liquid chromatography (HPLC) and gas chromatography (GC) instruments [63]. HPLC, along with diode array detection (DAD, which frequently requires an additional characterization period, is the most common and reliable method generally preferred over GC. To determine and separate the organic and inorganic solutes in a mixture, a method of HPLC is used in various samples such as food, pharmaceutical, industrial, environmental, and biological samples [57]. This technique separates substances based on the principle of their solid particle interactions with the solvent of the mobile phase inside a densely packed column. A solid non-polar phase (column C18) and a polar mobile phase, which is a mixture of water and solvent, are used in HPLC analysis. The analyte must be eluted down a column at a high pressure of up to 400 bars before passing through a diode array detector (DAD). To aid in analyte identification, a DAD examines the absorption spectra of the analytes. HPLC is excellent for detecting substances that cannot be vaporized or that break down at high temperatures, and it is a good complement to GC [64]. Chromatography is a separation technique for molecules based on their charge, size, and shape [65]. Chromatography involves dissolving analytes in solvent and passing them through a solid phase that acts as a sieve media. As the molecule passes through the molecular sieve, it is separated. Paper and thin layer chromatography (TLC) are chromatographic processes that rapidly produce qualitative data and enable the acquisition of quantitative data.

TABLE 7.4 Major Bioactive Chemical Constituents in Minor Millet

Millet Type	Analysis	Bioactive Constituents	References
Common millet	HPLC-DAD-ESI-MS	*p*-Hydroxybenzoic, chlorogenic, vanillic, gallic, gentisic, sinapic, and ferulic acids.	[22]
Foxtail millet	HPLC	Chlorogenic, ferulic, *p*-coumaric, syringic acids, caffeic.	[9]
Finger millet	GC-MS	Trans-cinnamic, gallic, *P*-hydroxybenzoic, vanillic, ferulic, sinapic, quercetin, caffeic, protocatechuic, syringic, *p*-coumaric acids.	[61]
Pearl millet	HPLC-DAD-ESI-MS	*p*-hydroxybenzoic and *p*-coumaric acids, ferulic, Methyl vanillate, apigenin, caffeic	[33]
Barnyard millet	HPLC/GC-MS	Gallic, *p*-Hydroxybenzoic, vanillic, sinapic, gentisic, chlorogenic, and ferulic acids.	[9]
Proso millet	RP-HPLC	Coumaric acid, ferulic, and chlorogenic acid, xanthophyll, caffeic, and syringic acid, zeaxanthin, and cryptoxanthin.	[22]
Kodo millet	GC-MS	9-octadecenoic acid methyl, 10-trans; 12-cisoctadecadienoate, N-5-hydroxynpentyl, pterin-6-carboxylic acid, N-(3,5-dinitropyridinyl) laspartic acid ester, stigmasterol, Çsitosterol, pregnenolone, campesterol.	[66]

7.5 INNOVATIVE PROCESSING METHODOLOGIES

Some novel extraction methods have been developed for the extraction of nutraceuticals from plants, including ultra-sonication extraction, solvent extraction, and supercritical fluid extraction (SCFE). Novel food processing techniques provide several advantages over traditional extraction methods, including being more environmentally friendly, reducing solvent usage, minimizing extraction time, and increasing yield [67]. Ultra-sonication is one such helpful methodology that, in the future, could lead to new economic extraction opportunities and methods. It does not necessitate the use of complicated instruments and is reasonably inexpensive. Ultrasonics enables the simultaneous extraction and encapsulation of compounds, as well as the selective hydroxylation of carotenoids and polyphenolics to enhance their bioactivity [68]. UAE has been employed to extract phenolics from wheat bran at the extraction temperature of 60°C and ethanol concentration of 64%, and time period of 25 minutes [69]. Ultrasound extraction yielded a total phenolic acid from 3.12 to 0.03 mgGAE/g bran, proving it to be as successful as solvent method, which yielded from 2.02–3.02 mg GAE/g outer layer after treatment of 15 hours. In another work, the traditional techniques of extraction and ultrasound-assisted method yielded equivalent yields of arabinoxylans (AXs) and phenolics-rich heteroxylans [58]. However, the UAE not only reduced the time necessary for traditional extraction from 60 minutes to 5 minutes, but also reduced the amount of sodium hydroxide applied in the process, constructing it potentially usable in cosmetics, pharmacy, and food industries. Some portions of bioactive extract have been utilized to extract oil from soybeans using high-intensity ultrasound-assisted extraction [60]. In comparison to non-treated samples, high oil yields obtain by using a mixed solvent with ultrasound. Gas chromatographic examination revealed that ultra-sonicated soybean oil has no substantial alterations in the saturated and unsaturated fatty acid composition. This technology has also been discovered to reduce the time it takes to extract edible oil, which could help commercial oil producers boost throughput. UAE has also been shown to be capable of extracting -glucans with a high molecular weight from a waxy barley genotype [70]. The findings have been attributed to the mechanical effects of ultrasonically produced cavitations, which increase the mass transfer limitations of the isolation process by breaking the cell walls of plants. According to study, the glucan concentration and molecular weight of targeted food items dictate the amount of energy required throughout the UAE process. Low-intensity treatment produced -glucans with a high molecular weight but low extraction yields, whereas high-intensity treatment increased -glucan yield while polymerizing the food sample [71].

UAE was utilized to extract flavonoids from wheat bran [68]. The extraction temperature and ethanol concentration required were 60°C and 64%, respectively, for a 25-minute time period. After 15 hours of treatment, ultrasound extraction gave a TPC of 3.12 ± 0.03 mg gallic acid equivalents (GAE)/g bran, demonstrating that it is as effective as Soxhlet extraction, which delivered 2.2–3.2 mg GAE/g bran. In another study, a similar yield of AXs and heteroxylans rich in phenolics was produced utilizing both traditional and ultrasonic aided extraction techniques [69]. The UAE, on the other hand, not only lowered the time required for typical extraction from 60 to 5 minutes, but also decreased the quantity of NaOH used in the process, making it potentially suitable for usage in the cosmetics, pharmaceutical, and food sectors. Apart from phenolics and polysaccharides, high-intensity ultrasound-assisted extraction was used to extract soybean oil using hexane, isopropanol, and a 3:1 combination of hexane and isopropanol [58]. In compared to untreated samples, ultrasound combined with a mixed solvent (isopropanol-hexane) resulted in significantly increased oil yields. No significant changes in the fatty acid content of ultra-sonicated soybean oil were detected using GC. Additionally, this approach was discovered to shorten the time required to extract edible oil, which might aid in increasing throughput in commercial oil production systems. Additionally, UAE has been demonstrated to be an efficient approach for extracting high molecular-glucans from a waxy barley genotype [60]. The energy (controlled by time and amplitude) required for the UAE process was determined to be dependent on the target product's properties, such as its -glucan concentration and molecular weight. Low-intensity treatment (250–425 kJ/L) generated -glucan with a high molecular weight (>400 kDa) but low extraction yields (less than 50%), whereas high-intensity treatment increased -glucan yield while depolymerizing the product, which may be desired in some cases.

The PLE technique is a unique approach for extracting bioactive compounds from natural sources. PLE is another term for rapid solvent extraction. PLE is a process for extracting solids from liquids that makes use of high-pressure solvents to allow solvents to be heated past their boiling temperatures. It was used to rapidly extract tocopherols and tocotrienols from cereals, with detection by electrospray ionization-mass spectrometry (MS) [72]. The best results were achieved when methanol was used at a temperature of 50°C, a pressure of 11 MPa, and a single extraction cycle with a 5-minute extraction (static) interval. Additionally, PLE has been used to extract polar and non-polar lipids such as phytosterols from ground corn kernels and rolled oats [39]. In both maize and oats, increasing the polarity of the solvent enhanced yield, and more lipids were extracted at 100°C than at 40°C for a static duration of 10 minutes at 7 MPa for each solvent. In all of these studies, the combination of high pressures (3.3–20.3 MPa) and temperatures (40–200°C) results in a more efficient extraction technique that minimizes solvent use while increasing automation and speed. High temperature and pressure enhance the solubility of analytes and their desorption kinetics from matrices. As a result, solvents for extraction such as water, which are useless at low temperatures, may be more successful at increased PLE temperatures.

7.6 POTENTIAL HEALTH BENEFITS OF MILLET GRAINS

Many scientists have discovered that diets high in plant or whole-grain materials can help prevent significant health problems such as Parkinson's disease and age-related ailments such as diabetes, cardiovascular disease, and cancer [73]. This might be because whole grains include vitamins, minerals, essential fatty acids, bioactive compounds, and fiber, all of which have been associated with increased health. Additionally to nutritive components, they are high in resistant starch, oligosaccharides, and bioactive substances. Millets are abundant in protein, minerals, vitamins, and dietary fiber, as well as antioxidants such as phenols and flavonoids, which have been found to protect the human body against carcinogenic and cardiovascular diseases, malignancies, hypertension, high cholesterol, and fat absorption [74].

7.6.1 ANTIOXIDANTS AND THEIR ACTIVITIES

Antioxidant chemicals are gaining prominence due to their primary functions as lipid stabilizers and suppressors of excessive oxidation, which is associated with cancer and aging [75]. Their stable radical intermediates protect a variety of dietary constituents, most notably fatty acids and oils, against oxidation [76]. The phenolic acids and their derivatives, flavonoids, and tannins found in millets' seed coats have a variety of functions, including acting as reducing agents (free radical terminators), metal chelators, and singlet oxygen quenchers [77]. The antioxidant activity of phenolic substances is due to their capacity to donate hydrogen atoms to electron-deficient free radicals through hydroxyl groups on benzene rings, forming a resonance-stabilized and less reactive phenoxyl radical. Natural antioxidants in the edible flours of tiny millets have been studied. The antioxidant capacity of finger, small, foxtail, and Proso millets was found to be greater, and their total carotenoids concentration ranged between 78 and 366 mg/100 g.

7.6.2 MILLET FOR DIABETICS

Diabetes mellitus is a chronic metabolic disease marked by hyperglycemia and abnormal carbohydrate, protein, and lipid metabolism. According to reports, hyperglycemia can result in non-enzymatic glycosylation of a variety of proteins, culminating in the development of chronic problems associated with diabetes [78]. As a result, regulating the postprandial blood glucose increase is crucial for managing diabetes and preventing chronic vascular problems [79], which may be accomplished by the consumption of a high-complex carbohydrate and high-fiber diet. Consumption of finger millet-based diets, for example, leads to considerably lower plasma glucose levels, which may be attributable to finger millet's greater fiber content than rice and wheat. As a result, millet grains may be beneficial in avoiding and treating diabetes.

7.6.3 MILLET AGAINST CARDIOVASCULAR DISEASE

Most of the countries on this list are confronted with high and rising rates of cardiovascular disease. Rats fed on a diet of native and processed starch from barnyard millet had the lowest blood glucose, serum cholesterol, and triglycerides levels when compared to rats fed rice or other minor millets [80]. Additionally, feeding proso millet protein improves adiponectin and high-density lipoprotein (HDL) cholesterol levels in genetically obese type-2 diabetic mice fed a high-fat diet [81]. Additionally, we evaluated the inhibitory effects of phenolic extracts from kodo, finger, proso, foxtail, little, and pearl millets on lipid peroxidation *in vitro* using copper-mediated human LDL cholesterol oxidation and several food model systems, including cooked comminuted pork, stripped corn oil, and a linoleic acid emulsion. Millet extracts reduce LDL cholesterol oxidation by 1 to 41% at a final concentration of 0.05 mg/ml. Millets also reduce lipid oxidation effectively in dietary systems [82].

7.6.4 MILLETS AGAINST CANCERS

Millet grains are known to be high in phenolic acids, tannins, and phytate, all of which act as "anti-nutrients" [83]. However, it has been demonstrated in animals that these antinutrients decrease the incidence of colon and breast cancer. Additionally, a recent study indicated that millet phenolics may be beneficial in preventing the beginning of cancer [16].

7.6.5 MILLETS AGAINST CELIAC DISEASE

The increased consumer desire for innovative, delicious, and "healthy" meals, along with the expanding number of celiac disease patients, has created a new market for cereal goods derived from grains other than wheat and rye. Oat, sorghum, and millet have developed a unique position in this competitive market [84]. Celiac disease is an immune-mediated enteropathy that is induced in genetically sensitive individuals by gluten consumption. It is one of the most prevalent chronic diseases globally. Historically, celiac disease was thought to be an uncommon condition that mostly affected children of European ancestry. Gluten-free foods and drinks are in high demand in industrialized nations due to the rising number of persons with celiac disease and other intolerances to wheat, barley, or rye. Millets, on the other hand, are gluten-free; they offer significant promise in meals and beverages suited for those with celiac disease. As a result, millet grains and their fractions have the potential to be beneficial in the production of gluten-free foods [82]. Table 7.5 summarizes the findings of numerous studies on the health benefits of various millets.

TABLE 7.5 Potential Health Benefits of Minor Millet

Sl. No.	Type of Millet	Potential Health Benefits	References
1.	Proso millet	Reduce heart diseases, reduction in glycemic index, anticancer properties.	[85]
2.	Kodo millet	Hypoglycemic activity, high antioxidant activity, HDL, and LDL management.	[20]
3.	Barnyard millet	Reduces blood glucose level, Fatty acids and serum, cholesterol, anticancer properties.	[86]
4.	Foxtail millet	Glycemic Index management, improve immune system; cholesterol management.	[9]
5.	Little millet	Reduces blood glucose level, Fatty acids and serum, cholesterol, anticancer properties.	[33]
6.	Finger millet	Prevent heart diseases, prohibition of cataracto-genesis, antioxidant properties and reduce of diabetes mellitus.	[61]
7.	Pearl millet	Antioxidant and anti-inflammatory properties, good heart health, reduce glucose release, slow down in inflammation diseases.	[33]

7.7 CONCLUSION

Minor millet and its derivatives are rich in bioactive chemicals that are good to human health. The amounts of bioactive chemicals in seed bran and germ are greater than in the rest of the seed. Methods for extracting and testing bioactive components from minor millets are compared in a comprehensive study. Using solvent extraction, ultrasonic technique, and pressured liquid extraction processes leads in larger yields of free and bound phenolic chemicals. GC and high-performance liquid chromatography (LC) techniques with mass spectrometers have been widely employed for separation, recognition, and limiting, but spectrophotometers are used for quick total quantification. Studying how processing impacts the concentration of bioactive chemicals in minor millets has attracted a lot of attention recently. In a limited way, NMR methods have been used to characterize and profile structural properties. Minor millet and minor millet products have not been studied extensively for their interactions with bioactive chemicals. Furthermore, there is also a paucity of research on how to make bound phenolics more bioavailable for their intended use.

KEYWORDS

- **antioxidants**
- **bioactive compounds**
- **cardiovascular disease**
- **celiac disease**
- **dietary fiber**
- **flavonoids**
- **minor millet**
- **quantitative analysis**

REFERENCES

1. Ravindran, G., (1991). Studies on millets: Proximate composition, mineral composition, and phytate and oxalate contents. *Food Chem.*, *39*, 99–107.
2. Cho, Y. I., Chung, J. W., Lee, G. A., Ma, K. H., Dixit, A., Gwag, J. G., & Park, Y., (2010). Development and characterization of twenty-five new polymorphic microsatellite markers in proso millet (*Panicum miliaceum L.*). *Genes Genomics*, *32*, 267–273.

3. Kalinova, J., & Moudry, J., (2006). Content and quality of protein in proso millet (*Panicum miliaceum L.*) varieties. *Plant Foods Hum Nutr.*, *61*, 43.
4. Upadhyaya, H. D., Pundir, R. P. S., Gowda, C. L. L., Gopal, R. V., & Singh, S., (2009). Establishing a core collection of foxtail millet to enhance the utilization of germplasm of an underutilized crop. *Plant Genet Resour.*, *7*, 177–184.
5. Amadou, I., Amza, T., Shi, Y. H., & Le, G., (2011). Chemical analysis and antioxidant properties of foxtail millet bran extracts. *Songklanakarin J. Sci. Technol.*, *33*, 509–515.
6. Chethan, S., & Malleshi, N., (2007). Finger millet polyphenols optimization of extraction and the effect of pH on their stability. *Food Chemi.*, *105*, 862–870. https://doi.org/10.1016/j. foodchem.2007.02.012.
7. Takan, J. P., Chipili, J., Muthumeenakshi, S., Talbot, N. J., Manyasa, E. O., Bandyopadhyay, R., et al., (2012). Magnaporthe oryzae populations adapted to finger millet and rice exhibit distinctive patterns of genetic diversity, sexuality and host interaction. *Mol Biotechnol.*, *50*, 145–158.
8. Anju, T., & Sarita, S., (2010). Suitability of foxtail millet (*Setaria italica*) and barnyard millet (*Echinochloa frumentacea*) for development of low glycemic index biscuits. *Malays J Nutr.*, *16,* 361.
9. Chethan, S., (2008). *Finger Millet (Eleusine Coracana) Seed Polyphenols and Their Nutraceutical Potential.* Thesis—Doctorate of Philosophy. University of Mysore, Mysore.
10. Rajasekaran, N. S., Nithya, M., Rose, C., & Chandra, T., (2004). The effect of finger millet feeding on the early responses during the process of wound healing in diabetic rats. *Biochim. Biophys. Acta.*, *1689*(3), 190–201. doi: 10.1016/j.bbadis.2004.03.004.
11. Halliwell, B., (2012). Free radicals and antioxidants: Updating a personal view. *Nutr. Rev.*, *70*(5), 257–265. doi: 10.1111/j.1753–4887.2012.00476.x.
12. Serrna-Saldivar, S., & Rooney, L., (2004). Structure and chemistry of sorghum and millets. In: Dendy, D. A. V., (ed.), *Sorghum and Millets: Chemis and Techno* (pp. 69–124).
13. Goldberg, A. L., (2003). Protein degradation and protection against misfolded or damaged proteins. *Nat.*, *426*(6968), 895–899.
14. Kris-Etherton, P. M., Hecker, K. D., Bonanome, A., Coval, S. M., & Binkoski, A., (2002). Bioactive compounds in foods: Their role in the prevention of cardiovascular disease and cancer. *Ameri J of Med.*, *13*(8), 234–256.
15. Chandrasekara, A., & Shahidi, F., (2011a). Determination of antioxidant activity in free and hydrolyzed fractions of millet grains and characterization of their phenolic profiles by HPLCDAD-ESI-MS. *J. Funct Foods*, *3,* 144–158.
16. Chandrasekara, A., & Shahidi, F., (2011c). Antiproliferative potential and DNA scission inhibitory activity of phenolics from whole millet grains. *J. Funct. Foods*, *3*, 159–170.
17. Rao, R. S. P., & Muralikrishna, G., (2006). Water soluble feruloyl arabinoxylans from rice and ragi: Changes upon malting and their consequence on antioxidant activity. *Phytochemistry.*, *67*(1), 91–99. doi: 10.1016/j.phytochem.2005.09.036.
18. Watanabe, M., (1999). Antioxidative phenolic compounds from Japanese barn yar D millet (*Echinochloa utilis*) grains. *J. Agr. Food Chem.*, *47*, 4500–4505. doi: 10.1021/jf990498s.
19. Devi, P. B., Vijayabharathi, R., Sathyabama, S., Malleshi, N. G., & Priyadarisini, V., (2014). Health benefits of finger millet (*Eleusine coracana* L.) polyphenols and

dietary fiber: A review. *J. Food Sci. Technol-Mysore*, *51*(6), 1021–1040. doi: 10.1007/s13197–011–0584–9.
20. Rao, M. V. S. S. T., & Muralikrishna, G., (2002). Evaluation of the antioxidant properties of free and bound phenolic acids from native and malted finger millet (Ragi, *Eleusine coracana* Indaf-15). *J. Agr. Food Chem.*, *50*(4), 889–892. doi: 10.1021/jf011210d.
21. Miyake, T., & Shibamoto, T., (1997). Antioxidative activities of natural compounds found in plants. *J. Agr. Food Chem.*, *45*, 1819–1822. doi: 10.1021/jf960620c.
22. Zhang, L. Z., & Liu, R. H., (2015). Phenolic and carotenoid profiles and antiproliferative activity of foxtail millet. *Food Chem.*, *174*, 495–501. doi: 10.1016/j.foodchem.2014.09.089.
23. Ralph, J., Bunzel, M., Marita, J. M., Hatfield, R. D., Lu, F., Kim, H., Schatz, P. F., et al., (2004). Peroxidase-dependent cross-linking reactions of P-hydroxycinnamates in plant cell walls. *Phytochem. Rev.*, *3*, 79–96. doi: 10.1023/B:PHYT.0000047811.13837.fb.
24. Castelluccio, C., Paganga, G., Melikian, N., Bolwett, G. P., Pridham, J., Sampson, J., & Rice-Evans, C., (1995). Antioxidant potential of intermediates in phenylpropanoid metabolism in higher plants. *FEBS Lett.*, *368*, 188–192. doi: 10.1016/0014–5793(95)00639-Q.
25. Ohta, T., Yamasaki, S., Egashira, Y., & Sanada, H., (1994). Antioxidant activity of corn bran hemicellulose fragments. *J. Agr. Food Chem.*, *42*, 653–656. doi:10.1021/jf00039a010.
26. Gulcin, I., (2012). Antioxidant activity of food constituents: An overview. *Arch. Toxicol.*, *86*, 345–391. doi: 10.1007/s00204–011–0774–2.
27. RiceEvans, C. A., Miller, N. J., & Paganga, G., (1996). Structure-antioxidant activity relationships of flavonoids and phenolic acids. *Free Radical Bio. Med.*, *20*(7), 933–956. doi: 10.1016/0891–5849(95)02227–9.
28. Stobiecki, M., (2000). Application of mass spectrometry for identification and structural studies of flavonoid glycosides. *Phytochemistry*, *54*, 237–256. doi: 10.1002/anie.201402958.
29. Brazier-Hicks, M., Evans, K. M., Gershater, M. C., Puschmann, H., Steel, P. G., & Edwards, R., (2009). The C-glycosylation of flavonoids in cereals. *J. Biol. Chem.*, *284*, 17926–17934. doi: 10.1074/jbc.M109.009258.
30. Duodu, K. G., & Awika, J., (2019). Phytochemical-related health promoting attributes of sorghum and millet. In: *Sorghum* and *Millets*. *Inter. J. Food Propert.*, *1659*, 225–258.
31. Hwang, S. L., Shih, P. H., & Yen, G., (2012). Neuroprotective effect of citrus flavonoids. *J of Agri and Food Chem.*, *60*, 877–885. https://doi.org/10.1021/jf204452y.
32. Pradeep, P. M., & Sreerama, Y., (2017). Soluble and bound phenolics of two different millet genera and their milled fractions: Comparative evaluation of antioxidant properties and inhibitory effects on starch hydrolyzing enzyme activities. *J of Functional Foods*, *35*, 682–693. https://doi.org/10.1016/j.jff.2017.06.033.
33. Banerjee, S., Sanjay, K. R., Chethan, S., & Malleshi, N., (2012). Finger millet (*Eleusine coracana*) polyphenols: Investigation of their antioxidant capacity and antimicrobial activity. *Afri. J. of Food Sci.*, *6*, 362–374. https://doi.org/10.5897/AJFS12.031.
34. Edge, M. S., Jones, J. M., & Marquart, L., (2005). A new life for whole grains. *J. of Ameri Dietetic Association*, *105*(12), 1856–1860. https://doi.org/10.1016/j.jada.2005.10.022.
35. Ekta, S., & Sarita, (2016). Potential functional implications of finger millet (*Eleusine coracana*) in nutritional benefits, processing, health and diseases: A review. *Inter J of Home Sci.*, *2*(1), 151–155.

36. Ahuja, K. G., Nath, P., & Swamy, K. R., (2013). Processing of millets. Paper presented national seminar on recent advances in processing, utilization and nutritional impact of small millets. *Foods and Nutrition* (1st edn.). New Delhi, India: Studium Press. Balasubramanian, S. Madurai Symposium, Thamukkam Grounds, Madurai.
37. Lestienne, I., Buisson, M., Lullien-Pellerin, V., Picq, C., & Trèche, S., (2007). Losses of nutrients and anti-nutritional factors during abrasive decortication of two pearl millet cultivars (*Pennisetum glaucum*). *Food Chemi.*, *100*, 1316–1323. https://doi.org/10.1016/j. foodchem.2005.11.02.
38. Upadhyaya, H. D., Ravishankar, C. R., Narasimhudu, Y., Sharma, N. D. R. K., Singh, S. K., Varshney, S. K., & Gowda, C. L., (2011). Identification of trait specific germplasm and developing a mini core collection for efficient use of foxtail millet genetic resources in crop improvement. *Field Crops Res.*, *124*, 459–467.
39. Moreau, R. A., Nystr€om, L., Whitaker, B. D., et al., (2018). Phytosterols and their derivatives: Structural diversity, distribution, metabolism, analysis, and health-promoting uses. *Progress in Lipid Res.*, *70*, 35–61.
40. Bhandari, S. R., & Lee, Y., (2013). The contents of phytosterols, squalene, and vitamin E and the composition of fatty acids of Korean landrace *Setaria italica* and *Sorghum bicolar* seeds. *Korean J. of Plant Resou.*, *26*, 663–672.
41. Ryan, E., Galvin, K., O'Connor, T. P., Maguire, A. R., & O'Brien, N. M., (2007). Phytosterol, squalene, tocopherol content, and fatty acid profile of selected seeds, grains, and legumes. *Plant Foods for Hum Nutri.*, *62*, 85–91.
42. Pang, M., He, S., Wang, L., Cao, X., Cao, L., & Jiang, S., (2014). Physicochemical properties, antioxidant activities and protective effect against acute ethanol-induced hepatic injury in mice of foxtail millet (*Setaria italica*) bran oil. *Food & Function, 5*, 1763–1770.
43. Ji, J., Liu, Y., Ge, Z., Zhang, Y., & Wang, X., (2019). Oleochemical properties for different fractions of foxtail millet bran. *J. of Oleo Science, 68*, 709–718.
44. Chandrasekara, A., & Shahidi, F., (2010). Content of insoluble bound phenolics in millets and their contribution to antioxidant capacity. *J. of Agric and Food Chem., 58*, 6706–6714.
45. Bouajila, A., Lamine, M., Rahali, F. Z., Melki, I., Prakash, G., & Ghorbel, A., (2020). Pearl millet populations characterized through *Fusarium* prevalence, morphological traits, phenolic content, and antioxidant potential. *J. of the Sci of Food and Agri., 100*, 4172–4181.
46. Ji, Z., Mao, J., Chen, S., & Mao, J., (2020). Antioxidant and anti-inflammatory activity of peptides from foxtail millet (*Setaria italica*) prolamins in HaCaT cells and RAW264. 7 murine macrophages. *Food Biosc., 36*, 100636.
47. Xiang, J., Apea-Bah, F. B., Ndolo, V. U., Katundu, M. C., & Beta, T., (2019a). Profile of phenolic compounds and antioxidant activity of finger millet varieties. *Food Chemi, 275*, 361–368.
48. Wu, Y. V., Sexson, K. R., Cavins, J. F., & Inglett, G. E., (1972). Oats and their dry-milled fractions. Protein isolation and properties of four varieties. *J. Agric. Food Chem., 20*, 757–761.
49. Holtekjølen, A. K., Kinitz, C., & Knutsen, S. H., (2006). Flavanol and bound phenolic acid contents in different barley varieties. *J. Agric. Food Chem., 54*, 2253–2260.

50. Ram, S., Narwal, S., Gupta, O. P., Pandey, V., & Singh, G., (2020). Anti-nutritional factors and bioavailability: Approaches, challenges, and opportunities. In: *Wheat and Barley Grain Biofortification* (pp. 101–128). Cambridge, USA: Elsevier.
51. Devisetti, R., Yadahally, S. N., & Bhattacharya, S., (2014). Nutrients and antinutrients in foxtail and proso millet milled fractions: Evaluation of their flour functionality. *LWT-Food Sci and Techno.*, *59*, 889–895.
52. Sharma, B., & Gujral, H., (2019b). Influence of nutritional and antinutritional components on dough rheology and in vitro protein& starch digestibility of minor millets. *Food Chemi.*, *299*, 125115.
53. Panwar, P., Dubey, A., & Verma, A., (2016). Evaluation of nutraceutical and antinutritional properties in barnyard and finger millet varieties grown in the Himalayan region. *J. of Food Sci. and Techno.*, *53*, 2779–2787.
54. Gull, A., Prasad, K., & Kumar, P., (2016). Evaluation of functional, antinutritional, pasting and microstructural properties of millet flours. *J. of Food Measu and Charac.*, *10*, 96–102.
55. Shyam, R., & Singh, R., (2018). Evalution of nutritional value and anti-nutritional factors of kodo millet (*Paspalum scrobiculatum L.*). Germplasm grown in the eastern. *Plant Archives*, *18*, 247–250.
56. Herrero, M., Cifuentes, A., & Ibañez, E., (2006). Sub- and supercritical fluid extraction of functional ingredients from different natural sources: Plants, food by-products, algae and microalgae: A review. *Food Chemi.*, *98*, 136–148.
57. Hancock, W. S., (1990). *High Performance Liquid Chromatography in Biotechnology.* Wiley-Interscience, New Jersey, USA.
58. Li, H., Pordesimo, L., & Weiss, J., (2004). High intensity ultrasound-assisted extraction of oil from soybeans. *Food Res. Int.*, *37*, 731–738.
59. Wang, L., & Weller, C. L., (2006). Recent advances in extraction of nutraceuticals from plants. *Trends Food Sci. Technol.*, *17*, 300–312.
60. Benito-Román, Ó., Alonso, E., & Cocero, M., (2013). Ultrasound-assisted extraction of β-glucans from barley. *LWT-Food Sci. Technol.*, *50*, 57–63.
61. Pradeep, P. M., & Sreerama, Y., (2015). Impact of processing on the phenolic profiles of small millets: Evaluation of their antioxidant and enzyme inhibitory properties associated with hyperglycemia. *Food Chem.*, *169*, 455–463.
62. Zhang, L., Liu, R., & Niu, W., (2014). Phytochemical and antiproliferative activity of proso millet. *PLoS One*, *9*(8), e104058. doi: 10.1371/journal.pone.0104058.
63. Lv, J., Lu, Y., Niu, Y., Whent, M., Ramadan, M. F., Costa, J., Luthria, D., & Yu, L., (2013). Effect of genotype, environment, and their interaction on phytochemical compositions and antioxidant properties of soft winter wheat flour. *Food Chemi.*, *138*, 454–462.
64. Katz, E. D., (1995). *High Performance Liquid Chromatography; Principle and Methods in Biotechnology (Separation Science Series)*. John Wiley & Sons, New Jersey, USA.
65. Heftmann, F., (1992). *Chromatography: Fundamentals and Application of Chromatographic and Electrophoretic Techniques* (5th edn.). Elsevier, Amsterdam, The Netherlands.
66. Sharma, S., Saxena, D. C., & Riar, C., (2018). Changes in the GABA and polyphenols contents of foxtail millet on germination and their relationship with in vitro antioxidant activity. *Food Chem.*, *245*, 863–870.

67. Vilkhu, K., Mawson, R., Simons, L., & Bates, D., (2008). Applications and opportunities for ultrasound assisted extraction in the food industry—A review. *Innov. Food Sci. Emerg. Technol.*, *9*, 161–169.
68. Wang, J., Sun, B., Cao, Y., Tian, Y., & Li, X., (2008). Optimization of ultrasound-assisted extraction of phenolic compounds from wheat bran. *Food Chem.*, *106*, 804–810.
69. Hromádková, Z., Košt'álová, Z., & Ebringerová, A., (2008). Comparison of conventional and ultrasound-assisted extraction of phenolics-rich heteroxylans from wheat bran. *Ultrason. Sonochem.*, *15*, 1062–1068.
70. Bustamante-Rangel, M., Delgado-Zamarreno, M., Sánchez-Pérez, A., & Carabias-Martínez, R., (2007). Determination of tocopherols and tocotrienols in cereals by pressurized liquid extraction–liquid chromatography–mass spectrometry. *Anal. Chim. Acta*, *587*, 216–221.
71. Thakur, M., & Tiwari, P., (2019). Millets: The untapped and underutilized nutritious functional foods. *Plant Archives*, *19*, 875–883.
72. Gupta, R. K., Gangoliya, S. S., & Singh, N., (2015). Reduction of phytic acid and enhancement of bioavailable micronutrients in food grains. *J of Food Sci and Techno.*, *52*, 676–684.
73. Nithiyanantham, S., Kalaiselvi, P., Mahomoodally, M. F., Zengin, G., Abirami, A., & Srinivasan, G., (2019). Nutritional and functional roles of millets—A review. *J of Food Biochemi.*, *43*, e12859.
74. Muthamilarasan, M., Dhaka, A., Yadav, R., & Prasad, M., (2016). Exploration of millet models for developing nutrient-rich graminaceous crops. *Plant Sci.*, *242*, 89–97.
75. Namikii, M., (1990). Antioxidants/antimutagens in food. *Criti Reviews in Food Sci and Nutri.*, *29*, 273–300.
76. Maillard, M. N., Soum, M. H., Boivia, P., & Berset, C., (1996). Antioxidant activity of barley and malt: Relationship with phenolic content. *LWT Food Sci and Techno.*, *3*, 238–244.
77. Sripriya, G., Chandrasekharan, K., Murty, V. S., & Chandra, T., (1996). ESR spectroscopic studies on free radical quenching action of finger millet (*Eleusine coracana*). *Food Chemis.*, *57*(4), 537–540.
78. Lebovitz, H. E., (2001). Effect of the postprandial state on nontraditional risk factors. *Americ. J. of Cardiolo.*, *88*, 204–205.
79. Bravo, L., (1998). Polyphenols: Chemistry, dietary sources, metabolism, and nutritional significance. *Nutrit. Reviews.*, *56*(11), 317–333.
80. Kumari, S. K., & Thayumanavan, B., (1997). Comparative study of resistant starch from minor millets on intestinal responses, blood glucose, serum cholesterol and triglycerides in rats. *J. Sci. Food Agricul.*, *75*, 296–302.
81. Park, K. O., Ito, Y., Nagasawa, T., Choi, M. R., & Nishizawa, N., (2008). Effects of dietary Korean proso-millet protein on plasma adiponectin, HDL cholesterol, insulin levels and gene expression in obese type 2 diabetic mice. *Biosci. Biotechn. Biochemi.*, *72*(11), 2918–2925.
82. Chandrasekara, A., & Shahidi, F., (2012). Antioxidant phenolics of millet control lipid peroxidation in human LDL cholesterol and food systems. *J. of the Ameri Oil Chemists' Soci.*, *89*(2), 275–285.
83. Thompson, L. V., (1993). Potential health benefits and problems associated with antinutrients in foods. *Food Resea. Internat.*, *26*, 131–149.

84. Angioloni, A., & Collar, C., (2013). Suitability of oat, millet and sorghum in breadmaking. *Food and Biopro Techno.*, *6*(6), 1486–1493.
85. Slama, A., Cherif, A., Sakouhi, F., Boukhchina, S., & Radhouane, L., (2019). Fatty acids, phytochemical composition, and antioxidant potential of pearl millet oil. *J. of Consumer Prot and Food Safety*. *15*, 145–151.
86. Sharma, S., Saxena, D. C., & Riar, C., (2016). Analyzing the effect of germination on phenolics, dietary fibers, minerals, and c-aminobutyric acid contents of barnyard millet (*Echinochloa frumentaceae*). *Food Biosc.*, *13*, 60–68.

CHAPTER 8

Technological and Analytical Aspects of Bioactive Compounds and Nutraceuticals from Coarse Grain Cereal Sources

NISAR AHMAD MIR,[1] BASHARAT YOUSUF,[2] KHALID GUL,[3] MOHD. AAQIB SHEIKH,[4] CHARANJIT S. RIAR,[4] and SUKHCHARN SINGH[3]

[1]*Department of Biotechnology Engineering and Food Technology, University Institute of Engineering (UIE), Chandigarh University, Mohali, Punjab, India*

[2]*Department of Food Science and Technology, University of Kashmir, Hazratbal, Srinagar, Jammu and Kashmir, India*

[3]*Department of Food Process Engineering, National Institute of Technology, Rourkela, Odisha, India*

[4]*Department of Food Engineering and Technology, Sant Longowal Institute of Engineering and Technology, Longowal, Punjab, India*

ABSTRACT

Coarse grains and *pseudo*cereals are gaining tremendous attention due to their excellent nutritional profile and the presence of different types of bioactive compounds. Nutritionally, they are superior to major cereals like wheat and rice due to high content of proteins with more balanced amino acid profile. Bioactive compounds present in coarse grains possess several health benefits due to their anticarcinogenic, hypolipidemic, antidiabetic, antihypertension, and immunomodulatory properties. Extraction of bioactive compounds from coarse grains and *pseudo* cereals are usually carried out by

Bioactive Compounds and Nutraceuticals from Plant Sources: Extraction Technology, Analytical Techniques, and Potential Health Prospects. Charanjit Singh Riar, Parmjit S. Panesar (Eds.)

using supercritical fluid extraction (SCFE), ultrasound-assisted extraction, enzymatic-assisted extraction, solvent extraction, and pulsed electric field technology. The selection of the technique for the extraction of bioactive compounds from these coarse grains and *pseudo*cereals depends upon the nature, chemical structure, and compatibility of the bioactive compound with the methodology employed. Apart from this, different techniques are also used for the qualitative and quantitative analysis of the extracted bioactive compounds, including ultra-high-performance liquid chromatography (LC) and spectrophotometric techniques. These methods accurately define the concentration and type of bioactive compound present in different coarse grains and *pseudo*cereals.

8.1 INTRODUCTION ABOUT COARSE GRAINS

Coarse grains are short-duration warm-weather grains and mainly include sorghum, maize, oats, barley, buckwheat, and some minor millet species like pearl millet, finger millet, foxtail millet, and kodo millet but do not include rice and wheat [1]. They have been widely used in food applications, fodder for animals, fuel, and in the preparation of fast-food products. Coarse cereal grains particularly grown in semi-arid tropical regions of Africa and Asia with rain-fed farming systems under little external inputs. The grain yield is comparatively lower than that of traditional cereal crops and is often less than 1 tons/ha [2]. In India, these coarse grains are mainly grown in poor agro-climatic regions, with high temperatures particularly in rain-fed areas of the country, and due to their unique agro-climatic conditions, they are sometimes referred to as dry land crops because they are mainly grown in areas with 50–100 cm rainfall. They are less sensitive to soil deficiencies and can be grown in an inferior alluvial or loamy soil. From the last few years, India has constantly increased its market share in the production of coarse cereal grains in the international market. India has exported about 1,31,416.51 MT of cereals to the world for the worth of Rs. 438.11 crores/61.12 USD million during the year 2019–2020. Coarse cereals are famous as they provide high dietary energy and also contain vitamins, several minerals (especially micronutrients such as iron and zinc), and insoluble dietary fiber. They are also considered an excellent source of bioactive compounds like polyphenols, phytosterols, beta-glucans, fagopyritols, and flavonoids. Phenolic compounds present in coarse grains are secondary metabolites because they are believed to be the first line of defense of plants and are categorized based on structural properties. The three main sub-groups which are found usually in plants include phenolic acids, flavonoids, and non-flavonoids. Flavonoids are the largest class of phenolic compounds and contain more than 8,000 metabolites and most abundant among the free phenolic fractions. Insoluble bound phenolic acids are mainly linked to dietary fiber (e.g., cell wall polysaccharides) by hydrophobic interactions, hydrogen bonds, or covalent bonds [3, 4]. Phenolic compounds are designated as antioxidants because of their potential ability to reduce the potential threat of free radicals and reactive oxygen and nitrogen species. They also possess a number of biological properties that are associated with abiotic and biotic stress defense mechanisms [5]. Phenolic compounds also contain a number of potential health-related bioactivities like anti-oxidative aging, anti-cancer, anti-inflammatory, hepatoprotective, anti-hypertension, antitumor, anti-inflammatory, anti-diabetic, nerve protection, lowering cholesterol, improving cognition, and antibacterial properties [6]. Coarse grains also impart beneficial effects on the intestinal microflora. Numerous studies have proved that consumption of coarse cereals in daily diet can effectively improve the probiotic intestinal microflora and can further lower the incidence of chronic diseases [7]. Bioactive compounds from coarse grains are extracted by using different extraction techniques. Qualitative and quantitative characteristics of the extracted bioactive depend mainly upon the type of extraction technique employed [8, 9]. Extraction of bioactive compounds from different plant sources is generally referred to as sample preparation techniques, and, in most cases, this part is usually neglected or carried out carelessly, which ultimately results in loss of the activity of different bioactive compounds. Accurate methodology, trained research personnel, and following standard operating procedures are some of the prerequisites for the extraction of bioactive compounds from coarse grains. For extraction of bioactive compounds, conventional methods like Soxhlet extraction, maceration, and hydrodistillation are commonly used and non-conventional methods like pulsed electric field, ultrasound-assisted extraction, enzyme-assisted extraction, PLE, microwave assisted-extraction, and supercritical fluid extraction (SCFE) are often used. In this chapter, we will be discussing the major bioactive compounds present in different coarse grains and the methodology for their extraction. In addition to this, the major impact of these bioactive compounds on human health would also be discussed in detail. The images of different coarse grains and their plants are presented in Table 8.1.

TABLE 8.1 Images of Different Coarse Grain Seeds and Their Plants

Name of Coarse Grain and *Pseudo*cereal	Image of Plant	Image of Seed
Maize		
Sorghum		
Oats		
Barley		

TABLE 8.1 *(Continued)*

Album		
Quinoa		
Buckwheat		
Amaranth		
Chia		

8.2 MAJOR BIOACTIVE COMPOUNDS OF COARSE GRAINS CEREALS AND *PSEUDO*CEREALS AND THEIR POTENTIAL IMPACT ON HEALTH

Coarse grains are a powerhouse of several bioactive compounds. Bioactive compounds are secondary metabolites and obtained from plants in subsequent growth stage. Although they have no direct role in growth, they possess a multitude of health benefits. Coarse grains have typical chemical structure and are often synthesized as mixtures of closely related members of a chemical family. Tiaz & Zeiger [10] proposed four different pathways responsible for the production of bioactive compounds which are; Malonic acid pathway, Shikimic acid pathway, Mevalonic acid pathway, and non-mevalonate (MEP) pathway. The major bioactive compounds present in coarse grains are usually in the form of polyphenols. Polyphenols in coarse grains are present in the form of phenolic acids, flavonoids, and their derivatives like lignans, tannins, stilbenoids, fagopyrin, triterpenoids, steroids, and phytosterols [11]. The main bioactive compounds present in coarse grains, their chemical characteristics, and their potential impact on health are discussed in subsections.

8.2.1 POLYPHENOLS

They are found in abundant quantities in coarse grains, and their concentration ranges from 347.26–552.98 mg/100 g. Polyphenols present in coarse grains are further divided into four types viz flavonoids, phenolic acid, stilbenoids, and lignans. Polyphenols possess the ability to scavenge free radicals which are otherwise responsible for cancer, diabetes, atherosclerosis, and reproductive endocrine dysfunction. Phenolic acids mainly feline acid has antioxidant potential [12], anticancer properties [13], and ability to cure type 2 diabetes, and also possess antihypertension and antihyperglycemic activity [14]. They also possess the ability to cure cardiovascular diseases [15]. In addition to this, caffeic acid possesses immunostimulatory properties [16] and on the other hand, chlorogenic acid is responsible for the decline of visceral adiposity content [17].

8.2.2 FLAVONOIDS

They are considered polyphenol-type secondary metabolites consisting of an aromatic ring that contains one hydroxyl group. Flavonoids are in the pericarp and are also the important antioxidants in human health [18]. Coarse grains contain several flavonoids like apigenin (Compound CID: 5280443) which is mostly present as a flavone and is usually found in oats. In addition to this, quercetin (Compound CID: 5280343), a type of flavonoid, is present as a flavonol and is found in maize. Quercetin is also present in other coarse cereals like oats [19], barley [20], and buckwheat [21]. It plays a crucial role in the reduction of atherosclerotic lesions by changing gut microflora. Hong & Piao [22] reported that quercetin is responsible for modifying dextran sodium sulfate-induced mice colitis by controlling oxidative related stress and intestinal microflora. Flavonoids present in coarse grains are further classified into numerous subgroups like flavonols, flavones, flavanones, flavanols, anthocyanins, fagopyrins, proanthocyanidins, isoflavones, and flavonolignans. The bioactivity characteristics of flavonoids was reported in 1936 for its potential impact on the capillary vessels of humans. Quercetin also plays a critical role in reducing Alzheimer's disease by different pathways [23].

8.2.3 PHENOLIC ACIDS

Phenolic compounds belong to the second class of polyphenols and are well known for their protective mechanism against biotic and abiotic stresses [24]. Phenolic acids are classified into two types, namely hydroxycinnamic and hydroxyl benzoic acids. The potential antioxidant property of phenolic acids is due to aromatic ring-type structure that accelerates donation and transfer of hydrogen atoms to free radicals [25]. Coarse cereals contain several phenolic compounds like chlorogenic, caffeic, ferulic, gallic, p-hydroxybenzoic, protocatechuic, syringic, p-coumaric, sinapic, and vanillic acids which are mainly found in buckwheat, maize oats, and barley. In coarse cereals like buckwheat, the concentration of phenolics mainly p-hydroxybenzoic, caffeic, chlorogenic, and protocatechuic acid is maximum in bran in free form and lower in bound form. On the other hand, phenolic acids like ferulic acid, p-cumaric, and protocatechuic acid are present in lower amounts in leaves, stems, roots, seeds, and so on in the case of common buckwheat. Hole, Grimmer, Jensen, & Sahlstrom [26] reported that phenolic acids possess antibacterial activity due to the presence of pyragallol group in their ring structure. In addition, phenolic acids are frequently absorbed in the GI tract and contribute significantly to the systematic circulation of different compounds [27]. Zhao et al. [28] reported that caffeic acid present in barley, maize, and oats is responsible for reducing the incidence of diabetes by controlling gut microfllora like Bifidobacterium, Prevotella_9, and Blautia. Caffeic acid also displays antitumor and anticancer properties by regulating different pathways. Apart from this, caffeic acid shows an ameliorative effect against Alzheimer's disease and also decreases the incidence of oxidative stress-related disorders [29].

8.2.4 LIGNANS

Coarse cereals like oats, rye, barley, and triticale are important sources of lignans. Lignans belong to the group of phytoestrogens and their concentration in coarse grains ranges from 8 to 299 μg/100 g [30]. Qu et al. [31] reported that after ingestion, lignans are converted into mammalian based lignans and help in the reduction of hormone-induced cancers. The most common lignin found in coarse cereals is Secoisolariciresinol (Compound CID: 65373). Among all the coarse cereals, barley and whole bran cereals contain large quantities of Secoisolariciresinol. Secoisolariciresino-rich flaxseed contain anti-oxidative, anti-inflammatory, and lipid-modulating activities. This activity promotes growth of gut micrflora like Bifidobacterium, *Akkermansia muciniphila*, and *Prevotella* spp. [32]. Imran et al. [33] reported that secoisolariciresinol has great potential against several human disorders because it contains anti-inflammatory, antioxidant, anti-mutagenic, anti-microbial, and neuroprotective properties.

8.2.5 PHYTOSTEROLS

Phytosterols are said to be widely distributed in coarse cereals and *pseudo*-cereals. They exist free or esters of fatty acid. In general, barley and oats contain approximately 55–76 mg/100 g and 35–46 mg/100 g of phytosterols [34]. Phytosterols are basically dietary phytochemicals with multitude of chemo-preventive properties. In the last few years, these phytosterols have gained tremendous attention due to their protective effect on different types of cancers [35]. On the other hand, beta-sitosterol and Daucosterol, the two main phytosterols, are responsible for controlling the cell cycle at the G2/M phase [36]. Phytosterols are also responsible for regulating fragmentation of DNA and protein expression [37, 38]. A detailed list of major bioactive compounds present in different coarse grains and *pseudo*cereals is presented in Table 8.2.

TABLE 8.2 Major Coarse Grains and Type of Bioactive Compounds Present

Coarse Grain Type	Major Bioactive Compounds Present	References
Maize	p-coumaric acid, ferulic acid	[39]
Barley	β-glucan, lignans	[40]
Oats	Flavnols, quercitin	[19]
Buckwheat	Stilbenoids, Reseratrol	[41]
Sorghum	Phytosterols	[41]
Chenopodium album	Flavonoids, saponins, lignans, alkaloids,	[42–45]
Chenopodium quinoa	Kaempferol and quercetin oligomeric glycosides	[46]
Chia	Tocopherols, phytosterols, carotenoids, and phenolic compounds	[47]
Amaranth	Betacyanins, polyphenols, flavonoids, phenolic acids, anthocyanins, tannins, and phytosterols	[48]

8.3 EXTRACTION OF BIOACTIVE COMPOUNDS FROM COARSE GRAINS AND *PSEUDO* CEREALS

Traditional cereals and millets are of primary economic importance due to the presence of excellent nutritional profile. Cereal grains contain a variety of compounds, especially those that are biologically active, such as antioxidants, phytoestrogens, dietary fiber, and lignins. The studies related to the quantitative and qualitative analysis of different bioactive compounds are dependent upon the selection of proper extraction methods. However, extraction is the main step for analysis of medicinal plants. Extraction of chemical compounds from different sources is usually followed by qualitative and quantitative analysis. Most cereals and *pseudo* cereals contain bran, germ, and endosperm, and the bioactive compounds are randomly distributed in these fractions. It has been reported that maximum content of bioactive compounds are located in germ and bran [49]. It has also been reported that phenolic acids usually exist in the bound insoluble form (80%) as compared to the free soluble form. Therefore, the extraction of phytochemicals from the cereals and *pseudo*cereals is quite a tricky process. Optimized extraction of phytochemicals from the plant materials is significant for accurate quantification of the phytochemical contents in foods. Considering the diverse bioactive compounds and large diversity in cereals and *pseudo*cereals, it is important to develop a novel approach to extract the compounds with proven health benefits. Techniques used for the extraction of bioactive compounds have some common objectives viz; (i) extraction of the bioactive compound of interest from a complex plant matrix; (ii) increasing the efficiency and selectivity of these extraction techniques; (iii) converting the bioactive compound of interest into a more valuable form for detection and separation; and (iv) providing a robust and accurate method that is independent of variations in the sample matrix [50]. A schematic of the extraction of bioactive compounds from cereals and *pseudo*cereals is shown in Figure 8.1.

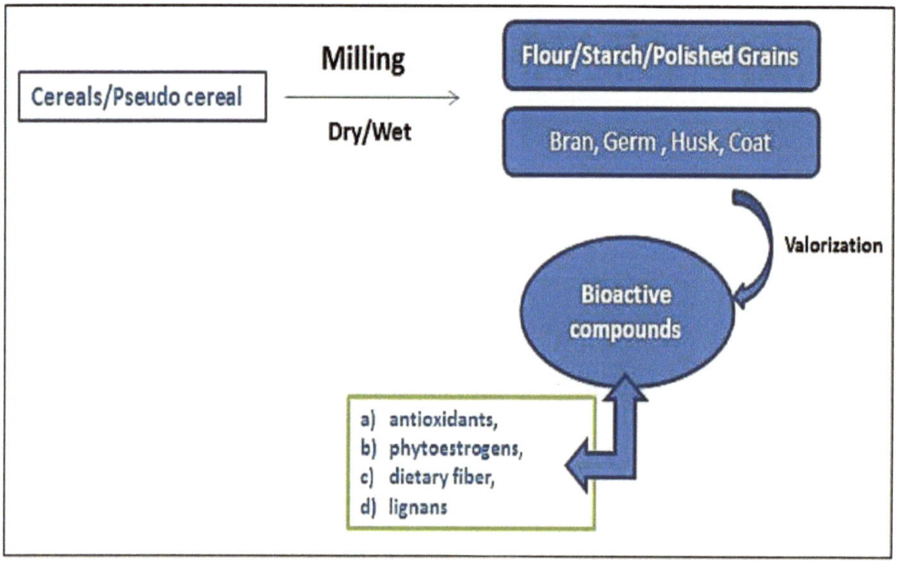

FIGURE 8.1 Schematic presentation for bioactive compound isolation.

8.3.1 DIFFERENT METHODS FOR THE EXTRACTION OF BIOACTIVE COMPOUNDS FROM COARSE CEREALS AND PSEUDOCEREALS

The biologically active compounds from cereals and other plant materials are often extracted by using different conventional methods (Table 8.3). Most of these methods are based on the polarity of a solvent that determines its extraction power and the application of temperature and/or mixing. The main traditional extraction techniques for the extraction of bioactive compounds from cereals include – (i) Soxhlet extraction; (ii) maceration; and (iii) hydrodistillation. Soxhlet extraction is one of the oldest means of extraction, primarily designed for the extraction of lipids, but is now widely used for the extraction of valuable bioactive compounds from different sources. This process still serves as a model for designing new and robust extraction processes. Yang et al. [51]; and Abdel et al. [52] reported the extraction for anthocyanins from corn and purple wheat, respectively using Soxhlet extraction. Similar kinds of extraction procedures have been reported in the literature by different authors. Oufnac et al. [53] reported the extraction of antioxidants from wheat bran using a conventional solvent extraction process.

TABLE 8.3 Some Extraction Methods Used for the Extraction of Bioactive Compounds from Coarse Grain Cereals and *Pseudo*cereals

Extraction Process	Commodity	Target Compound(s)	References
Soxhlet extraction	Corn	Anthocyanins	[51]
	Purple wheat	Anthocyanins	[52]
Solvent extraction	Barley	Free phenolics	[54]
Solvent extraction	Maize	Flavonoids and anthocyanins	[39]
Solvent extraction	Chenopodium album	Phytochemicals	[55]
Solvent extraction	Oat-buckwheat	Phenolic compounds	[56]
Solvent extraction	Colored rice bran	Anthocyanin, proanthocy-anidins, phenolics	[57]
Solvent extraction	Black, purple rice	Procyanidins, carotenoids, oryzanol	[58]
Solvent extraction	Rice	Bioactive compounds (phenolics, tocopherol, oryzanol)	[59]
Solvent extraction	Wheat, rice, and corn	Phytochemicals	[60]
Soxhlet extraction	White maize	Free and bound phenolics	[60]
Pressurized liquid extraction	Chenopodium album	Tropolone derivatives	[61]
Pressurized liquid extraction	Barley	Free phenolics	[54]
Solid-phase extraction	Wheat-Tartary buckwheat products	Polyphenolic compounds	[62]
Matrix solid-phase dispersion	Buckwheat sprouts	Flavonoids from buckwheat sprouts	[62]

In this process, the sample is placed in a thimble, which in turn is placed in the distillation flask, which is filled with a solvent. At the desired temperature, the solvent mixes with the contents, boils, condenses, and the process is then repeated until the extraction is completed. Maceration is a process widely used in homes to extract desired compounds from herbs and other plant materials. It has since long become an inexpensive way for the extraction of different essential oils and important bioactive compounds from a number of plant sources. Maceration for a small number of materials consists of different steps. In this process, the particle size of plant materials is reduced to a significant amount in order to increase the surface area, which in turn can facilitate the efficient mixing with the solvent. The solvent is then added to a closed vessel, and then the liquid is strained off, but the solid residue, marc, is pressed to recover a large number of occluded solutions. The liquid and the resultant strain are mixed and separated from impurities by filtration. During maceration, occasional shaking of the contents is carried out to facilitate extraction by increasing the diffusion and by removing the concentrated solution from the sample surface to bring new solvent to the menstruum, thus helping in increased extraction. Hydrodistillation, another traditional method used for the extraction of bioactive compounds and essential oils from plants, does not involve the use of organic solvents. Hydrodistillation is performed before dehydration of plant materials. Water distillation, water, and steam distillation, and direct steam distillation are three different types of hydrodistillation processes used in the food industry [63].

Traditional extraction methods have a number of limitations. They are time-consuming and also generate a large number of residual wastes. The advantages and disadvantages of modern extraction technologies are presented in Table 8.4.

TABLE 8.4 Some Advantages and Disadvantages of Modern Extraction Technologies

Method of Extraction	Advantages	Disadvantages
Ultrasound-assisted extraction	• Rapid extraction of compounds. • Requires less solvent. • Low additional operating costs.	• Non-selective process (too many compounds extracted). • Limited sample can be used.
Supercritical fluid extraction	• Rapid extraction of compounds. • Environmentally friendly solvents are used. • No solvent residue. • Thermally labile compounds are preserved. • Selective extraction (small number of compounds extracted).	• High installation cost. • Requires technical knowledge.

These traditional techniques, however, cannot meet the current trend of analysis that focuses on the use of very few samples and simplification of the process. Reduction of the sample and the simpler sample-preparation techniques lead to the low consumption of experimental resources reduces or prevents waste, quick analysis, and energy efficiency. Novel food processing methods are advantageous in being environmentally friendly, shortening extraction time, reducing solvent consumption, and efficiency is also higher than conventional methods of extraction methods [64]. UAE is one of the potential technologies that can offer new commercial extraction opportunities and processes, and it does not involve the use of any complex instruments. While UAE primarily helps enhance the extraction efficiency and rate, the simultaneous extraction and encapsulation of extracted compounds and targeted hydroxylation of polyphenols and carotenoids can also be achieved to enhance the bioavailability of the extracts [65]. Pressurized liquid extraction (PLE) or accelerated solvent extraction (ASE) is a relatively new technology and is used for the extraction of different phytochemicals. PLE is a solid-liquid extraction process that uses a solvent at high pressures, permitting the application of temperature to the solvents beyond their boiling points. PLE is used as a rapid method for the extraction of tocopherols and tocotrienols from cereals and cereal-based products [66]. SFE involves the use of supercritical fluids (SFs) for the extraction of biologically active compounds. SFE using CO_2 as a solvent for the extraction of oil from millet bran [67], oil and lipids from oat [68], and rice bran [69] is reported in the literature. The products obtained from SC-CO_2 are completely free of solvents, as opposed to conventionally extracted products. SFE procedures are accomplished without exposure to light and air, thus reducing the oxidation processes. However, due to the application of high pressure, the cost of the equipment and the method on an industrial scale is high.

8.4 NOVEL EXTRACTION METHODS FOR THE EXTRACTION OF BIOACTIVE COMPOUNDS FROM COARSE GRAINS AND *PSEUDO*CEREALS

8.4.1 SUPERCRITICAL FLUID EXTRACTION (SFE)

The method of extraction of bioactive compounds from different plant sources by the application of different supercritical fluids is known as SFE. For the selective extraction of compounds, SFE is considered as a "green" technology. Extraction is usually carried from a solid matrix. SFE provides mild extraction conditions, and for solvent recovery, it has fewer energy requirements. SFE consists of a high-pressure stainless-steel extraction vessel and a reciprocating pump to compress and introduce carbon dioxide into the extraction system (Figure 8.2). SFE assembly consists of a pump which is used for increasing the pressure of solvent above its supercritical pressure. It also consists of a heater, which is used to maintain the temperature of the carbon dioxide. This extraction technique has high selectivity and potential in preventing the extracted materials from oxidation [70]. The time and amount of solvent required for extraction are less than the conventional extraction techniques. This technique is ideal for extracting natural pigments because of all of these advantages [71]. The principle of SFE states that the solubility of a compound in a supercritical fluid (SF) increases with the increase in density [72]. The concept of critical temperature and pressure, i.e., when temperature and pressure are increased above the critical values of temperature of pressure, the solvent possesses both transport and diffusion properties, is an important parameter in the SFE process [80]. Carbon dioxide is frequently used SF for the extraction of heat-labile biomolecules [71]. This process is most suited for heat-labile bioactive compounds as heat-treatment can cause thermal degradation of bioactive compounds. It also removes non-polar substances, as non-polar substances could otherwise can create problems in extraction process. Solvent used in supercritical extraction process plays a crucial role in the extraction process because its activity particularly depends on the polarity of extracted bioactive compounds [70].

Apart from CO_2, some polar solvents can also be used for the extraction of bioactive compounds like polyphenols [80]. The solvent residue obtained in this technique can be removed in the later phases by using some novel purification methods [71]. Carbon dioxide is pressurized and then passed into the extraction vessel, and extraction of bioactive compound takes place from the sample matrix. Recovery of solvent in SFE process can be achieved by reducing the pressure and it is then recycled for the next run [73]. Although higher temperature are suitable for the isolation of polyphenols, but in this case temperature was maintained below 40°C because higher temperatures can lead to the degradation of bioactive [74].

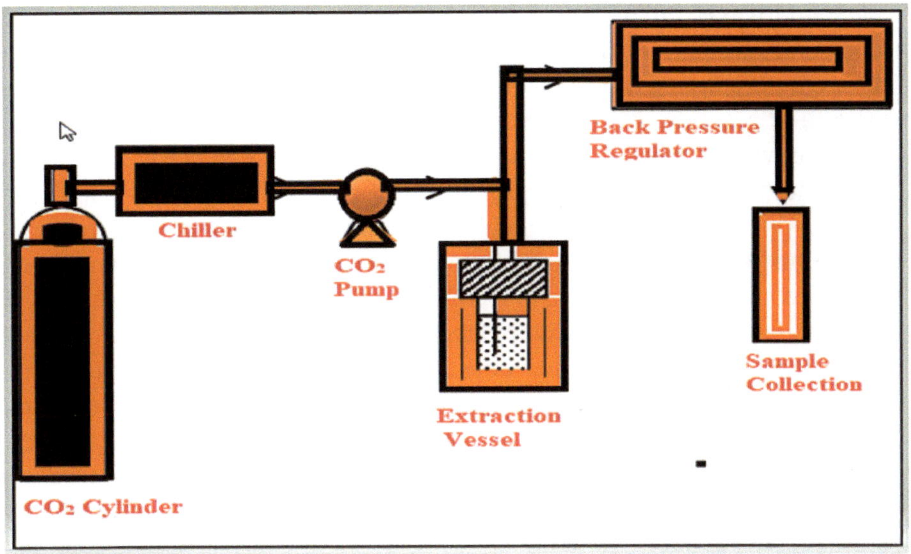

FIGURE 8.2 Schematic representation of supercritical fluid extraction technique.

8.4.2 ULTRASOUND-ASSISTED EXTRACTION (UAE)

UAE extraction is a novel approach for the isolation of important bioactive which involves application of ultrasound waves to rupture plant tissues which in turn releases bioactive compounds [75]. Ultrasonic waves have a frequency ranging between 20 kilohertz (kHz) and several gigahertz (GHz) generally, more significant than a hearing range of humans [76]. Ultrasound frequency in the power range of 20–25 kHz results in the production of mechanical, chemical, and biochemical action due to which breakdown of bubbles takes place which in turn increases the size and then finally bursts because of the increase in internal pressure [77]. These waves cause cavitation in sample tissues which leads to physical and chemical changes in the material and hence result in better extraction [78]. During the cavitation phenomenon, bubbles are formed. These bubbles, when burst, result in the generation of high pressure and temperature, and hence provide enough energy required for the reaction [75]. It has been observed that the diffusion process results in the increase in the size of the bubble cavity. This is followed by the collapse of bubbles, making it unstable at higher values of temperatures and pressures. Higher shear forces are then released, which causes rupturing of the surface, and this mechanism is responsible for the dissolution of the compounds.

Sound waves, after their entry into the liquid medium, cause compressions and rarefactions, resulting in the rise of temperature, which aids in different mass transfer processes [77]. An increase in temperature increases the rate of reaction, which in turn increases the rate of diffusion efficiency of bioactive compounds. The possible reason for this effect is mainly due to an increase in internal energy, which decreases the viscosity coefficient [79]. UAE yielded maximum number of anthocyanins were extracted in the fast diffusion stage of extraction, and the rest bioactive compounds were extracted at a slow transfer rate. Optimization of reaction conditions in UAE is essential as high temperature and pressure generated during acoustic cavitation may degrade the desirable components in the sample [80]. UAE is carried out in ultrasonic electronic equipment, A flask containing the desired amount of sample (from which compound is to be extracted), solvent, flask, and water bath. After the extraction process is completed, the contents of the flask are cooled. The mixture is filtered, and the amount of desired compound present in the extract is determined. Considering the results obtained, around 10 minutes are required to extract about 80% of the desired compound from the sample [81]. UAE is efficient as it has high mass transfer rates, lower liquid-solid ratios, and shorter extraction times.

8.4.3 PULSED ELECTRIC FIELD ASSISTED EXTRACTION (PEFAE)

It is a non-thermal processing technique involving the use of a short high-voltage pulse for a minimal time. The intensity of the high voltage pulses varies from 0.1 to 80 kV/cm. The pulses are conducted through electrodes to the material, which may lead to disruption of cell integrity and softening of tissues which in turn causes the formation of reversible or irreversible pores (Figure 8.3). The experiment is carried out in a chamber containing multiple electrodes. This process is called electroporation or electropermeabilization. PFE works by puncturing the cell membrane. This process of pore formation results in more efficient diffusion of solvent in the sample tissues; as a result, mass transfer of intracellular components is enhanced [82]. PEF is also considered a cold extraction process. Extraction selectivity of the target material (bioactive compounds) can be achieved by controlling the pore formation, which depends on the product size, shape, and electric field orientation. It also depends on the intensity of the electric field, treatment time, pulse duration, etc. [83].

PEF assisted extraction, being environment friendly in nature, has received significant attention. It decreases the extraction cost, processing time and improves extraction yields. It also results in the production of a purified extract with less degradation of heat-sensitive components [84]. The presence of an electric field leads to the plasmolysis of chloroplast due to osmolytic disequilibrium, which results in ease in the diffusion of bioactive compounds from the sample. PEF-assisted extraction has been successfully used for the extraction of lycopene from tomatoes. Since the bioactive compounds are present in the intracellular membrane of cells so, during the extraction process, barriers like cell walls and cell membranes preserve it from extraction. This problem is found mostly in conventional extraction processes, which cannot disrupt the cell wall properly for efficient extraction of bioactive compounds. Due to incomplete cell wall degradation, the efficiency of bioactive compounds extraction is also low. However, PEF leads to the electroporation of cell walls, which leads to enhanced extraction efficiency.

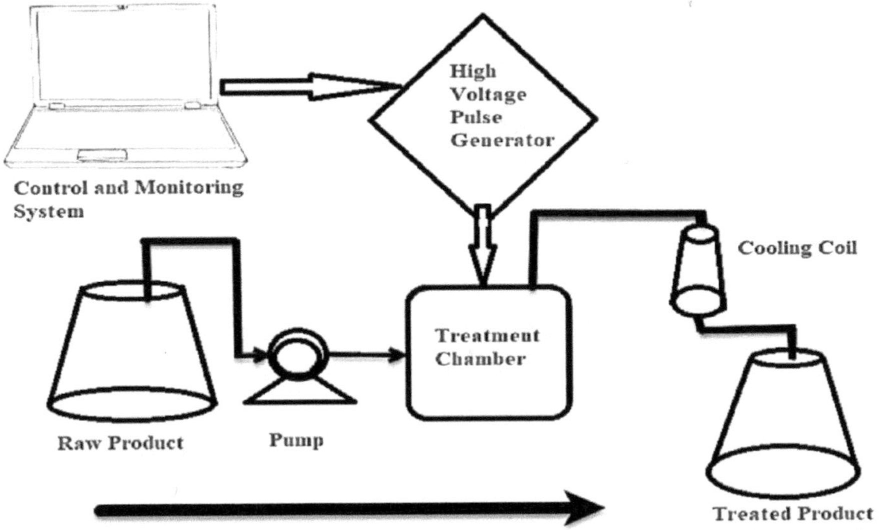

FIGURE 8.3 Schematic representation of pulsed electric field assisted extraction.

8.4.4 *ENZYME ASSISTED EXTRACTION (EAE)*

Enzyme-assisted extraction (EAE) is considered an eco-friendly technology for the extraction of bioactive compounds from different plant sources because enzymes mainly solubilize the bioactive compounds and then finally lead to their extraction from the sample matrix. Enzymes mainly target the cell wall by disrupting the cells or tissues where bioactive compounds are usually located or trapped [85]. These enzymes degrade cell wall components, thus enhancing the liberation of the intracellular constituents present in the material. To achieve this, different enzyme mixtures are used for this purpose [86]. For extraction of bioactive compounds, different enzyme mixtures like pectinase, cellulases, etc., are commercially available [87]. There are two ways by which we can apply this technique, the first one is enzyme-assisted aqueous extraction, and the second one is enzyme-assisted cold pressing (EACP); the former one is primarily used for the extraction of oil from different oil seeds. Moreover, this technique employs water as a primary solvent instead of chemicals, which makes it eco-friendly. It has been observed that pretreatment of the enzyme with optimized or appropriate conditions improves the extraction of bioactive compounds [88].

8.4.5 SOLVENT EXTRACTION

It is a conventional extraction technique involving maceration or thermal extraction using organic or inorganic solvents at ambient (low) pressure and boiling temperature providing selective extraction of targeted compounds [74]. It gives the highest recovery of compounds of interest and is used for bulk extraction [89]. The solvent saturated with the extracted compounds is collected in the flask, and fresh solvent contacts the material present in the thimble and re-extracts the compounds (Figure 8.4). The selection of a suitable solvent is of prime importance for selectively and efficiently extracting lycopene with greater purity [75]. The polarity of solvent is an important parameter which is needed for the dissolution of the compound. In addition to these, acidified solvents are prepared with the addition of acids, e.g., HCl in ethanol, methanol, and water [90]. Solvent extraction has a few limitations that include high consumption of solvent and less purity and efficiency [91]. To overcome this problem, a multistage extraction process is preferred for the extraction of bioactive compounds. ASE or PLE involves extraction at high pressure and temperature. The high pressure causes an increase in cell permeability and intermolecular interaction, resulting in increased solvent penetration and increased mass transfer rates [75]. The high temperature used will reduce the viscosity of the solvent used, facilitating the diffusion of solvent in the sample. Temperature, strain, and extraction time must all be carefully controlled in this phase.

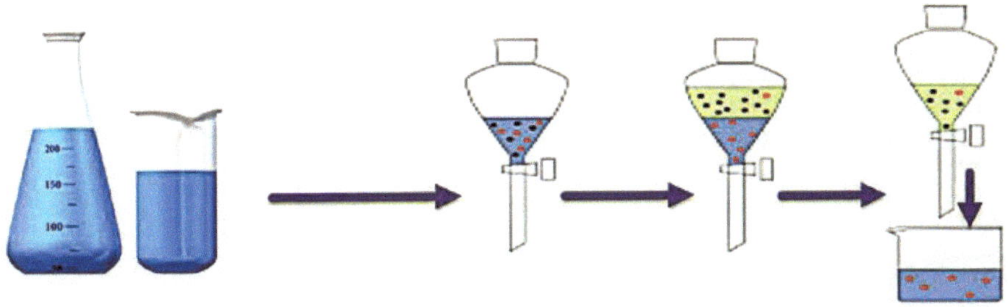

FIGURE 8.4 Schematic representation of solvent extraction process.

8.5 ANALYTICAL TECHNIQUES FOR QUALITATIVE AND QUANTITATIVE ANALYSIS OF BIOACTIVE COMPOUNDS OBTAINED FROM COARSE GRAINS AND *PSEUDO*CEREALS

This section will emphasize analytical techniques for the detection/quantification of bioactive compounds obtained from coarse grain and *pseudo*cereals. Bioactive components may be present in various parts of the cereal plant. However, the content of these bioactive components may vary with different factors such as plant growth stage, plant part being considered, species, growing season, and area [92]. As such, grains are also rich in bioactive compounds. Several techniques have been employed for the detection/quantification of bioactive compounds. Tables 8.5 and 8.6 show various detection/quantification techniques for the different bioactives present in cereals. Literature studies report the use of high-performance liquid chromatography (HPLC), gas chromatography-mass spectroscopy (GC-MS), near-infrared reflectance, and electron impact-mass spectrometer (EI-MS). Chromatography techniques are most frequently being used for this purpose. There are various modes of chromatography for the analysis of bioactives in foods.

HPLC is a common and most predominant technique used in separating and quantifying flavonoids. Various kinds of stationary phases and eluents are used for this purpose. Most of the chromatographic columns are made up of C18 or C8 stationary phases and on the other hand mobile phase is a combination of methanol and water or acetonitrile. Mass spectrometry (MS) has a number of applications, including the detection of polyphenolic compounds. HPLC has been used to detection/quantification bioactive compounds in maize [93], sorghum [94], oats [95], barley [96]. Przybylska-Balcerek,

TABLE 8.5 Detection/Quantification of Bioactive Compounds from Coarse Grain Cereals (Maize, Sorghum, Oats, Barley)

Class of Bioactive Compound	Name of Bioactive Compound	Obtained From	Detection/Quantifications Technique	Quantity (if any)	References
Anthocyanins	Anthocyanin	Colored maize (*Zea mays* L.) kernels	Ultrahigh performance chromatography (U-HPLC)	2.50 to 696.07 mg CGE/kg d.m.	[98]
Avenanthramides	Avenanthramide A, avenanthramide B and avenanthramide C	Oats	HPLC-DAD	1.1 and 2 mg per 40 g	[101]
Carotenoids	Carotenoids	Colored barley	Column reversed-phase HPLC	2.98–3.85 μg g^{-1}	[102]
–	Lutein, zeaxanthin, β-carotene	White corn	RP-HPLC	(5.73, 6.01, 4.92) μg/100 g of dry weight	[103]
Flavonoids	Catechin, kaempferol, naringenin, quercetin, rutin	Red sorghum	UPLC-PDA	(3.61, 0.33, 0.58, 0.17, 0.42) mg/kg	[97]
–	Catechin, kaempferol, naringenin, quercetin, rutin	White sorghum	UPLC-PDA	(4.57, 0.43, 1.11, 0.49, 1.61) mg/kg	[97]
–	Total flavanols	Different barley varieties	HPLC-DAD-MS(n)	325 to 527 μg/g	[104]
Phenolic acids	Total phenolic acids (4-hydroxybenzoic acid, vanillic acid, caffeic acid, 4-hydroxybenzaldehyde, syringic acid, p-coumaric acid vanillin, ferulic acid, and sinapic acid)	Oats	HPLC-DAD	15.79 to 25.05 mg per 40 g	[101]
–	Ferulic acid	White corn	RP-HPLC	120 mg/100 g of dry weight	[103]

TABLE 8.5 *(Continued)*

Class of Bioactive Compound	Name of Bioactive Compound	Obtained From	Detection/Quantifications Technique	Quantity (if any)	References
–	Total phenolic acid	Different barley varieties	HPLC-DAD-MS(n)	604 mg/g and 1,346 mg/g	[104]
Stilbene	Trans-piceid	Red sorghum	RP-HPLC-APCI(+)-HRMS and -HRMS/MS	0.4–1 mg/kg	[105]
Phytosterols	Phytosterols	Sorghum	HPLC	4.13 to 24.45 μg/g, dry weight basis	[106]
Tannins	Procyanidins	Sorghum	HPLC with fluorescence detection	33 mg/g	[94]
Vitamins	α-tocopherol	Barley	HPLC	4.92 ± 0.84 mg/kg dry matter	[107]

Abbreviations: CGE: cyanidine-3-glucoside equivalent; dm: dry matter; mg: milligrams; g: gram; HPLC: high performance liquid chromatography; UPLC-PDA: ultra-performance liquid chromatograph coupled with a photodiode array detector; RP-HPLC: reverse phase high-performance liquid chromatography; and HPLC-DAD-MS(n): high-performance liquid chromatography coupled with diode-array detection and electrospray ionization tandem mass spectrometry.

TABLE 8.6 Detection/Quantification of Bioactive Compounds from *Pseudo*cereals (Album, Quinoa, Chia, Buckwheat, Amaranthus)

Class of Bioactive Compound	Name of Bioactive Compound	Obtained From	Detection/Quantifications Technique	Quantity (if any)	References
Anthocyanins	Anthocyanins	*Amaranthus cruentus* v. Aztek	Spectrophotometric analysis.	103.6 ± 10.4 g CGE 100 g^{-1} dm	[108]
–	Anthocyanins	*Chenopodium quinoa*	Spectrophotometric analysis.	120.4 ± 7.2 g CGE 100 g^{-1} dm	[108]
Anthraquinones	Emodin	Tartary buckwheat	HPLC-DA	1.72 to 2.71 mg/kg	[109]
Betalains	Betacyanins and betaxanthins	Colored quinoa seeds	Spectrophotometric multiple-component method.	0.15–6.10 mg/100 g	[110]
Carotenoids	Lutein	Amaranth seeds	HPLC and LC-MS	3.55–4.44 μg g^{-1} of seed	[111]
–	Zeaxanthin	Amaranth seeds	HPLC and LC-MS	0.14–0.32 μg g^{-1} of seed	[111]
Fagopyrins	Fagopyrin	*Fagopyrum tataricum* (Tartary buckwheat)	HPLC	68 μg g^{-1} of product dry matter	[112]
Flavonoids	Rutin, quercetin	*Fagopyrum tataricum* (Tartary buckwheat)	HPLC	(13, 0.05) mg/g of product dry matter	[112]
Phenolic acids	Protocatechuic acid	Chenopodium album	PLC-DAD	0.41–0.57 mg/100 g	[113]
–	Vanillic acid	Chenopodium album	PLC-DAD	7.11–12.24 mg/100 g	[113]
Phenylpropanoid glycosides	1,3,6,6′-tetra-feruloyl sucrose; 3,6-di-p-coumaroyl-1,6′-di-feruloyl sucrose; 1,6,6′-tri-feruloyl-3-p-coumaroyl sucrose	Tartary buckwheat	LC-PDA/LTQ-FTICR-MS	–	[114]
Stilbene	Trans-resveratrol	Common buckwheat	RP-HPLC	0.98–1.68 mg/k	[115]

TABLE 8.6 *(Continued)*

Class of Bioactive Compound	Name of Bioactive Compound	Obtained From	Detection/Quantifications Technique	Quantity (if any)	References
Vitamins	α-tocopherol	Quinoa	HPLC	47.42 ± 13.95 mg/kg dry matter	[107]
–	Vitamin A	Chenopodium album	Chemical analysis	4.89 mg/100 g to 6.61 mg/100 g	[116]
–	Vitamin C	Chenopodium album	Chemical analysis	10.2 mg/100 g to 12.8 mg/100 g	[116]
–	γ-tocopherol	Tartary buckwheat	HPLC	107.69 mg/100 g of lipids	[117]

Abbreviations: CGE: cyanidine-3-glucoside equivalent; dm: dry matter; mg: milligrams; g: gram; HPLC: high-performance liquid chromatography; UPLC-PDA: ultra-performance liquid chromatograph coupled with a photodiode array detector; and RP-HPLC: reverse phase high performance liquid chromatography.

Frankowski, & Stuper-Szablewska [97] employed UPLC to quantify various bioactive components in red sorghum. Based on this evaluation, it was stated that sorghum grain is a rich source of bioactive compounds. Žilić et al. [98] used U-HPLC for characterization of anthocyanins in maize kernels.

Gas chromatography (GC) is a type of separation technique in which components are distributed between stationary and mobile phase. After this process, samples are then vaporized and carried by the carrier gas through the column, where analyte separation may be based on adsorption or solubility with the stationary phase. Dora, Sarada, & Polumahanthi [99] evaluated the phytochemical composition of sorghum using the GC-MS technique. Sorghum contains maximum content of polyphenols and bioactive lipids as compared to other cereal grains [100]. Among all the polyphenols, ferulic acid and its derivatives are present mostly in sorghum, constituting about 90% of the bound phenolic acids in sorghum. In addition to this, sorghum also contains higher levels of monomeric and/or polymeric flavonoids.

8.6 CONCLUSION

In conclusion, coarse grains and *pseudo*cereals, besides providing basic nutritional requirements, also possess the ability to supply an adequate amount of important bioactive compounds. These bioactive compounds have tremendous health benefits, which have been proven by clinical studies. Adequate extraction technologies employed for the extraction of bioactive compounds from coarse grains and *pseudo*cereals will also preserve their bioactivity and will open new directions for proper utilization in a diverse range of food products. Apart from providing basic dietary requirements, the excellent bioactive profile of coarse grains and *pseudo*cereals is an added advantage to the farmers. Wheat, rice, and maize require favorable climatic conditions for their optimal growth. Coarse grains and *pseudo*cereals do not require favorable climatic conditions for their optimal growth and development. They can thrive in extreme climatic conditions and can also tolerate drought conditions. This property of coarse grains can be used for maximizing global food security and additionally can be used to combat the hidden hunger crisis. Moreover, the need of the hour is to create extensive research and development programs about the possibilities for efficient utilization of important nutrients from different coarse grains and *pseudo*cereals. In addition, genetic engineering programs for the release of new varieties should also be placed in line, which will surely increase the popularity of these important crops among farmers.

KEYWORDS

- **bioactive compounds**
- **coarse grains**
- **enzyme assisted extraction**
- **extraction**
- **high-performance liquid chromatography**
- **novel extraction methods**
- **spectroscopy**

REFERENCES

1. Kumar, S. A., Singh, R., Subramani, R., Kumar, R., & Wankhede, P. D., (2016). Molecular approaches to understand nutritional potential of coarse cereals. *Current Genomics, 17*(3), 177–192.
2. Rai, K. N., Gowda, C. L. L., Reddy, B. V. S., & Sehgal, S., (2008). Adaptation and potential uses of sorghum and pearl millet in alternative and health foods. *Comprehensive Reviews in Food Science and Food Safety, 7*(4), 320–396.
3. Li, Q., Yang, S., Li, Y., Huang, Y., & Zhang, J., (2019). Antioxidant activity of free and hydrolyzed phenolic compounds in soluble and insoluble dietary fibers derived from hulless barley. *LWT-Food Science and Technology, 111*, 534–540.
4. Martínez-Subirà, M., Romero, M. P., Puig, E., Macià, A., Romagosa, I., & Moralejo, M., (2020). Purple, high β-glucan, hulless barley as valuable ingredient for functional food. *LWT- Food Science and Technology, 131*, 109582.
5. Cheynier, V., Comte, G., Davies, K. M., Lattanzio, V., & Martens, S., (2013). Plant phenolics: Recent advances on their biosynthesis, genetics, and ecophysiology. *Plant Physiology. and Biochemistry, 72*, 1–20. https://doi.org/10.1016/j.plaphy.2013.05.009.
6. Shahidi, F., & Yeo, J., (2018). Bioactivities of phenolics by focusing on suppression of chronic diseases: A review. *International Journal of Molecular Sciences, 19*(6), 1573. https://doi.org/10.3390/ijms19061573.
7. David, L. A., Maurice, C. F., Carmody, R. N., Gootenberg, D. B., Button, J. E., Wolfe, B. E., et al., (2014). Diet rapidly and reproducibly alters the human gut microbiome. *Nature, 505*, 559–563.
8. Smith, R. M., (2002). Extractions with superheated water. *Journal of Chromatography A, 975*(1), 31–46.
9. Sasidharan, S., Chen, Y., Saravanan, D., Sundram, K. M., & Latha, L. Y., (2011). Extraction, isolation, and characterization of bioactive compounds from plants' extracts. *African Journal of Traditional, Complementary and Alternative Medicines, 8*(1).
10. Tiaz, L., & Zeiger, E., (2006). Secondary metabolites and plant defense; Chapter 13. In: *Plant Physiology* (4th edn., pp. 283–308). Sinauer Associates, Inc., Sunderland, Massachusetts.

11. Kawabata, K., Yoshioka, Y., & Terao, J., (2019). Role of intestinal microbiota in the bioavailability and physiological functions of dietary polyphenols. *Molecules, 24*, E370.
12. Urias-Lugo, D. A., Heredia, J. B., Muy-Rangel, M. D., Valdez-Torres, J. B., Serna-Saldívar, S. O., & Gutiérrez-Uribe, J. A., (2015). Anthocyanins and phenolic acids of hybrid and native blue maize (*Zea mays* L.) extracts and their antiproliferative activity in mammary (MCF7), liver (HepG2), colon (Caco2 and HT29) and prostate (PC3) cancer cells. *Plant Foods for Human Nutrition, 70*(2), 193–199. https://doi.org/10.1007/s11130–015–0479–4.
13. De Oliveira, S. E., & Batista, R., (2017). Ferulic acid and naturally occurring compounds bearing a feruloyl moiety: A review on their structures, occurrence, and potential health benefits. *Comprehensive Reviews in Food Science and Food Safety, 16*(4), 580–616. https://doi.org/10.1111/1541–4337.12266.
14. Ranilla, L. G., Apostolidis, E., Genovese, M. I., Lajolo, F. M., & Shetty, K., (2009). Evaluation of indigenous grains from the Peruvian Andean region for anti-diabetes and antihypertension potential using in vitro methods. *Journal of Medicinal Food, 12*(4), 704–713.
15. Lao, F., Sigurdson, G. T., & Giusti, M. M., (2017). Health benefits of purple corn (*Zea mays* L.) phenolic compounds. *Comprehensive Reviews in Food Science and Food Safety, 16*(2), 234–246. https://doi.org/10.1111/1541–4337.12249.
16. Pandey, R., Singh, A., Maurya, S., Singh, U. P., & Singh, M., (2013). Phenolic acids in different preparations of maize (*Zea mays*) and their role in human health. *International Journal of Current Microbiology and Applied Sciences, 2*(6), 84–92.
17. Chaiittianan, R., Sutthanut, K., & Rattanathongkom, A., (2017). Purple corn silk: A potential anti-obesity agent with inhibition on adipogenesis and induction on lipolysis and apoptosis in adipocytes. *Journal of Ethnopharmacology, 201*, 9–16. https://doi.org/10.1016/j.jep.2017.02.044.
18. Ginter, E., & Simko, (2012). Plant polyphenols in prevention of heart disease. *Bratislavské Lekárske Listy, 113*, 476–480.
19. Chen, C., Wang, L., Wang, R., Luo, X., Li, Y., Li, J., & Chen, Z., (2018). Phenolic contents, cellular antioxidant activity, and antiproliferative capacity of different varieties of oats. *Food Chemistry, 239*, 260–267.
20. Quan, M., Li, Q., Zhao, P., & Tian, C., (2018). Chemical composition and hepatoprotective effect of free phenolic extract from barley during malting process. *Scientific Reports, 8*, 4460.
21. Zhu, H., Liu, S., Wang, Yao, L., & Li, (2019). Free and bound phenolics of buckwheat varieties: HPLC characterization, antioxidant activity, and inhibitory potency towards α- glucosidase with molecular docking analysis. *Antioxidants, 8*, 606.
22. Hong, Z., & Piao, M., (2018). Effect of quercetin monoglycosides on oxidative stress and gut microbiota diversity in mice with dextran sodium sulphate-induced colitis. *BioMed Research International, 2018*, 1–7.
23. Zaplatic, E., Bule, M., Shah, S. Z., Uddin, M. S., & Niaz, K., (2019). Molecular mechanisms underlying protective role of quercetin in attenuating Alzheimer's disease. *Life Sciences, 224*, 109–119.
24. Călinoiu, L. F., & Vodnar, D. C., (2018). Whole grains and phenolic acids: A review on bioactivity, functionality, health benefits and bioavailability. *Nutrients, 10*, 36–46.
25. Sevgi, K., Tepe, B., & Sarikurkcu, C., (2015). Antioxidant and DNA damage protection potentials of selected phenolic acids. *Food and Chemical Toxicology, 77*, 12–21.

26. Hole, A. S., Grimmer, S., Jensen, M. R., & Sahlstrom, S., (2012). Synergistic and suppressive effects of dietary phenolic acids and other phytochemicals from cereal extracts on nuclear factor kappa B activity. *Food Chemistry*, *133*, 969–977.
27. Neacsu, M., McMonagle, J., Fletcher, R. J., Hulshof, T., Duncan, S. H., Scobbie, L., Duncan, G. J., et al., (2017). Availability and dose-response of phytosterols from a wheat bran-rich cereal product in healthy human volunteers. *Molecular Nutrition & Food Research*, *61*, 1600202.
28. Zhao, X., Guo, S., Lu, Y., Hua, Y., Zhang, F., Yan, H., Shang, E., et al., (2020). *Lycium barbarum* L. leaves ameliorate type 2 diabetes in rats by modulating metabolic profiles and gut microbiota composition. *Biomedicine and Pharmacotherapy*, *121*, 109559.
29. Habtemariam, S., (2017). Protective effects of caffeic acid and the Alzheimer's brain: An update. *Mini-Reviews in Medicinal Chemistry*, *17*, 667–674.
30. Buri, R. C., Von, R. W., & Gavin, M. H., (2004). Description and characterization of wheat aleurone. *Cereal Foods World*, *49*, 274–281.
31. Qu, H., Madl, R. L., Takemoto, D. J., Baybutt, R. C., & Wang, W., (2005). Lignans involved in the anti-tumor activity of wheat bran in colon cancer SW480 cells. *Journal of Nutrition*, *135*, 598–602.
32. Parikh, M., Maddaford, T. G., Austria, J. A., Aliani, M., Netticadan, T., & Pierce, G. N., (2019). Dietary flaxseed as a strategy for improving human health. *Nutrients*, *11*, 117.
33. Imran, M., Ahmad, N., Anjum, F. M., Khan, M. K., Mushtaq, Z., Nadeem, M., et al., (2015). Potential protective properties of flax lignan secoisolariciresinol diglucoside. *Nutrition Journal*, *14*, 71.
34. Kaur, K. D., Jha, A., Sabikhi, L., & Singh, A. K., (2014). Significance of coarse cereals in health and nutrition: A review. *Journal of Food Science and Technology*, *51*(8), 1429–1441.
35. Ramprasath, V. R., & Awad, A. B., (2015). Role of phytosterols in cancer prevention and treatment. *Journal of AOAC International*, *98*, 735–738.
36. Rajavel, T., Mohankumar, R., Archunan, G., Ruckmani, K., & Pandima, D. K., (2017). Beta-sitosterol and Daucosterol (phytosterols identified in *Grewia tiliaefolia*) perturb the cell cycle and induce apoptotic cell death in A549 cells. *Scientific Reports*, *7*, 3418.
37. Le, C., Kailaivasan, T. H., Chow, S., Abdullah, Z., Ling, S., & Fang, C., (2017). Phytosterols isolated from *Clinacanthus nutans* induce immunosuppressive activity in murine cells. *International Immunopharmacology*, *44*, 203–210.
38. Shahzad, N., Khan, W., Shadab, Ali, A., Saluja, S. S., Sharma, S., & Alghamdi, S. S., (2017). Phytosterols as a natural anticancer agent: Current status and future perspective. *Biomedicine & Pharmacotherapy*, *88*, 786–794.
39. Zhang, S., Ji, J., Zhang, S., Xiao, W., Guan, C., Wang, G., & Wang, Y., (2020). Changes in the phenolic compound content and antioxidant activity in developmental maize kernels and expression profiles of phenolic biosynthesis-related genes. *Journal of Cereal Science*, *96*, 103113.
40. Bai, Y. P., Zhou, H. M., Zhu, K. R., & Li, Q., (2021). Effect of thermal treatment on the physicochemical, ultrastructural, and nutritional characteristics of whole grain highland barley. *Food Chemistry*, *346*, 128657.
41. Fu, J., Zhang, Y., Hu, Y., Zhao, G., Tang, Y., & Zou, L., (2020). Concise review: Coarse cereals exert multiple beneficial effects on human health. *Food chemistry*, *325*, 126761.

42. Rahiminejad, M. R., & Gornall, R. J., (2004). Flavonoid evidence for allopolyploidy in the *Chenopodium album* aggregate (*Amaranthaceae*). *Plant Systematics and Evolution, 246*(1), 77–87.
43. Lavaud, C., Voutquenne, L., Bal, P., & Pouny, I., (2000). Saponins from *Chenopodium* album. *Fitoterapia, 71*(3), 338–340.
44. Cutillo, F., DellaGreca, M., Gionti, M., Previtera, L., & Zarrelli, A., (2006). Phenols and lignans from *Chenopodium* album. *Phytochemical Analysis: An International Journal of Plant Chemical and Biochemical Techniques, 17*(5), 344–349.
45. Cutillo, F., D'Abrosca, B., DellaGreca, M., & Zarrelli, A., (2004). Chenoalbicin, a novel cinnamic acid amide alkaloid from *Chenopodium* album. *Chemistry & Biodiversity, 1*(10), 1579–1583.
46. Dini, I., Tenore, G. C., & Dini, A., (2004). Phenolic constituents of Kancolla seeds. *Food Chemistry, 84*(2), 163–168.
47. Firtin, B., Yenipazar, H., Saygun, A., & Şahin, Y. N., (2020). *Encapsulation of Chia Seed Oil with Curcumin and Investigation of Release Behavior & Antioxidant Properties of Microcapsules During in Vitro Digestion Studies.*
48. Martinez-Lopez, A., Millan-Linares, M. C., Rodriguez-Martin, N. M., Millan, F., & Montserrat-De La Paz, S., (2020). Nutraceutical value of kiwicha (*Amaranthus caudatus* L.). *Journal of Functional Foods, 65*, 103735.
49. Luthria, D. L., Lu, Y., & John, K. M., (2015). Bioactive phytochemicals in wheat: Extraction, analysis, processing, and functional properties. *Journal of Functional Foods, 18*, 910–925.
50. Smith, R. M., (2003). Before the injection—Modern methods of sample preparation for separation techniques. *Journal of Chromatography A, 1000*(1, 2), 3–27.
51. Yang, Z., Chen, Z., Yuan, S., Zhai, W., Piao, X., & Piao, X., (2009). Extraction and identification of anthocyanin from purple corn (*Zea mays* L.). *International Journal of Food Science & Technology, 44*(12), 2485–2492.
52. Abdel-Aal, E. S. M., Abou-Arab, A. A., Gamel, T. H., Hucl, P., Young, J. C., & Rabalski, I., (2008). Fractionation of blue wheat anthocyanin compounds and their contribution to antioxidant properties. *Journal of Agricultural and Food Chemistry, 56*(23), 11171–11177.
53. Oufnac, D. S., Xu, Z., Sun, T., Sabliov, C., Prinyawiwatkul, W., & Godber, J. S., (2007). Extraction of antioxidants from wheat bran using conventional solvent and microwave-assisted methods. *Cereal Chemistry, 84*(2), 125–129.
54. Bonoli, M., Marconi, E., & Caboni, M. F., (2004). Free and bound phenolic compounds in barley (*Hordeum vulgare* L.) flours: Evaluation of the extraction capability of different solvent mixtures and pressurized liquid methods by micellar electrokinetic chromatography and spectrophotometry. *Journal of Chromatography A, 1057*(1, 2), 1–12.
55. Jan, R., Saxena, D. C., & Singh, S., (2018). Comparative study of raw and germinated *Chenopodium* (*Chenopodium* album) flour on the basis of thermal, rheological, minerals, fatty acid profile and phytocomponents. *Food Chemistry, 269*, 173–180.
56. Baczek, N., Jarmułowicz, A., Wronkowska, M., & Haros, C. M., (2020). Assessment of the glycaemic index, content of bioactive compounds, and their *in vitro* bioaccessibility in oat-buckwheat breads. *Food Chemistry, 330*, 127199.
57. Huang, Y. P., & Lai, H. M., (2016). Bioactive compounds and antioxidative activity of colored rice bran. *Journal of Food and Drug Analysis, 24*(3), 564–574.

58. Pereira-Caro, G., Watanabe, S., Crozier, A., Fujimura, T., Yokota, T., & Ashihara, H., (2013). Phytochemical profile of a Japanese black–purple rice. *Food Chemistry, 141*(3), 2821–2827.
59. Moongngarm, A., & Saetung, N., (2010). Comparison of chemical compositions and bioactive compounds of germinated rough rice and brown rice. *Food Chemistry, 122*(3), 782–788.
60. Félix-Medina, J. V., Gutiérrez-Dorado, R., López-Valenzuela, J. A., López-Ángulo, G., Quintero-Soto, M. F., Perales-Sánchez, J. X. K., & Montes-Ávila, J., (2021). Nutritional, antioxidant and phytochemical characterization of healthy ready-to-eat expanded snack produced from maize/common bean mixture by extrusion. *LWT, 142*, 111053.
61. Ma, Q. G., Wei, R. R., Zhang, X. D., Sang, Z. P., Dong, J. H., Lu, Q. X., & Jiang, L., (2020). Tropolone derivatives with hepatoprotective and antiproliferative activities from the aerial parts of *Chenopodium album* Linn. *Fitoterapia, 146*, 104733.
62. Gawlik-Dziki, U., Dziki, D., Baraniak, B., & Lin, R., (2009). The effect of simulated digestion in vitro on bioactivity of wheat bread with Tartary buckwheat flavones addition. *LWT-Food Science and Technology, 42*(1), 137–143.
63. Vankar, P. S., (2004). Essential oils and fragrances from natural sources. *Resonance, 9*(4), 30–41.
64. Wang, L., & Weller, C. L., (2006). Recent advances in extraction of nutraceuticals from plants. *Trends in Food Science & Technology, 17*(6), 300–312.
65. Vilkhu, K., Mawson, R., Simons, L., & Bates, D., (2008). Applications and opportunities for ultrasound assisted extraction in the food industry—A review. *Innovative Food Science & Emerging Technologies, 9*(2), 161–169.
66. Bustamante-Rangel, M., Delgado-Zamarreño, M. M., Sánchez-Pérez, A., & Carabias-Martínez, R., (2007). Determination of tocopherols and tocotrienols in cereals by pressurized liquid extraction–liquid chromatography–mass spectrometry. *Analytica Chimica Acta, 587*(2), 216–221.
67. Devittori, C., Gumy, D., Kusy, A., Colarow, L., Bertoli, C., & Lambelet, P., (2000). Supercritical fluid extraction of oil from millet bran. *Journal of the American Oil Chemists' Society, 77*(6), 573.
68. Fors, S. M., & Eriksson, C. E., (1990). Characterization of oils extracted from oats by supercritical carbon dioxide. *Lebensmittel-Wissenschaft & Technologie, 23*(5), 390–395.
69. Kuk, M. S., & Dowd, M. K., (1998). Supercritical CO_2 extraction of rice bran. *Journal of the American Oil Chemists' Society, 75*(5), 623–628.
70. Pereira, C. G., & Meireles, M. A. A., (2010). Supercritical fluid extraction of bioactive compounds: Fundamentals, applications, and economic perspectives. *Food and Bioprocess Technology, 3*(3), 340–372.
71. Roh, M. K., Uddin, M. S., & Chun, B. S., (2008). Extraction of fucoxanthin and polyphenol from *Undaria pinnatifida* using supercritical carbon dioxide with co-solvent. *Biotechnology and Bioprocess Engineering, 13*(6), 724–729.
72. Da Silva, R. P., Rocha-Santos, T. A., & Duarte, A. C., (2016). Supercritical fluid extraction of bioactive compounds. *TrAC Trends in Analytical Chemistry, 76*, 40–51.
73. Brunner, G., (2013). *Gas Extraction: An Introduction to Fundamentals of Supercritical Fluids and the Application to Separation Processe*s (Vol. 4). Springer Science & Business Media.
74. Rodríguez-Solana, R., Salgado, J. M., Domínguez, J. M., & Cortés-Diéguez, S., (2015). Comparison of Soxhlet, accelerated solvent and supercritical fluid extraction techniques

for volatile (GC–MS and GC/FID) and phenolic compounds (HPLC–ESI/MS/MS) from *Lamiaceae* species. *Phytochemical Analysis*, *26*(1), 61–71.
75. Cai, Z., Qu, Z., Lan, Y., Zhao, S., Ma, X., Wan, Q., Jing, P., & Li, P., (2016). Conventional, ultrasound-assisted, and accelerated-solvent extractions of anthocyanins from purple sweet potatoes. *Food Chemistry*, *197*, 266–272.
76. Vinatoru, M., (2001). An overview of the ultrasonically assisted extraction of bioactive principles from herbs. *Ultrasonics Sonochemistry*, *8*(3), 303–313.
77. Paniwnyk, L., (2017). Applications of ultrasound in processing of liquid foods: A review. *Ultrasonics Sonochemistry*, *38*, 794–806.
78. Lianfu, Z., & Zelong, L., (2008). Optimization and comparison of ultrasound/microwave assisted extraction (UMAE) and ultrasonic assisted extraction (UAE) of lycopene from tomatoes. *Ultrasonics Sonochemistry*, *15*(5), 731–737.
79. Cong-Cong, X. U., Bing, W. A. N. G., Yi-Qiong, P. U., Jian-Sheng, T. A. O., & Zhang, T., (2017). Advances in extraction and analysis of phenolic compounds from plant materials. Chinese *Journal of Natural Medicines*, *15*(10), 721–731.
80. Silva, S., Costa, E. M., Calhau, C., Morais, R. M., & Pintado, M. E., (2017). Anthocyanin extraction from plant tissues: A review. *Critical Reviews in Food Science and Nutrition*, *57*(14), 3072–3083. https://doi.org/10.1080/10408398.2015.1087963.
81. Ranveer, R. C., Gatade, A. A., Kamble, H. A., & Sahoo, A. K., (2015). Microencapsulation and storage stability of lycopene extracted from tomato processing waste. *Brazilian Archives of Biology and Technology*, *58*, 953–960.
82. Raso, J., Frey, W., Ferrari, G., Pataro, G., Knorr, D., Teissie, J., & Miklavčič, D., (2016). Recommendations guidelines on the key information to be reported in studies of application of PEF technology in food and biotechnological processes. *Innovative Food Science & Emerging Technologies*, *37*, 312–321.
83. Vorobiev, E., & Lebovka, N., (2008). *Electrotechnologies for Extraction from Food Plants and Biomaterials* (Vol. 5996). New York: Springer.
84. Roohinejad, S., Everett, D. W., & Oey, I., (2014). Effect of pulsed electric field processing on carotenoid extractability of carrot purée. *International Journal of Food Science & Technology*, *49*(9), 2120–2127.
85. Brglez, M. E., Knez, H. M., Škerget, M., Knez, Ž., & Bren, U., (2016). Polyphenols: Extraction methods, antioxidative action, bioavailability and anticarcinogenic effects. *Molecules*, *21*(7), 901.
86. Surin, S., You, S., Seesuriyachan, P., Muangrat, R., Wangtueai, S., Jambrak, A. R., Phongthai, S., et al., (2020). Optimization of ultrasonic-assisted extraction of polysaccharides from purple glutinous rice bran (*Oryza sativa* L.) and their antioxidant activities. *Scientific Reports*, *10*(1), 1–10.
87. Buchert, J., Koponen, J. M., Suutarinen, M., Mustranta, A., Lille, M., Törrönen, R., & Poutanen, K., (2005). Effect of enzyme-aided pressing on anthocyanin yield and profiles in bilberry and blackcurrant juices. *Journal of the Science of Food and Agriculture*, *85*(15), 2548–2556.
88. Çinar, I., (2005). Effects of cellulase and pectinase concentrations on the color yield of enzyme extracted plant carotenoids. *Process Biochemistry*, *40*(2), 945–949.
89. Myong-Kyun, R., Min-Hee, J., Jin-Nam, M., Woi-Sook, M., Sun-Mee, P., & Jae-Suk, C., (2013). A simple method for the isolation of lycopene from *Lycopersicon esculentum*. *Botanical Sciences*, *91*(2), 187–192.

90. Ali, O. H., Al-Sayed, H., Yasin, N., & Afifi, E., (2016). Effect of different extraction methods on stability of anthocyanins extracted from red onion peels (*Allium cepa*) and its uses as food colorants. *Bulletin of the National Nutrition Institute of the Arab Republic of Egypt*, *47*(2), 1–24.
91. Poojary, M. M., & Passamonti, P., (2015). Optimization of extraction of high purity all-trans-lycopene from tomato pulp waste. *Food Chemistry*, *188*, 84–91.
92. Matsui, K., & Walker, A. R., (2019). Biosynthesis and regulation of flavonoids in buckwheat. *Breeding Science*, 1–11.
93. Paucar-Menacho, L. M., Martínez-Villaluenga, C., Dueñas, M., Frias, J., & Peñas, E., (2017). Optimization of germination time and temperature to maximize the content of bioactive compounds and the antioxidant activity of purple corn (*Zea mays* L.) by response surface methodology. *LWT-Food Science and Technology*, *76*, 236–244.
94. Awika, J. M., Dykes, L., Gu, L., Rooney, L. W., & Prior, R. L., (2003). Processing of sorghum (*Sorghum bicolor*) and sorghum products alters procyanidin oligomer and polymer distribution and content. *Journal of Agricultural and Food Chemistry*, *51*, 5516–5521.
95. Aparicio-García, N., Martínez-Villaluenga, C., Frias, J., & Peñas, E., (2021). Sprouted oat as a potential gluten-free ingredient with enhanced nutritional and bioactive properties. *Food Chemistry*, *338*, 127972.
96. Irakli, M., Lazaridou, A., Mylonas, I., & Biliaderis, C. G., (2020). Bioactive components and antioxidant activity distribution in pearling fractions of different Greek barley cultivars. *Foods*, *9*, 783.
97. Przybylska-Balcerek, A., Frankowski, J., & Stuper-Szablewska, K., (2019). Bioactive compounds in sorghum. *European Food Research and Technology*, *245*, 1075–1080.
98. Žilić, S., Serpen, A., Akıllıoğlu, G., Gökmen, V., & Vančetović, J., (2012). Phenolic compounds, carotenoids, anthocyanins, and antioxidant capacity of colored maize (*Zea mays* L.) kernels. *Journal of Agricultural and Food Chemistry*, *60*, 1224–1231.
99. Dora, S. V., Sarada, M. N., & Polumahanthi, S., (2015). Phytochemical composition of wild Sorghum by GC-MS. *International Journal of Pharmaceutical Science Invention*, *4*, 1–6.
100. Girard, A. L., & Awika, J. M., (2018). Sorghum polyphenols and other bioactive components are functional and health-promoting food ingredients. *Journal of Cereal Science*, *84*, 112–124.
101. Soycan, G., Schär, M. Y., Kristek, A., Boberska, J., Alsharif, S. N. S., Corona, G., Shewrya, P. R., & Spencer, J. P., (2019). Composition and content of phenolic acids and avenanthramides in commercial oat products: Are oats an important polyphenol source for consumers? *Food Chemistry*, *X3*, 100047.
102. Suriano, S., Iannucci, A., Codianni, P., Fares, C., Menga, V., Russo, M., Marciello, U., & Troccoli, A., (2020). Carotenoids and tocols content in genotypes of colored barley. *Journal of Cereal Science*, *96*, 103110.
103. Parra, D. L. C., Serna, S. S. O., & Liu, R. H., (2007). Effect of processing on the phytochemical profiles and antioxidant activity of corn for production of masa, tortillas, and tortilla chips. *Journal of Agricultural and Food Chemistry*, *55*, 4177–4183.
104. Holtekjølen, A. K., Kinitz, C., & Knutsen, S. H., (2006). Flavanol and bound phenolic acid contents in different barley varieties. *Journal of Agricultural and Food Chemistry*, *54*, 2253–2260.

105. Bröhan, M., Jerkovic, V., & Collin, S., (2011). Potentiality of red sorghum for producing stilbenoid-enriched beers with high antioxidant activity. *Journal of Agricultural and Food Chemistry, 59*, 4088–4094.
106. Chung, I. M., Yong, S. J., Lee, J., & Kim, S. H., (2013). Effect of genotype and cultivation location on β-sitosterol and α-, β-, γ-, and δ-tocopherols in sorghum. *Food Research International, 51*, 971–976.
107. Granda, L., Rosero, A., Benešová, K., Pluháčková, H., Neuwirthová, J., & Cerkal, R., (2018). Content of selected vitamins and antioxidants in colored and nonpigmented varieties of quinoa, barley, and wheat grains. *Journal of Food Science, 83*, 2439–2447.
108. Paśko, P., Bartoń, H., Zagrodzki, P., Gorinstein, S., Fołta, M., & Zachwieja, Z., (2009). Anthocyanins, total polyphenols and antioxidant activity in amaranth and quinoa seeds and sprouts during their growth. *Food Chemistry, 115*, 994–998.
109. Peng, L. X., Wang, J. B., Hu, L. X., Zhao, J. L., Xiang, D. B., Zou, L., & Zhao, G., (2013). Rapid and simple method for the determination of emodin in Tartary buckwheat (*Fagopyrum tataricum*) by high-performance liquid chromatography coupled to a diode array detector. *Journal of Agricultural and Food Chemistry, 61*, 854–857.
110. Abderrahim, F., Huanatico, E., Segura, R., Arribas, S., Gonzalez, M. C., & Condezo-Hoyos, L., (2015). Physical features, phenolic compounds, betalains and total antioxidant capacity of colored quinoa seeds (*Chenopodium quinoa* Willd.) from Peruvian Altiplano. *Food Chemistry, 183*, 83–90.
111. Tang, Y., Li, X., Chen, P. X., Zhang, B., Liu, R., Hernandez, M., Draves, J., Marcone, M., & Tsao, R., (2016). Assessing the fatty acid, carotenoid, and tocopherol compositions of amaranth and quinoa seeds grown in Ontario and their overall contribution to nutritional quality. *Journal of Agricultural and Food Chemistry, 64*, 1103–1110.
112. Stojilkovski, K., Glavač, N. K., Kreft, S., & Kreft, I., (2013). Fagopyrin and flavonoid contents in common, Tartary, and cymosum buckwheat. *Journal of Food Composition and Analysis, 32*, 126–130.
113. Laghari, A. H., Memon, S., Nelofar, A., Khan, K. M., & Yasmin, A., (2011). Determination of free phenolic acids and antioxidant activity of methanolic extracts obtained from fruits and leaves of *Chenopodium album. Food Chemistry, 126*(4), 1850–1855.
114. Ren, Q., Wu, C., Ren, Y., & Zhang, J., (2013). Characterization and identification of the chemical constituents from Tartary buckwheat (*Fagopyrum tataricum* Gaertn) by high performance liquid chromatography/photodiode array detector/linear ion trap FTICR hybrid mass spectrometry. *Food Chemistry, 136*, 1377–1389.
115. Němcová, L., Zima, J., Barek, J., & Janovská, D., (2011). Determination of resveratrol in grains, hulls and leaves of common and Tartary buckwheat by HPLC with electrochemical detection at carbon paste electrode. *Food Chemistry, 126*, 374–378.
116. Pachauri, T., Lakhani, A., & Kumari, K. M., (2017). Nutritional and antinutritional characterization of *Chenopodium* album seeds: A neglected wild species. *International Journal of Nutrition and Agriculture Research, 4*, 9–21.
117. Dziedzic, K., Kurek, S., Mildner-Szkudlarz, S., Kreft, I., & Walkowiak, J., (2020). Fatty acids profile, sterols, tocopherol and squalene content in *Fagopyrum tataricum* seed milling fractions. *J. of Cereal Sci., 96*, 103118.

CHAPTER 9

Technological and Analytical Aspects of Bioactive Compounds and Nutraceuticals from Cereal Processing Byproducts

REN-YOU GAN,[1,2] KHALID GUL,[3] LOKESWARI RAMIREDDY,[3] NISAR AHMAD MIR,[4] and FARHANA MEHRAJ ALLAI[5]

[1]Key Laboratory of Coarse Cereal Processing (Ministry of Agriculture and Rural Affairs), Sichuan Engineering and Technology Research Center of Coarse Cereal Industrialization, School of Food and Biological Engineering, Chengdu University, Chengdu, China

[2]Research Center for Plants and Human Health, Institute of Urban Agriculture, Chinese Academy of Agricultural Sciences, Chengdu, China

[3]Department of Food Process Engineering, National Institute of Technology, Rourkela, Odisha, India

[4]Department of Food Technology, Chandigarh University, Mohali, Punjab, India

[5]Department of Food Technology, Islamic University of Science and Technology, Awantipora, India

ABSTRACT

Cereals and cereal processing byproducts have received wide attention in recent years due to the associated health benefits. Cereals are rich in bioactive compounds which have been directly linked to the prevention of different cancers and cardiovascular diseases and to the proper physiological functions in the human body. Cereal and cereal products are rich sources

Bioactive Compounds and Nutraceuticals from Plant Sources: Extraction Technology, Analytical Techniques, and Potential Health Prospects. Charanjit Singh Riar, Parmjit S. Panesar (Eds.)

of biologically active compounds such as tocopherols, tocotrienols, dietary fiber, (arabinoxylans, b-glucans, cellulose, lignin, and lignans), sterols, phenolic acids, vitamins, and minerals. Cereals are milled before processing and the rich byproducts have, however, remained unexplored and the extraction and re-processing of these physiologically beneficial components are, therefore, vital and need of the hour. Different extraction processes including solvent extraction, maceration, supercritical fluid extraction (SFE), etc., are widely used to fractionate the biologically active compounds. This chapter details different bioactive compounds in cereals and their byproducts, associated health benefits, and the extraction processes used.

9.1 INTRODUCTION

The rapid increase in global population, decrease in cultivable land, and exhaustion of renewable sources creates the necessity to think about saving food resources and utilizing the raw materials effectively. The waste in the food chain starts from harvesting, throughout post-harvesting, product manufacturing, to consumption. It's been evaluated that 1.3 billion tons of food waste were generated from all stages of food processing [1]. However, these food wastes are the organic residues of processed raw materials which have a negative impact on the environment due to the issues like disposal, pollution, and sustainability [2]. Moreover, they exhibit certain attributes like low biological stability, high water activity, rapid oxidation, and high enzymatic activity which makes it difficult to standardize the food waste components. The food industries started to concentrate on waste management, recycling the waste by developing value-added products with commercial applications [3]. This idea led to the concept of waste volarization, which proposes a wide range of alternatives for the sustainable management of food waste. It offers the opportunity of reusing the potential nutrient and non-nutrient health-beneficial components from the waste. This will generate economic benefits for the food processing industries, help to reduce nutritional problems and reduce environmental consequences that can arise from mishandling of waste [4, 5]. Currently, the industries are focusing on novelties to acquire zero waste, where the generated waste is utilized as a raw ingredient to produce bioactive and nutraceutical compounds.

Among the food groups, cereals are considered a staple food for most of the world population and are recommended in all dietary guidelines. They comprise almost 70% carbohydrates and 6–15% proteins, contribute more than 56% of the energy and 50% of protein needed worldwide [6]. Cereals belong to the *Gramineae* family and are present as nine species namely barley, corn, rice, rye, oat, wheat, sorghum, millet, and triticale (a hybrid of wheat and rye). Despite their nutritional properties, many cereals are rich in one component and deficient in others. To overcome this, pseudocereals have been introduced. Pseudocereals belong to the non-grasses family and have similar end-uses as cereals. Major pseudocereals are Amaranth, buck-wheat, quinoa, and others like chia and breadnut. Unlike cereals, these are dicotyledonous but their composition and functions resemble true cereals so, they are called false or pseudocereals [7, 8]. The cereals and pseudocereals are processed further to a range of products such as flours, breakfast cereals, bread, pasta, biscuits, extruded snacks, fermented beverages, etc. [9]. These products are energy-dense foods packed with carbohydrates, proteins, fiber, lipids, and micronutrients.

The processing of cereal grains generates a large amount of wastage. A 12.9% of the total generated food wastes are produced from the cereal-based manufacturing industries. The byproducts exerted from cereals are turned into feed, substrates for bio-refineries, or waste. The cereal industries are giving importance to finding alternative ways to reduce the volume of wastage. Simultaneously, looking for novel solutions and applications to valorize them by reprocessing to extract the valuable components present in them or converting them chemically, enzymatically, or microbiologically for new applications [2]. Cereal byproducts represent an unexploited package of components with high nutritional benefits. The cereals generate valuable byproducts like aleurone, germ, husk, pericarp, spent grain, testa, bran, germ meal during milling, pearling, malting, etc., which have high market value. The obtained byproducts can be subjected to valorization by extracting the components like flavonoids, polyphenols, dietary fibers (DFs), vitamins, minerals [10].

The cereal's bioactive compounds exert antioxidant, anti-inflammatory, and anti-cancer activities in the human body. Several *in vitro* studies and randomized control human trials have shown that these compounds have shown a positive impact on various diseases like coronary heart diseases, cancer, type 2 diabetes, and neurodegenerative diseases. DFs and polyphenols are key nutrients of a healthy diet. The fibers like β-glucan, arabinoxylans (AXs) from barley, oat, and wheat have been known for their cholesterol-lowering properties and their intake has shown a positive impact on colon cancer [11]. Dietary fiber consumption is also associated with the decreased risk of diabetes, gastrointestinal disorders, obesity, and hypertension. It may influence energy intake and body weight by stimulating the production of gut hormones or peptides, which results in a delay in gastric emptying and helps to reduce weight [12]. The polyphenols are accountable for the antioxidant activity. The intake of phenolic compounds reduces oxidative stress and decreases the risk of cancer, inflammation, CVD. Moreover, the phenolic compounds can effectively target the signaling pathways in cancer, which is helpful in reducing the risk of cancer [13]. The functional groups of flavonoids enhance the plasma antioxidant and anti-inflammatory ability [14]. Moreover, Phytosterols like sitosterol, stigmasterol, and campesterol reduce the absorption of LDL cholesterol in the intestines and lower the risk of CVD [15].

9.1.1 BIOACTIVE COMPOUNDS AND NUTRACEUTICALS IN CEREAL BYPRODUCTS

Bioactive compounds are substances that possess biological activities, the ability to regulate metabolic processes and promote health benefits. They exhibit anti-inflammatory, antioxidant, and anti-microbial activities on human beings. However, the characteristics of these compounds differ based on several criteria like chemical structure, bioavailability, amount, and form of intake. The cereal husks, bran, germ are the low-value waste materials in the cereal processing units due to their inedible nature. However, they have been associated with important nutrients such as DFs, polyphenols, tocopherols, and tocotrienols.

9.1.1.1 DIETARY FIBERS (DFS)

Dietary fibers (DFs) are the principal and major component of cereal byproducts and could be represented as non-starch carbohydrate compounds. DFs are the edible part of grains that resist digestion and undergo partial or complete fermentation in the large intestines. These are primarily composed of plant cell walls and produced at various stages of plant development [16]. The byproducts like cereal bran and husk, brewers spent grain, distiller grains are rich in DFs. It is present in both soluble and insoluble forms. Some of the major DFs present in the cereal byproducts are mentioned in succeeding sections.

9.1.1.2 ARABINOXYLANS (AXS)

Arabinoxylan (AX) is a branched heteroglycan molecule composed of pentose sugars, xylose, and arabinose. The AX is present in the bran, endosperm cell walls, and aleurone layer of the cereal grains and found abundant in the rye and wheat followed by barley, maize, rice, oats, sorghum, and other pseudocereals. It is made of (β1-4) D-Xylopyranosyl backbone chain linked with ∝-L arabinofuranosyl by ∝-1,2 and ∝-1,3 glycosidic bonds [17, 18]. The C-5 position of ∝-L arabinofuranosyl is esterified with ferulic acid compound, allowing the formation of covalent cross-link bond between AX via phenolic oxidative coupling of ferulic acid present in adjacent AX. The residues of ferulic acid also form bonds between AX and other components like cellulose, lignin, glucose, β glucan, proteins, reducing the solubility of certain AX structures. The water-soluble nature of AX can be determined by their molecular weight. The AX with molecular weight from 10 to 10,000 kDa can be extracted by water, and molecular weight beyond 10,000 kDa cannot be extracted by water [19]. AX appears to be in different forms at various tissues of the cereal grains. The bran contains higher branched AX compared to other lignified tissues. This structural heterogeneity of AX is observed on the degree of polymerization (DP) of the xylan backbone, as well as on the ratio of Arabinose to Xylan residues [17]. The ratio normally varies from 0.4 to 1.2 in cereal bran, 0.5 to 0.8 in cereal flours, and lower than 0.3 in lignified husks and straws [20]. AX aids in a plethora of biological activities like immune enhancement, antioxidant properties, prebiotic nature, cholesterol-lowering, and blood sugar modifying characteristics. Apart from nutritional significance, the AX has potential food application as a thickening and stabilizing agent due to its high viscosity and water retention nature [17].

9.1.1.3 B-GLUCANS

β-glucan is a bioactive polysaccharide compound profoundly present in the bran, endosperm, and aleurone layer of cereal grains. Cereal β-glucans are unbranched homopolysaccharide of D-glucopyranosyl monomers with a, β(1,4) and (1,3) linkages. The structural pattern resembles a staircase, which is composed of one β(1,3) linkage for a consecutive three or four β(1,4) linkage [21]. The chemical alignment of β-glucans differs in terms of trisaccharide to tetra-saccharide ratio based on plant sources. The existence of β(1,3) linkages allows β-glucans to exert high solubility, water binding capacities, and viscosity [22, 23]. Barley (5–11%) and oats (3–7%) are the major sources of β-glucans followed by wheat (~0.5%), rye (~2%), and other grains.

β-glucans possess health benefit properties like control diabetics, reducing the circulating blood cholesterol, eventually lowers the risk of cardiovascular diseases, improves gut health [24]. The β-glucans can hydrate and form gels that impart stickiness to the dough. It tends to improve water or oil binding capacity, emulsifying capacity of the product, which may help in developing pseudoplastic, thixotropic suspensions [24]. These properties determine the suitability of β-glucan as an alternative to thickening agents like guar gum, alginates, pectin.

9.1.1.4 CELLULOSE AND LIGNIN

The cellulose and lignin are considered cellulosic components of cereal cell walls. Cellulose is a linear polysaccharide of β(1,4) D-glucopyranosyl units at different degrees of polymerization (500–14,000), linked with hydrogen bonding and formed in myofibrils. It accounts for nearly 20–30% of the primary cell wall and 50% of secondary cell wall components. The cellulose molecules connect each other strongly by hydrogen bond and result in compact aggregates named microfibrils [25]. These microfibrils are encased by other components like lignin, which reduces the soluble nature of cellulose. Lignin comprises almost 30% of the secondary cell walls of the grains. Cellulose is highly available in rice bran (16%), barley hull (19%), oat hull (19%), corn distiller grains (10%), and low amounts in other cereal byproducts [26].

Lignin is a non-carbohydrate polysaccharide composed of 40 oxygenated phenylpropanoid units of coniferyl, sinapyl alcohols, and p-coumaryl derivatives guaiacyl, syringyl, and *p*-hydroxyphenyl, respectively [27]. These are cross-linked with R-O-R' (ether) and C-C bonds in an irregular 3D pattern. The bran portion of the cereal is highly lignified compared to other parts. In cereals, the lignin cross-link bonds are facilitated by ferulates that are bonded with AX. Lignin is the component that mostly occurs in the cell walls of grains it is widely distributed among the cereal grains, wheat bran (5%), brewers spent grain (12–18%), oat bran (2%), barley (4.8%), corn bran (3%), rye bran (6%), barley hulls (11%), oat hulls (14%), sorghum distiller grains (9%) [26]. These cellulosic components are hydrophobic in nature and poorly fermented by gut microbes. The lignin acquires antioxidant activity and may alleviate DNA damage in relation to oxidative stress [26]. These cellulosic components exhibit antioxidant and prebiotic properties, which have a wide range of health benefits.

9.1.1.5 POLYPHENOLIC COMPOUNDS

Polyphenolic compounds are naturally occurring plant secondary metabolites which contribute astringency, flavor, bitterness, color, and oxidative stability in foods. These are non-nutritive molecules with the phenolic hydroxyl structures present in a wide range of foods [28]. Polyphenols are majorly present in the cortical layer of grains and are either soluble in water/oil or insoluble in nature. Some polyphenols are phenolic acids, alkylresorcinols, tocopherols, tocotrienols, and flavonoids [29]. Almost 95% of grain phenolic compounds are either ester or ether-linked to polysaccharides and form intra/intermolecular networks. These compounds are known as dietary fiber-phenolic compounds [30].

9.1.1.6 PHENOLIC ACIDS

The phenolic acids are present in the outer layers of the grains and exhibit strong antioxidant properties. These molecules are composed of a carboxylic group majorly categorized as hydrobenzoic and hydrocinnamic acids [31]. The cinnamic acid, cholregenic acid, ferulic acid, sinnapic acid, *p*-coumaric acids are derived from the hydrocinnamic acids. Hydroxybenzoic acid is a derivative of benzoic acid, composed of a C_6–C_1 structure. Benzoic acids, p-hydroxybenzoic acids, syringic acid, vanillic acid, gallic acid, and protocatechuic acid are the derivatives of hydrobenzoic acid [30]. The common phenolic acids present in cereal grains are ferulic acid, coumaric acid, vanillic acid, syringic acid, and caffeic acid.

The phenolic compounds are present in the outer layers of the kernels and mostly in the bran, hull, and endosperm regions. The phenolic acids are undesirable in some of the products like beer which often causes clouding in the end product [32]. The various phenolic compounds present in the bran of cereals and pseudocereals were mentioned in Table 9.1. These exhibit excellent radical scavenging and anti-bacterial properties. The cinnamic acid exhibits high antioxidant properties followed by gallic, caffeic, benzoic, sinapic, ferulic, p-coumaric, vanillic, chlorogenic, and 4-hydroxy benzoic acids, respectively.

9.1.1.7 FLAVONOIDS

Flavonoids are often known as dietary bioactive compounds with potential antioxidant and anti-inflammatory properties. These are low molecular weight phenolic compounds that are widely distributed in the plant kingdom. Flavonoids comprise aromatic ring structures associated with phenolic hydroxyl compounds by carbon bridge. These are further majorly categorized into flavanones, anthocyanins, flavones, flavonols, anthocyanidins, and chalcones. These are predominantly located in the pericarp and germ portion of the grains [45]. Compared with fruits and vegetables, cereal grains compose fewer amounts of flavonoids.

TABLE 9.1 Phenolic Acids in the Various Cereal Brans

Cereal Bran	Phenolic Acids	References
Wheat	Gallic acid, protocatechuic acid, chlorogenic acid, coffee acid, syringic acid, ferulic acid, vanillic acid, *p*-coumaric acid, *p*-hydroxybenzoic acid.	[30]
Rice	Ferulic acid, coffee acid, syringic acid, protocatechuic acid, chlorogenic acid, coumaric acid, salicylic acid, coumaric acid, syringol, 2-4-dihydroxybenzaldehyde, sinapic acid, 2-hydroxycinnamic acid, ellagic acid.	[33, 34]
Corn	Ferulic acid, caffeic acid, gallic acid, *p*-coumaric acid, *o*-coumaric acid, protocatechuic acid, P-hydroxybenzoic acid.	[35]
Oats	Chlorogenic acid, syringic acid, vanillic acid, *p*-coumaric acid, *p*-hydroxybenzoic acid, 2-4-dihydroxybenzaldehyde, sinapic acid, chlorogenic, ferulic acid, p-hydroxyphenyl acetic acid.	[36]
Rye	Vanillic acid, p-coumaric acid, ferulic acid, synigric acid, sinapic acid, *p*-hydroxybenzoic acid, caffeic acid	[37, 38]
Sorghum	Ferulic acid, p-coumaric acid, vanillic acid, cinnamic acid, gallic acid, p-hydroxybenzoic acid, gentisic acid, protocatechuic acid.	[39]
Barley	Synigric acid, salicylic acid, ferulic acid, caffeic acid, *p*-hydroxybenzoic acid, sinapic acid, *m*-coumaric acid, gallic acid.	[40]
Buckwheat	Vanillic acid, p-coumaric acids, p-hydroxybenzoic acid, caffeic acid, protocatechuic acid, gallic acid, vanillic acid.	[41]
Amaranth seeds	Ferulic acids, caffeic acid, p-hydroxybenzoic acid, protocatechuic acid, salicylic acid, syringic acid, gallic acid, p-coumaric acid.	[42, 43]
Quinoa	Gallic acid, p-hydroxybenzoic acid, vanillic acid, caffeic acid, cinnamic acid, rutin, p-coumaric acid, chrysin.	[43, 44]

9.1.1.8 PHYTOSTEROLS

Phytosterols are the key components that monitor the permeability and fluidity of phospholipid bilayers. Their chemical structure contains a tetracyclic structure and a side chain at the C17 position. These are further classified into Δ5-sterols (plant sterols) and 5α-sterols (plant stanols). The plant stanols contain a saturated sterol ring, whereas plant sterols contain a double bond at the 5th position of the sterol ring. The most common phytosterols present in cereal grains are sitosterol, sitostanol, campesterol, campestanol, and stigmasterol [46]. The considerable amounts of phytosterols are present in rice bran, oat hull, wheat bran, and germ. These occur in both free and conjugated forms. These are known to reduce total and LDL cholesterol in the body, which aids in decreasing the risk of cardiovascular diseases and exhibits anti-inflammatory and antioxidative properties. A high amount of phytosterols is present in the rye (80–90 mg/100 g), barley (70–80 mg/100 g), wheat (80 mg/100 g), and oats (45–50 mg/100 g). The cereal phytosterols form complexes with carbohydrates, phytic acids, and phenolic acids which reduce their bioavailability in humans [47].

9.1.1.9 TOCOLS

Tocols are the chemical derivatives of vitamin E, including tocopherols (T) and tocotrienols (T3). These are lipophilic in nature and exhibit high antioxidant activities. The basic structure of vitamin E contains a long isoprenoid side chain associated with a 6-hydroxylchroman group (polar chromanol) [48]. The tocopherol comprises a phytyl side-chain whereas tocotrienols comprise the geranylgeranyl side chain. These occur in eight congeners namely α-T, β-T, γ-T and δ-T tocopherols and α-T3, β-T3, γ-T3, and δ-T3 tocotrienols based on the position of methylated chroman group. Among all forms of tocols, a-tocopherol exhibits the greatest vitamin E activity, αnd a-tocotrienol exhibits better antioxidant and anti-inflammatory properties. The wheat (27.8 μg) and rye (27.7 μg) are a good source of tocols followed by quinoa (24.7 μg), amaranth (15.4 μg) barley (18.7 μg), and oats (11.5 μg) [11, 49].

9.1.1.10 OTHER PHYTOCHEMICAL COMPOUNDS

Apart from the above-mentioned bioactive compounds, avenanthramides, lignans, alkyl resorcinols, and γ-oryzanol are other major bioactive compounds present in the cereal byproducts. Avenanthramides are polyphenolic compounds present in the oat bran (13 μg/g) and flakes (26–27 μg/g).

These are a group of 20–25 structurally related low molecular weight hydrophilic compounds, Aventhramide-1,3,4 is majorly noticed aventhramides in oats. These are easily available in the body and exhibit anti-atherogenic, antioxidant, and anti-inflammatory activities [11, 50]. Lignans are phytoestrogen compounds present in the cereals like barley, triticale, oat, wheat, and rye, in the range of 8–299 μg/100 g. Syringaresinol, secoisolariciresinol, lariciresinol, pinoresinol, and matairesinol are the common plant lignans [11]. Among them, matairesinol and secoisolariciresinol can be converted to mammalian lignans enterolactone and enterodiol, respectively, with the aid of colon microbial enzymes. They possess antioxidant properties and are capable of reducing the risk of cancer, heart diseases [51]. Alkyl resorcinols are the polyphenols majorly present in the bran of rye (575–1,008 μg/g) and wheat (339–759 μg/g) [51]. These are derivatives of 1,3-dehydrobenzene with the presence of an odd alkyl chain at C-5. These are amphiphilic in nature and have *in vitro* antioxidant, anti-microbial, and anti-fungal activities [52]. γ-oryzanol is a ferulic acid ester of phytosterols and triterpene alcohols. These are also known as steryl ferulates. Though these are identified in wheat, rye, tritacle, barley, and maize, rice bran oil is the richest source. The γ-oryzanol exhibit strong antioxidant activity and have anti-inflammatory, antihyperlipidemic, anti-allergic, and antidiabetic effects [53, 54].

9.2 TECHNOLOGICAL METHODS FOR PRODUCTION OF BIOACTIVE COMPOUNDS FROM CEREAL BYPRODUCTS

The re-utilization of the byproducts that are evolved in cereal processing can give origin to a plethora of new products with functional compounds which has potential benefit for both industries from an economic point of view and consumers from health benefits that can replace synthetic additives. A sequence of separation techniques is used to concentrate and purify the individual bioactive compound. The methodology of production of bioactive compounds majorly includes isolation, extraction, and identification of the compounds from the byproducts. The isolation of the compound can be done by centrifugation, filtration, drying, etc. The isolated compound is further extracted using various extraction techniques. The conventional techniques like solvent extraction, percolation, distillation, maceration, etc., have disadvantages like wastage of organic solvents, low selectivity, extraction yields, extended extraction times, chances of degradation of the compounds compared to novel techniques. These processes are time-consuming and unprofitable to produce on a large scale. Some of the green extraction techniques like supercritical fluid extraction (SFE), microwave-assisted extraction (MAE), ultrasonic-assisted extraction, pulsed electric field-assisted extraction, enzyme-assisted extraction, Deep eutectic solvent extraction, pressurized liquid extraction (PLE), and their combination.

9.2.1 CONVENTIONAL TECHNIQUES FOR EXTRACTION OF BIOACTIVE COMPOUNDS

9.2.1.1 MACERATION

Maceration, also known as a steady-state extraction, is a low-cost technology widely adopted at household levels to extract active compounds and essential oils. It is a combination of a series of steps which include reducing particle size, mixing with solvent followed by filtration. The powdered sample was mixed with the solvent (menstruum) and kept in a closed vessel for a certain time (mostly 3–7 days) as shown in Figure 9.1(a). The maceration softens and breaks the cell wall and helps in releasing the bioactive compounds. The sample mixture is agitated frequently to facilitate the diffusion of compounds in the extract. Finally, the mixture is strained by decantation or filtration. The separated from the solid residues are pressed to recover the solution. Subsequently, the active compounds are separated by evaporating the solvent. This is mainly used to extract thermolabile compounds. The increase in processing temperature improves the solute solubility and promotes diffusion resulting in increased extraction [55]. The extraction efficiency of phenolic compounds, anthocyanins, and flavonoid compounds by this method is very less compared to other techniques [56].

9.2.1.2 PERCOLATION

The percolation is used in the production of bioactive compound fluid extracts. It is an exhaustive extraction, where all soluble components in the sample are completely removed by using the solvent. The percolator (A narrow cone-shaped glass apparatus with an opening at two sides) is used for the percolation process. It can be a continuous process in which the saturated solvent always passes downwards through the sample bed to extract bioactive compounds. The lower valve of the percolator is opened to allow the extract to drip slowly. The extraction process is carried out with the aid of gravitational force. The percolation extraction process is presented in Figure 9.1(b). The obtained wet extract is subjected to evaporation to obtain the desired concentration of the sample. This extraction can be used for the isolation of polyphenolic compounds, essential oils from the sample [57, 58]. Percolation extraction can be applied in shallow bed-type extractors where the bed heights are lower than 1 m, and the extraction process is carried out in such a way that the sample and solvent move in the opposite direction (counter-current) for maximum extraction [58]. The major disadvantage of this extraction is that large volumes of solvent are required, and it can be a time-consuming process.

9.2.1.3 SOXHLET EXTRACTION (CONTINUOUS HOT PERCOLATION)

Soxhlet extraction is a widely used technique for the extraction of bioactive compounds that are thermally stable. In this process, the finely ground sample was packed in the porous bag/thimble and placed in a distillation flask that contains the suitable solvent. The solvent is continuously circulated through the sample by boiling and condensation, where the active compounds are collected in the hot solvent. The Soxhlet apparatus and the extraction were represented in Figure 9.1(c). These extracts further needed to be cleaned up from the solvent residues and concentrated. Depending on the size of the sample, it can be used as a batch processor that can be converted to a continuous process [59]. This technique can be successfully employed in combination with novel extraction techniques (high pressure, microwave-assisted, or ultrasound-assisted extraction) to enhance the extraction efficiencies. This can be used to extract the lipids from the cereal brans/germ and also helpful for extraction of phytosterols, tocotrienols [60].

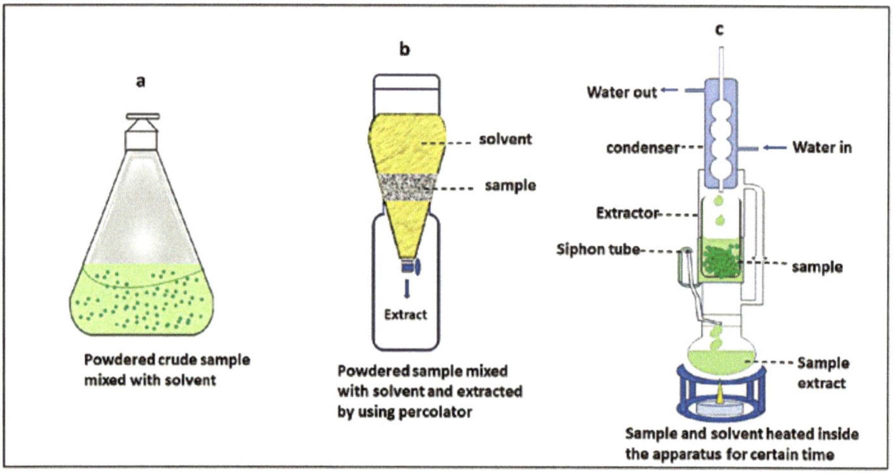

FIGURE 9.1 The conventional extraction techniques (a) maceration; (b) percolation; and (c) Soxhlet apparatus.

9.2.1.4 HYDRODISTILLATION

Hydro distillation is a process used to extract flavonoids and bioactive compounds with the help of water or steam instead of organic solvents. The main processes in hydro distillation are hydro-diffusion, hydrolysis, and decomposition by heat. The hot water/steam aids as the critical factors to free bioactive compounds from the cell matrix. The condensed mixture flows through the condenser to the separator, where the lipids and bioactive compounds separate from the water. This method is mostly used to extract oils from the bran or germ of the cereals. Exposure of sample to high temperature during extraction limits the usage of this technique [61].

The extraction efficiency of these conventional techniques depends majorly on the solvent used and the polarity of the compound, temperature, time, and type of extraction. The most widely used solvents for the extraction process are hexane, methanol, petroleum ether, ethanol, chloroform, ether, acetonitrile, and benzene at various ratios with water. These solvents are used to extract both polar and nonpolar bioactive compounds like phenolic acids, DFs, phytosterols, and flavonoids.

9.3 GREEN TECHNOLOGIES FOR EXTRACTION OF BIOACTIVE COMPOUNDS

As conventional techniques are time-consuming and associated with certain risks, and most of the bioactive compounds are thermally labile, the novel green technologies are gaining more interest. The key objective for the development of non-conventional techniques is to reduce the security risks, time of extraction, solvent consumption and to improve the yield, stability of the compound.

9.3.1 MICROWAVE-ASSISTED EXTRACTION (MAE)

The microwaves are electromagnetic waves in the wavelength of 0.001 m to 1 m with oscillating electric and magnetic fields in the perpendicular direction [62]. When these waves pass through the sample, the electromagnetic energy is absorbed and converted to thermal energy [63]. The schematic diagram of MAE is presented in Figure 9.2. The energy transfer takes place by the ionic conduction and dipole rotation principles. The ionic conduction helps in the generation of homogenous heat in the media through the resistance offered by the solvent for ionic migration. Dipole rotation leads to the reorientation of the water molecules in the direction of the electric field by creating thermal agitation. The rapid rise in localized pressure and temperature of the sample makes the moisture present in the sample evaporate and allows the solvent to penetrate the cell membrane. Moreover, the dipole rotation of the organic molecules resulting in the destruction of hydrogen bonds facilitates the extraction efficiency of bioactive compounds. The selection of appropriate solvent can be done based on the parameters like solubility, dielectric constant, and dissipation factors. Solvents with high dielectric constant and dissipation factor can absorb a high amount of microwave energy and increase the recovery of bioactive compounds [64].

The MAE offers a wide range of advantages like reducing the solvent consumption and time of the extraction, increasing the extraction efficiency of the compounds. This technique can be used to extract polysaccharide compounds like AXs, β-glucans, phenolic compounds, antioxidants, and flavonoids for increased extraction efficiencies [65].

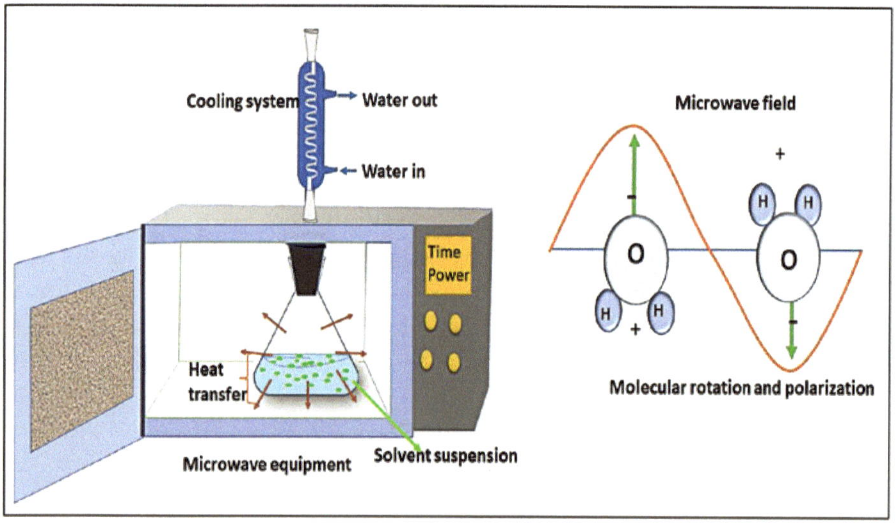

FIGURE 9.2 Schematic diagram of lab-scale microwave-assisted extraction equipment and molecular rotation mechanism of water molecules under applied microwave field.

9.3.2 *ULTRASOUND-ASSISTED EXTRACTION (UAE)*

Ultrasound-assisted extraction provides high recovery of bioactive compounds with low solvent consumption. The principle of acoustic cavitation facilitates increased extraction efficiency. The UAE is widely used to extract carbohydrate compounds, proteins, fiber components [66]. Ultrasounds are sound waves with frequencies ranging from 20 kHz to 100 MHz, transmitted through the medium by generating continuous compressions and rarefactions. The high-frequency waves induce the hotspots at the macroscopic level and produce high temperatures and stress within the medium by causing bubble cavitation. The cavitation creates turbulence and agitation in the medium and inner particle collisions in the sample [61]. The stimulation of acoustic cavitation in the solvent elevates the extraction of the desired compounds. The cavitation caused on the surface of the sample increases the contact area across the solid (sample)-liquid (solvent) phases resulting in rapid diffusion of the solute components into the solvent [67]. The shear forces increase the mass transfer of the sample by acoustic streaming and turbulent mixing. The factors like frequency, time, temperature, and pressure controls the extraction rate.

The ultrasound can be applied to the sample by both direct and indirect methods. In the direct method, the probe system is used, where it provides high-intensity waves, i.e., approx. 100 times greater. The schematic representation of UAE using probe system is presented in Figure 9.3. In the case of an indirect system, the water bath is used as the medium for transferring the sound waves to the sample [68]. This increases the yield of bioactive compounds by elevating the rate of mass transfer and kinetics. The UAE is faster than MWE and helps to recover the heat liable bioactive compounds [69].

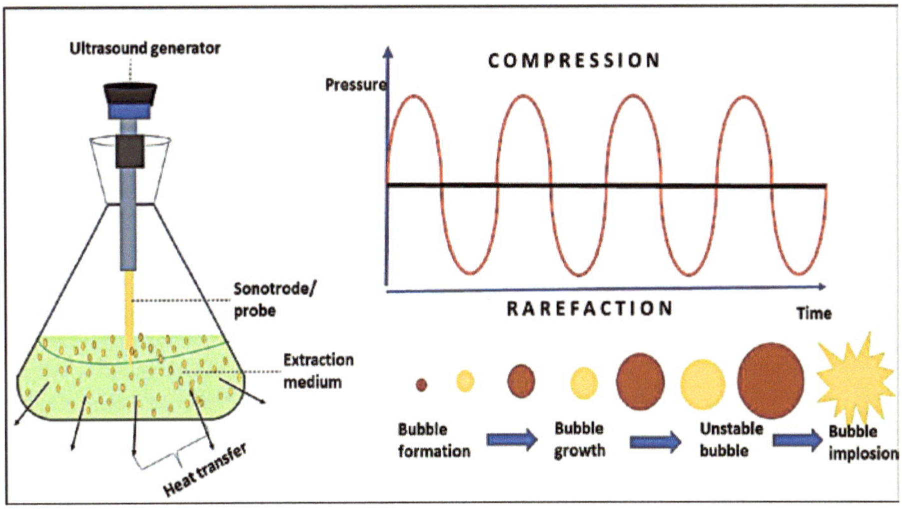

FIGURE 9.3 Ultrasound-assisted extraction of the sample using the probe and the cavitation mechanism of the air bubbles.

9.3.3 PULSED ELECTRIC FIELD ASSISTED EXTRACTION (PFE)

The pulsed electric field is a promising non-thermal technology with the application of high voltage electricity (100–300 V/cm, 20–80 kV/cm in batch mode and continuous mode, respectively). The PEF is mainly used for microbial destruction in liquid foods. The efficiency of the PEF majorly depends on the type of the solution and the treatment parameters. The PEF produces transmembrane potential within the cell membrane, which leads to a rise in conductivity and permeability of the cell [70]. This causes reversible or irreversible damage to the cell membrane and increases the mass transfer rate. The high-intensity electric pulses create electroporation of cell membrane and improve diffusion efficiency of bioactive compounds at ambient temperature.

It has been employed in wine industries to hasten the aging process and to extract flavor compounds. The PEF can be used to extract lipids, polysaccharides, alkaloids, polyphenol compounds, and polyphenols, flavonoid compounds [71]. The effectiveness of extraction majorly depends on the electric strength, shape, width, and number of pulses, specific energy, and frequency of pulses. The schematic diagram of PFE and various electric pulses used for extraction was presented in Figure 9.4.

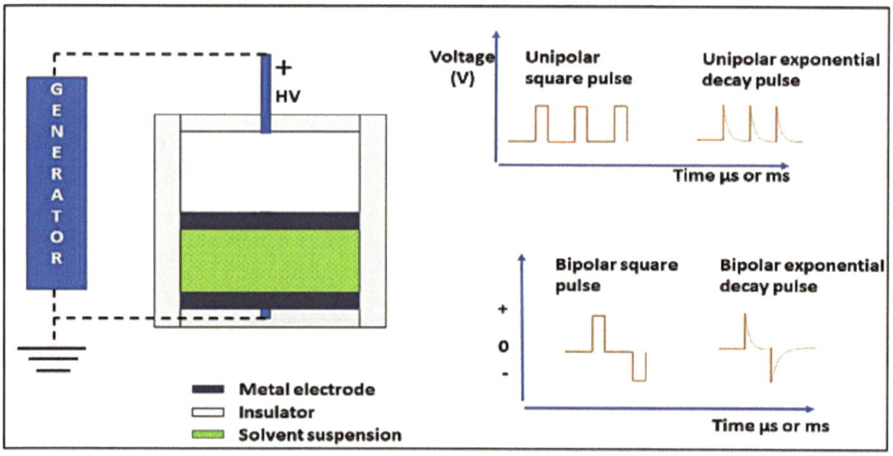

FIGURE 9.4 Experimental set-up of pulsed electric field-assisted extraction and different kinds of electric pulses.

9.3.4 ENZYME-ASSISTED EXTRACTION (EAE)

Enzymes are globular proteins derived from various sources like microorganisms, animals, or plant origin. These are categorized as oxidation-reduction enzymes, hydrolyzing enzymes, carboxylation enzymes, isomerizing enzymes, ligases, and group transfer enzymes based on their catalytic properties. Enzymes aid in the extraction of compounds by destroying cell walls. The enzymes disrupt the cell wall, subsequently improving the extraction rate of the bioactive compounds. The more often used enzymes are cellulases, β-glucosidase, hemicelluloses for the extraction of polysaccharide compounds from the cereal brans. The EAE conditions like reaction time and temperature, pH, particle size, and enzyme concentration are major influential factors [72]. The compounds like flavonoids that have poor solubility in the solvents like ethanol or water can be successfully extracted using this technology [73]. The EAE can be used in combination with other techniques like the US, MW, SFE to improve the extraction rates and to reduce solvent usage [74].

9.3.5 SUPERCRITICAL FLUID EXTRACTION (SFE)

Supercritical fluid extraction (SFE) is used to extract the bioactive compounds by employing fluids at the temperature and pressure above their critical point. Supercritical fluids (SF) form a mesophase because of their liquid and gas-like properties. They have a density equal to liquids, viscosity like gas, and diffusivity in between the liquids and gas. These properties, like low viscosity and high density, facilitate SF to penetrate faster and deeper into the sample [75]. Though there are a lot of solvents such as propane, freon, isobutane, nitrous oxide, dimethyl ether, and water, carbon dioxide is the most used solvent because of its low cost, non-flammable, and reusable nature, available in abundance, and compatible with all food ingredients. The optimum conditions for supercritical CO_2 are above 31°C and 74 bar temperature and pressure, respectively [76].

The degradation of bioactive compounds is little as the extraction process is carried out at low temperatures and in the absence of light and oxygen, which prevents oxidation of the compounds [77]. The solubility of the selected bioactive compound depends on the molecular interaction at the supercritical phase, operating pressure, temperature, and solute molecular weight, structure, vapor pressure, dielectric constant, and polarity. Since CO_2 is a low polar solvent, it is less effective in extracting high polar bioactive compounds. Thus, the co-solvents or modifiers are used to enhance the extraction rate. Some of the co-solvents include methanol, ethanol, hexane, isopropanol, and dichloromethane. The SFE can be used in combination with high-pressure processing (HPP) for increased extraction efficiency of phenolic acids. The compounds with high solubility in SFs like phytosterols, phenolic acids, tocols, γ-oryzanol, and carotenoids can be extracted by this method [64].

9.3.6 PRESSURIZED LIQUID EXTRACTION (PLE)

Pressurized liquid extraction (PLE) is also known as enhanced or accelerated solvent extraction (ASE), it is partially like SFE. The PLE uses the high temperatures (50–200°C) and pressure (4–20 MPa) without attaining the critical point of the solvents to retain the solvents in the liquid state and to expedite the extraction kinetics of the compounds. The high temperatures and pressure reduce the surface tension and viscosity of the solvent and increase the mass transfer ratio. The high-temperature aids solubilization of the bioactive compound in the solvent, pressure helps to penetrate the solvent into the sample. PLE consumes a very low amount of solvents and requires less time for extraction in comparison with other techniques [78]. At elevated temperatures, the dielectric constant and polarity of the solvent reduce. Hence, the operating temperature can be determined according to the polarity of the solvent and the solubility of the bioactive compounds [79]. The common solvents used for the extraction process are water or ethanol. The solvents like propane and dimethyl ether can also be used to increase the recovery of non-polar compounds from the sample. The PLE is mainly used for the extraction of polysaccharides, polyphenolic compounds, and flavonoids from the sample. The PLE can be used in combination with US or SFE for the high recovery of polyphenols [80].

9.4 ANALYTICAL TECHNIQUES FOR QUALITATIVE AND QUANTITATIVE ANALYSIS OF THE BIOACTIVE COMPOUNDS

The type of extraction influences the qualitative and quantitative analysis of the compounds. It plays a key role in the quality of the final compound. Since the cereal extracts are composed of a range of bioactive compounds with various polarities, the isolation and identification of a particular component is a challenging task [81]. Advanced, high-resolution analytical techniques are employed for analyzing the extracted compounds. Chromatographic techniques are one of the most used techniques for analyzing bioactive compounds. They enable the complex process of separation, identification, and quantification of multiple food components easy with high efficiency [82]. Some of the most used chromatographic techniques are high-performance liquid chromatography (HPLC), GC, and liquid chromatography (LC).

9.4.1 HIGH PERFORMANCE LIQUID CHROMATOGRAPHY (HPLC)

High-performance liquid chromatography (HPLC) is the most used technique for the identification and quantification of these bioactive compounds. Apart from that, LC, GC, capillary electrophoresis (CEP) chromatography, supercritical fluid chromatography (SFC) are other popular techniques. Quantification of bioactive compounds is accomplished by using retention rates in the given column and mobile phase [83]. High chromatographic resolution can be achieved by optimizing the experimental parameters like stationary phase, mobile phase, the temperature of column, and gradient elution. However, quantification of the compounds is majorly dependent on the choice of the column, stationary, and mobile phase [84]. The detector helps to identify the compound; UV detectors (UVD) are highly preferred because of their sensitivity, and the bioactive compounds exhibit absorbance of UV at certain wavelengths. The other detectors including fluorescence detector (FD), photodiode array detector (PDA), and mass spectrometer (MS) can be used in combination with HPLC and LC to acquire identification and structural elucidation of the bioactive compounds. PDA is used for the identification of phenolic compounds, flavanols, and flavanones, which can be detected at the wavelength of 240–285 nm and 350–365 nm, respectively. Whereas the FD is used for the identification of flavonoids like catechins, procyanidins, etc. Mass spectroscopy detectors rapid and accurate identification of the compounds [85].

American Chemical Association style HPLC can be employed for both identification and quantification of bioactive compounds. It can operate at pressures lower than 40 MPa whereas, ultra-pressure liquid chromatography (UPLC) can operate at 100 MPa pressure. Flavonoids and phenolic compounds are generally analyzed by a UV-visible diode array detector (DAD). The elution can be carried out by using binary solvents, an organic solvent (methanol or acetonitrile), and a polar solvent (acidified water a 0.05–5% of formic acid or acetic acid can be used). The pH of the mobile phase should be maintained between 2 and 4 in order to avoid the ionization of hydroxyl groups in the phenolic compounds and minimize the peak shape distortions. Usually, the identification and quantification of phytochemicals can be achieved by the isocratic system. The analysis of bioactive components is usually done by using C-8 or 18 based silica columns in a reverse phase mode (RPLC). The normal phase HPLC with octadecyl silica column is used to isolate and quantify the major cereal phytosterols. They are identified and quantified using various detectors like flame ionization detector (FID), UVD, evaporative light scattering detector (ELSD), and MS [85]. Although there are several organic solvents like acetic acid, tetrahydrofuran, isopropanol. Methanol and acetonitrile are widely used solvents in the mobile phase. The new columns like monolithic, superficially porous particles columns can be used along with advanced HPLC techniques like UHPLC, high-temperature liquid chromatography (HTLC), and two-dimensional liquid chromatography (LC X LC) can be used for quantification of the bioactive compounds.

9.4.2 GAS CHROMATOGRAPHY (GC)

Gas chromatography (GC) is a simple, versatile, and highly efficient technique primarily used for the quantification of non-polar bioactive compounds like lipids, phytosterols, and phenols with volatile nature [82]. The mobile phase of GC is composed of inert gasses like He or N_2. Flame ionization detectors (GC-FID) or a mass spectrometer (GC-MS) are widely used detectors for the quantification of bioactive compounds. The GC-FID is primarily used for the quantification of the compounds based on their retention time (RT). It is preferred for quantifying phenolic compounds and phytosterols. Contrastingly, GC-MS uses peaks for the identification and quantification of the compounds. GC-MS uses gradient/isothermal temperature modes that vaporize bioactive compounds, a uniform flow of carrier gas facilitates the separation of compounds within the GC column [86]. Apart from detectors, columns like capillary column offer more benefits by providing thermal stability, decreasing processing time, and enhancing peak resolution of the phytosterols [72].

9.4.3 CAPILLARY ELECTROPHORESIS (CEP)

Capillary electrophoresis (CE) is one of the novel methods used for the identification and quantification of various bioactive compounds in the sample. CE gained a lot of preference over other techniques due to high resolution, rapid analysis time, and high selectivity [87]. CE is composed of separating fused silica capillary column with high voltage power driving force. The variations in electrophoretic mobility of the components facilitate the quantification of bioactive compounds. Similar to other chromatographic techniques, CE can also couple with detectors like UVD, MS, evaporative light scattering detector (ESD) for analyzing various flavonoids, polyphenols, and other bioactive compounds [88].

9.5 CONCLUSION

The proper utilization of cereal byproducts can be beneficial in economic, environmental, and health aspects. The bioactive compounds extracted from these byproducts are gaining a lot of popularity in the food and pharmaceutical industries because of their nutritious and health-beneficial properties. Increasing demand for bioactive components like DFs, polyphenols, phytosterols, and other phytochemicals promotes the development of various techniques to increase extraction efficiency and safety. Since most of these compounds are generally present in bound with other cereal components, the complete extraction is difficult, and many components remain unidentified. However, the cost, safety, and scalability of the compounds are the major challenges that have to be addressed in order to produce industrially. Though a lot of new technologies were established for extraction, the purity of the extracted compounds is still questionable. The production of pure bioactive compounds would allow knowing in depth about the bioavailability of the compound in the body, their mechanism of action, and the benefits associated with their consumption. This will enhance the applications of bioactive compounds in food fortification and nutraceuticals production.

KEYWORDS

- **bioactive compounds**
- **byproducts**
- **capillary electrophoresis**
- **cereals**
- **extraction**
- **health benefits**
- **nutraceuticals**
- **supercritical fluid extraction**

REFERENCES

1. Alexander, C., Gregson, N., & Gille, Z., (2013). Food waste. In: *The Handbook of Food Research* (pp. 471–485).
2. Skendi, A., Zinoviadou, K. G., Papageorgiou, M., & Rocha, J. M., (2020). Advances on the valorization and functionalization of byproducts and wastes from cereal-based processing industry. *Foods*, *9*(9). 10.3390/foods9091243.

3. Dey, D., Richter, J. K., Ek, P., Gu, B. J., & Ganjyal, G. M., (2021). Utilization of food processing byproducts in extrusion processing: A review. *Front. Sustain. Food Syst, 4*, 1–18. 10.3389/fsufs.2020.603751.
4. Torres-León, C., et al., (2018). Food waste and byproducts: An opportunity to minimize malnutrition and hunger in developing countries. *Frot. Sustain. Food Syst.*, *2*(52), 10.3389/fsufs.2018.00052.
5. Anal, A. K., (2018). Food processing byproducts and their utilization: Introduction. *Food Processing Byproducts and their Utilization*, 1–10.
6. Laskowski, W., Górska-Warsewicz, H., Rejman, K., Czeczotko, M., & Zwolińska, J., (2019). How important are cereals and cereal products in the average polish diet? *Nutrients, 11*(3). 10.3390/nu11030679.
7. Alvarez-Jubete, L., Arendt, E. K., & Gallagher, E., (2010). Nutritive value of pseudocereals and their increasing use as functional gluten-free ingredients. *Trends Food Sci. Technol, 21*(2), 106–113. 10.1016/j.tifs.2009.10.014.
8. Mir, N. A., Riar, C. S., & Singh, S., (2018). Nutritional constituents of pseudo cereals and their potential use in food systems: A review. *Trends Food Sci. Technol, 75*, 170–180. 10.1016/j.tifs.2018.03.016.
9. Aisbitt, B., Caswell, H., & Lunn, J., (2008). Cereals – current and emerging nutritional issues. *Nutr. Bull., 33*(3), 169–185. 10.1111/j.1467–3010.2008.00704.x.
10. Saini, A., Panwar, D., Ps, P., & Mb, B., (2019). Bioactive compounds from cereal and pulse processing byproducts and their potential health benefits. *Austin J. Nutr. Metab, 6*(2), 1–7.
11. Gani, A., Wani, S. M, & Masoodi, F. A., (2012). Whole-grain cereal bioactive compounds and their health benefits: A review. *J. Food Process. Technol, 3*(3). 10.4172/2157–7110.1000146.
12. Gutzwiller, J. P., et al., (1999). Glucagon-like peptide-1: A potent regulator of food intake in humans. *Gut, 44*(1), 81–86. 10.1136/gut.44.1.81.
13. Xie, M., Liu, J., Tsao, R., Wang, Z., Sun, B., & Wang, J., (2019). Whole grain consumption for the prevention and treatment of breast cancer. *Nutrients, 11*(1769), 1–21. 10.3390/nu11081769.
14. Nignpense, B., Francis, Ed. N., Blanchard, C., & Santhakumar, A. B., (2021). Bio-accessibility and bioactivity of cereal polyphenols: A review. *Foods, 10*(7). 10.3390/foods10071595.
15. Bartłomiej, S., Justyna, R. K., & Ewa, N., (2012). Bioactive compounds in cereal grains – occurrence, structure, technological significance, and nutritional benefits: A review. *Food Sci. Technol. Int, 18*(6), 559–568. 10.1177/1082013211433079.
16. Roth, M., Jekle, M., & Becker, T., (2019). Opportunities for upcycling cereal byproducts with special focus on distiller's grains. *Trends Food Sci. Technol., 91*, 282–293. 10.1016/j.tifs.2019.07.041.
17. Bastos, R., Coelho, E., & Coimbra, M. A., (2018). Arabinoxylans from cereal byproducts: Insights into structural features, recovery, and applications. *Sustainable Recovery and Reutilization of Cereal Processing Byproducts*, 227–251.
18. Onipe, O. O., Jideani, A. I. O., & Beswa, D., (2015). Composition and functionality of wheat bran and its application in some cereal food products. *Int. J. Food Sci. Technol, 50*(12), 2509–2518. 10.1111/ijfs.12935.
19. Niño-Medina, G., Carvajal-Millán, E., Rascon-Chu, A., Marquez-Escalante, J. A., Guerrero, V., & Salas-Muñoz, E., (2010). Feruloylated arabinoxylans and arabinoxylan

gels: Structure, sources and applications. *Phytochem. Rev.*, *9*(1), 111–120. 10.1007/s11101–009–9147–3.
20. Zhang, Z., Smith, C., & Li, W., (2014). Extraction and modification technology of arabinoxylans from cereal byproducts: A critical review. *Food Res. Int.*, *65*, 423–436. 10.1016/j.foodres.2014.05.068.
21. Henrion, M., Francey, C., Lê, K. A., & Lamothe, L., (2019). Cereal B-glucans: The impact of processing and how it affects physiological responses. *Nutrients*, *11*(8), 1–14. 10.3390/nu11081729.
22. Hu, X., Zhao, J., Zhao, Q., & Zheng, J., (2015). Structure and characteristic of β-glucan in cereal: A review. *J. Food Process. Preserv*, *39*(6), 3145–3153. 10.1111/jfpp.12384.
23. Tiefenbacher, K. F., (2017). Technology of main ingredients—Water and flours. *The Technology of Wafers and Waffles*, 15–121.
24. Ahmad, A., Munir, B., & Abrar, M., (2012). Perspective of β-glucan as functional ingredient for food industry. *J. Nutr. Food Sci.*, *2*(2). 10.4172/2155–9600.1000133.
25. Huma, B. U. A., et al., (2019). Isolation and characterization of cereal cell walls. *Int. J. Food Prop.*, *22*(1), 130–137. 10.1080/10942912.2019.1573832.
26. Navarro, D. M. D. L., Abelilla, J. J., & Stein, H. H., (2019). Structures and characteristics of carbohydrates in diets fed to pigs: A review. *J. Anim. Sci. Biotechnol*, *10*(39), 1–17. 10.1186/s40104–019–0345–6.
27. Lu, Y., Lu, Y. C., Hu, H. Q., Xie, F. J., Wei, X. Y., & Fan, X., (2017). Structural characterization of lignin and its degradation products with spectroscopic methods. *J. Spectrosc.*, 1–15. 10.1155/2017/8951658.
28. Tian, S., Sun, Y., Chen, Z., Yang, Y., Wang, Y., & Trabelsi, N., (2019). Functional properties of polyphenols in grains and effects of physicochemical processing on polyphenols. *J. Food Qual, 2019.* 10.1155/2019/2793973.
29. Sidhu, J. S., Kabir, Y., & Huffman, F. G., (2007). Functional foods from cereal grains. *Int. J. Food Prop.*, *10*(2), 231–244. 10.1080/10942910601045289.
30. Călinoiu, L. F., & Vodnar, D. C., (2018). Whole grains and phenolic acids: A review on bioactivity, functionality, health benefits and bioavailability. *Nutrients*, *10*(11), 10.3390/nu10111615.
31. Kumar, N., & Goel, N., (2019). Phenolic acids: Natural, versatile molecules with promising therapeutic applications. *Biotechnol. Reports*, *24*, e00370. 10.1016/j.btre.2019.e00370.
32. Stuper-Szablewska, K., & Perkowski, J., (2019). Phenolic acids in cereal grain: Occurrence, biosynthesis, metabolism, and role in living organisms. *Crit. Rev. Food Sci. Nutr.*, *59*(4), 664–675. 10.1080/10408398.2017.1387096.
33. Zhou, Z., Robards, K., Helliwell, S., & Blanchard, C., (2004). The distribution of phenolic acids in rice. *Food Chem*, *87*(3), 401–406. 10.1016/j.foodchem.2003.12.015.
34. Goufo, P., & Trindade, H., (2014). Rice antioxidants: Phenolic acids, flavonoids, anthocyanins, proanthocyanidins, tocopherols, tocotrienols, c-oryzanol, and phytic acid. *Food Sci. Nutr.*, *2*(2), 75–104. 10.1002/fsn3.86.
35. Zhang, R., et al., (2017). Phenolic content and antioxidant activity of eight representative sweet corn varieties grown in South China. *Int. J. Food Prop.*, *20*(12), 3043–3055. 10.1080/10942912.2016.1270964.
36. Chen, C. Y., Milbury, P. E., Kwak, H. K., Collins, F. W., Samuel, P., & Blumberg, J. B., (2004). Avenanthramides and phenolic acids from oats are bioavailable and

act synergistically with vitamin C to enhance hamster and human LDL resistance to oxidation. *J. Nutr., 134*(6), 1459–1466. 10.1093/jn/134.6.1459.
37. Kulichová, K., et al., (2019). Phenolic compounds and biological activities of rye (*Secale cereale* L.) grains. *Open Chem., 17*(1), 988–999. 10.1515/chem-2019–0103.
38. Andreasen, M. F., Christensen, L. P., Meyer, A. S., & Hansen, Å., (2000). Content of phenolic acids and ferulic acid dehydrodimers in 17 rye (*Secale cereale* L.) varieties. *J. Agric. Food Chem., 48*(7), 2837–2842. 2000. 10.1021/jf991266w.
39. Kaur, K. D., Jha, A., Sabikhi, L., & Singh, A. K., (2014). Significance of coarse cereals in health and nutrition: A review. *J. Food Sci. Technol., 51*(8), 1429–1444. 10.1007/s13197–011–0612–9.
40. Šimić, G., Horvat, D., Lalić, A., Koceva, K. D., Abičić, I., & Zdunić, Z., (2019). Distribution of β-glucan, phenolic acids, and proteins as functional phytonutrients of hull-less barley grain. *Foods, 8*(12), 680.
41. Guo, X. D., Ma, Y. J., Parry, J., Gao, J. M., Yu, L. L., & Wang, M., (2011). Phenolics content and antioxidant activity of Tartary buckwheat from different locations. *Molecules, 16*(12), 9850–9867. 10.3390/molecules16129850.
42. Ogrodowska, D., Czaplicki, S., Zadernowski, R., Mattila, P., Hellström, J., & Naczk, M., (2012). Phenolic acids in seeds and products obtained from *Amaranthus cruentus. J. Food Nutr. Res., 51*(2), 96–101.
43. Paśko, P., Sajewicz, M., Gorinstein, S., & Zachwieja, Z., (2008). Analysis of selected phenolic acids and flavonoids in *Amaranthus cruentus* and *Chenopodium quinoa* seeds and sprouts by HPLC. *Acta Chromatogr., 20*(4), 661–672. 10.1556/AChrom.20.2008.4.11.
44. Al-qabba, M. M., El-mowafy, M. A., Althwab, S. A., & Alfheeaid, H. A., (2020). Phenolic profile, antioxidant activity, and ameliorating efficacy of *Chenopodium quinoa* sprouts against CCl_4-induced oxidative stress in rats. *Nutrients, 12*(10), 2904. 10.3390/nu12102904.
45. Suchowilska, E., Bieńkowska, T., Stuper-Szablewska, K., & Wiwart, M., (2020). Concentrations of phenolic acids, flavonoids and carotenoids and the antioxidant activity of the grain, flour and bran of *Triticum polonicum* as compared with three cultivated wheat species. *Agriculture, 10*(12), 1–19. 10.3390/agriculture10120591.
46. Tlais, A. Z. A., Fiorino, G. M., Polo, A., Filannino, P., & Di Cagno, R., (2020). High-value compounds in fruit, vegetable and cereal byproducts: An overview of potential sustainable reuse and exploitation. *Molecules, 25*(13), 2987. 10.3390/molecules25132987.
47. Piironen, V., Toivo, J., & Lampi, A. M., (2002). Plant sterols in cereals and cereal products. *Cereal Chem., 79*(1), 148–154. 10.1094/CCHEM.2002.79.1.148.
48. Borneo, R., & León, V. X., (2012). Whole grain cereals: Functional components and health benefits. *Food Funct., 3*(2), 110–119. 10.1039/c1fo10165j.
49. Alvarez-Jubete, L., Holse, M., Hansen, Å., Arendt, E. K., & Gallagher, E., (2009). Impact of baking on vitamin E content of pseudocereals amaranth, quinoa, and buckwheat. *Cereal Chem., 86*(5), 511–515. 10.1094/CCHEM-86–5-0511.
50. Sur, R., Nigam, A., Grote, D., Liebel, F., & Southall, M. D., (2008). Avenanthramides, polyphenols from oats, exhibit anti-inflammatory and anti-itch activity. *Arch. Dermatol. Res., 300*(10), 569–574. 10.1007/s00403–008–0858-x.
51. Dykes, L., & Rooney, L. W., (2007). Phenolic compounds in cereal grains and their health benefits. *Cereal Foods World, 52*(3), 105–111. 10.1094/CFW-52–3-0105.

52. Tian, S., Sun, Y., Chen, Z., & Zhao, R., (2020). Bioavailability and bioactivity of alkylresorcinols from different cereal products. *J. Food Qual., 2020*, 5781356. 10.1155/2020/5781356.
53. Eslami, S., Esa, N. M., Marandi, S. M., Ghasemi, G., & Eslami, S., (2014). Effects of gamma oryzanol supplementation on anthropometric measurements & muscular strength in healthy males following chronic resistance training. *Indian J. Med. Res., 139*, 857–863.
54. Lemus, C., Angelis, A., Halabalaki, M., & Skaltsounis, A. L., (2014). γ-oryzanol. An attractive bioactive component from rice bran. *Wheat and Rice in Disease Prevention and Health,* 409–430.
55. Lee, N. Y., Yunus, M. A. C., Idham, Z., Ruslan, M. S. H., Aziz, A. H. A., & Irwansyah, N., (2016). Extraction and identification of bioactive compounds from agarwood leaves. *IOP Conf. Ser. Mater. Sci. Eng., 162*(1), 012028. 10.1088/1757–899X/162/1/012028.
56. Abubakar, A. R., & Haque, M., (2020). Preparation of medicinal plants: Basic extraction and fractionation procedures for experimental purposes, *J. Pharm. Bioallied Sci., 12*(1), 1–10. 10.4103/jpbs.JPBS_175_19.
57. Sridhar, A., Ponnuchamy, M., Kumar, P. S., Kapoor, A., Vo, D. V. N., & Prabhakar, S., (2021). Techniques and modeling of polyphenol extraction from food: A review. *Environ. Chem. Lett., 17*, 1–35. 10.1007/s10311–021–01217–8.
58. Pramparo, M., Gregory, S., & Mattea, M., (2002). Immersion vs. percolation in the extraction of oil from oleaginous seeds. *JAOCS, J. Am. Oil Chem. Soc., 79*(10), 955–960. 10.1007/s11746–002–0586–3.
59. Baiano, A., (2014). Recovery of biomolecules from food wastes: A review. *Molecules, 19*(9), 14821–14842. 10.3390/molecules190914821.
60. Górnaś, P., Rudzińska, M., Raczyk, M., & Soliven, A., (2016). Lipophilic bioactive compounds in the oils recovered from cereal byproducts. *J. Sci. Food Agric., 96*(9), 3256–3265. 10.1002/jsfa.7511.
61. Soquetta, M. B., Terra, L. De. M., & Bastos, C. P., (2018). Green technologies for the extraction of bioactive compounds in fruits and vegetables. *CYTA – Journal of Food, 16*(1), 400–412. 10.1080/19476337.2017.1411978.
62. Angiolillo, L., Del Nobile, M. A., & Conte, A., (2015). The extraction of bioactive compounds from food residues using microwaves. *Curr. Opin. Food Sci., 5*, 93–98. 10.1016/j.cofs.2015.10.001.
63. Akhtar, I., Javad, S., Yousaf, Z., Iqbal, S., & Jabeen, K., (2019). Microwave-assisted extraction of phytochemicals is an efficient and modern approach for botanicals and pharmaceuticals. *Pak. J. Pharm. Sci., 32*(1), 223–230.
64. Joana Gil-Chávez, G., et al., (2013). Technologies for extraction and production of bioactive compounds to be used as nutraceuticals and food ingredients: An overview. *Compr. Rev. Food Sci. Food Saf., 12*(1), 5–23. 10.1111/1541–4337.12005.
65. Kim, H., Kim, Y., Lian, T., Song, S., & Bang, K., (2019). Effect of microwave assisted water extraction on insoluble phenolic compounds through bioactivity of *Fagopyrum esculentum. Korean J. Plant Resources, 32*(6), 723–729. 10.7732/kjpr.2019.32.6.723.
66. Bonifácio-Lopes, T., Teixeira, J. A., & Pintado, M., (2020). Current extraction techniques towards bioactive compounds from brewer's spent grain: A review. *Crit. Rev. Food Sci. Nutr., 60*(16), 2730–2741. 10.1080/10408398.2019.1655632.
67. Medina-Torres, N., Ayora-Talavera, T., Espinosa-Andrews, H., Sánchez-Contreras, A., & Pacheco, N., (2017). Ultrasound-assisted extraction for the recovery of phenolic compounds from vegetable sources. *Agronomy, 7*(3), 47. 10.3390/agronomy7030047.

68. Wen, L., Zhang, Z., Sun, D. W., Sivagnanam, S. P., & Tiwari, B. K., (2020). Combination of emerging technologies for the extraction of bioactive compounds. *Crit. Rev. Food Sci. Nutr.*, *60*(11), 1826–1841. 110.1080/10408398.2019.1602823.
69. Alc, C., Žugˇ, T., Jambrak, A. R., Lorenzo, J. M., Ganato, D., & Barba, F. J., (2020). Effects of ultrasound-assisted extraction and solvent on the phenolic profile, bacterial growth, and anti-inflammatory/antioxidant activities of Mediterranean olive and fig leaves extracts Cristina. *Molecules*, *25*, 1718.
70. Lakka, A., Bozinou, E., Makris, D. P., & Lalas, S. I., (2021). Evaluation of pulsed electric field polyphenol extraction from *Vitis vinifera*, *Sideritis scardica* and *Crocus sativus*. *Chem. Engineering*, *5*(2), 25. 10.3390/chemengineering5020025.
71. Ranjha, M. M. A. N., et al., (2021). A critical review on pulsed electric field: A novel technology for the extraction of phytoconstituents. *Molecules*, *26*(16), 1–23. 10.3390/molecules26164893.
72. Cheng, X., Bi, L., Zhao, Z., & Chen, Y., (2015). Advances in enzyme assisted extraction of natural products. *International Conference on Material, Mechanical and Manufacturing Engineering*, IC3ME, 371–375. 10.2991/ic3me-15.2015.72.
73. Chen, S., Xing, X. H., Huang, J. J., & Xu, M. S., (2011). Enzyme-assisted extraction of flavonoids from *Ginkgo biloba* leaves: Improvement effect of flavonol transglycosylation catalyzed by *Penicillium decumbens* cellulase. *Enzyme Microb. Technol*, *48*(1), 100–105. 10.1016/j.enzmictec.2010.09.017.
74. Nadar, S. S., Rao, P., & Rathod, V. K., (2018). Enzyme assisted extraction of biomolecules as an approach to novel extraction technology: A review. *Food Res. Int*, *108*(2017), 309–330. 10.1016/j.foodres.2018.03.006.
75. Ibáñez, E., Mendiola, J. A., & Castro-Puyana, M., (2015). Supercritical fluid extraction. *Encycl. Food Heal.*, *8*(2), 227–233. 10.1016/B978–0-12–384947–2.00675–9.
76. Khaw, K. Y., Parat, M. O., Shaw, P. N., & Falconer, J. R., (2017). Solvent supercritical fluid technologies to extract bioactive compounds from natural sources: A review. *Molecules*, *22*(7), 1186. 10.3390/molecules22071186.
77. Díaz-Reinoso, B., Moure, A., Domínguez, H., & Parajó, J. C., (2006). Supercritical CO_2 extraction and purification of compounds with antioxidant activity. *J. Agric. Food Chem*, *54*(7), 2441–2469. 10.1021/jf052858j.
78. Camel, V., (2001). Recent extraction techniques for solid matrices - Supercritical fluid extraction, pressurized fluid extraction and microwave-assisted extraction: Their potential and pitfalls. *Analyst*, *126*(7), 1182–1193. 10.1039/b008243k.
79. Rodríguez De, L. S. L., Ramírez-Garza, R. E., & Serna, S. S. O., (2020). Environmentally friendly methods for flavonoid extraction from plant material: Impact of their operating conditions on yield and antioxidant properties. *Sci. World J*, *2020*. 10.1155/2020/6792069.
80. Abhari, K., & Mousavi, K. A., (2020). Alternative extraction techniques to obtain, isolate and purify proteins and bioactive from aquaculture and byproducts. *Adv. Food Nutr. Res.*, *92*, 35–52. 10.1016/bs.afnr.2019.12.004.
81. Azmir, J., et al., (2013). Techniques for extraction of bioactive compounds from plant materials: A review. *J. Food Eng.*, *117*(4), 426–436. 10.1016/j.jfoodeng.2013.01.014.
82. Coskun, O., (2016). Separation techniques: Chromatograph. *North. Clin. Istanbul.*, *3*(2), 156–160. 10.14744/nci.2016.32757.

83. Sasidharan, S., Chen, Y., Saravanan, D., Sundram, K. M., & Latha, L. Y., (2011). Extraction, isolation, and characterization of bioactive compounds from plants' extracts. *African J. Tradit. Complement. Altern. Med.*, *8*(1).
84. Câmara, J. S., et al., (2021). Food bioactive compounds and emerging techniques for their extraction: Polyphenols as a case study. *Foods*, *10*(1), 37. 10.3390/foods10010037.
85. Uddin, M. S., et al., (2018). Techniques for the extraction of phytosterols and their benefits in human health: A review. *Sep. Sci. Technol. 53*(14), 2206–2223. 10.1080/01496395.2018.1454472.
86. Khakimov, B., Bak, S., & Engelsen, S. B., (2014). High-throughput cereal metabolomics: Current analytical technologies, challenges and perspectives. *J. Cereal Sci*, *59*(3), 393–418. 10.1016/j.jcs.2013.10.002.
87. Przybylska, A., Gackowski, M., & Koba, M., (2021). Application of capillary electrophoresis to the analysis of bioactive compounds in herbal raw materials. *Molecules*, *26*(8), 1–24. 10.3390/molecules26082135.
88. Ma, H., Bai, Y., Li, J., & Chang, Y. X., (2018). Screening bioactive compounds from natural product and its preparations using capillary electrophoresis. *Electrophoresis*, *39*(1), 260–274. 10.1002/elps.201700239.

Index

I

J

K

L

M

N

O

P